Bearings

Basic Concepts and Design Applications

Bearings

Basic Concepts and Design Applications

Maurice L. Adams, Jr.

CRC Press is an imprint of the
Taylor & Francis Group, an informa business

CRC Press
Taylor & Francis Group
6000 Broken Sound Parkway NW, Suite 300
Boca Raton, FL 33487-2742

Printed on acid-free paper

International Standard Book Number-13: 978-1-138-04908-6 (Hardback)

Library of Congress Cataloging-in-Publication Data

Names: Adams, Maurice L., Jr., author.
Title: Bearings : basic concepts and design applications / Maurice L. Adams.
Description: Boca Raton : CRC Press, 2017. | Includes bibliographical references.
Identifiers: LCCN 2017049075| ISBN 9781138049086 (hardback : acid-free paper)
| ISBN 9781315169798 (e-book)
Subjects: LCSH: Bearings (Machinery)
Classification: LCC TJ1061 .A33 2017 | DDC 621.8/22--dc23
LC record available at https://lccn.loc.gov/2017049075

Visit the Taylor & Francis Web site at
http://www.taylorandfrancis.com

and the CRC Press Web site at
http://www.crcpress.com

eResource material is available for this title at https://www.crcpress.com/Bearings-Basic-Concepts-and-Design-Applications/Adams/p/book/9781138049086

This book is dedicated to my late parents and late brother

Maury, Libby and George

And to my late wives

Heidi and Kathy

And to my four mechanical engineering sons

Maury, Professor Dr. Mike, RJ and Nate.

Contents

Preface

All machines and mechanisms have parts connected to each other that are in relative motion. The components that connect those parts in relative motion are called *bearings*, and thus are at the heart of all machines and mechanisms in the modern world. One widely recognized machine type is the *automotive engine*, which has crankshaft main bearings, rod bearings, wrist-pin bearings, and valve/camshaft bearings. Vehicles also need bearings for the wheels, transmission internal components, steering mechanism, and suspension components. Similarly, a wide variety of bearings are also needed in aircraft, ships, power turbines, pumps, manufacturing machine tools, and home appliances.

In performing their operating function, bearings often must reliably sustain significant interactive forces between the parts they connect. With those significant connection forces, bearings often also simultaneously sustain significantly high relative sliding velocities between the parts they connect. In the presence of simultaneous significant sliding velocities and connection forces, sliding friction between the mating bearing parts naturally results. As is well known, rubbing friction between parts in relative sliding motion may cause wear, often an important bearing design selection factor. The sliding friction forces also convert mechanical energy into heat. So, bearings may have to sustain elevated temperatures from that friction heat production. Furthermore, when bearing components heat up they may experience significant thermal expansion and distortion. In some bearing types, that thermal expansion can be comparable sizewise to critical bearing dimensions, for example, journal-bearing radial clearance. Sliding friction also is an energy loss from the overall energy performance of a machine. That energy efficiency factor often plays a major role in determining the optimum bearing type for a given application.

When bearings wear out the need to be replaced, so the ease of replacement can be a major design consideration, for example, labor costs accrued in auto repairs and industrial machinery maintenance. Related to this, the bearing cost and the need to have replacement bearings as part of a spare parts inventory are important factors.

Providing both *depth and breadth of fundamental and practical knowledge for bearings* is the goal of this book. To that end, the book's 18 chapters are divided into three sequential Sections: Section I: *Bearing Basic Technology*, Section II: *Bearing Design and Applications*, and Section III: *Troubleshooting Case Studies*.

Acknowledgments

Truly qualified technologists invariably acknowledge the shoulders upon which they stand. I am unusually fortunate in having worked for several expert caliber individuals during my formative 14 years of industrial employment prior to becoming a professor in 1977, especially my 4 years at the Franklin Institute Research Laboratories followed by my 6 years at the Westinghouse Corporate R&D Center's Mechanics Department. I am also highly appreciative of many subsequent rich interactions with other technologists. I here acknowledge those individuals, many of whom have unfortunately passed away over the years. They were members of a *now extinct breed of giants* who unfortunately have not been replicated in today's industrial workplace environment.

My work in rotating machinery began (1963–1965) at the Allis-Chalmers Hydraulic Products Division in my hometown of York, PA. There I worked on hydroelectric turbine design. That was followed by employment (1965–1967) at Worthington's Advanced Products Division (APD) in Harrison, NJ. There I worked under two highly capable European-bred engineers, Chief Engineer Walter K. Jekat (German) and his assistant John P. Naegeli (Swiss). John Naegeli later returned to Switzerland and eventually became general manager of Sulzer's Turbo-Compressor Division and later general manager of their Pump Division. The APD general manager was Igor Karassik, the world's most prolific writer of centrifugal pump articles, papers, and books and an energetic teacher to all the then-young recent engineering graduates at APD like me. My first assignment at APD was to be "thrown into the deep end" of a new turbomachinery development for the U.S. Navy that even today would be considered highly challenging. That new product was comprised of a 42,000-rpm rotor having an overhung centrifugal air compressor impeller at one end and an overhung single-stage impulse steam turbine powering the rotor from the other end, with inboard water-lubricated turbulent fluid-film bearings. Worthington sold several of these units to the U.S. Navy over a period of many years.

I seized upon an opportunity to work (1967–1971) for an internationally recognized group at the Franklin Institute Research Laboratories in Philadelphia. I am eternally indebted to several FIRL technologists for the knowledge I gained from them and for their encouragement for me to pursue graduate studies part-time, which led to my master's degree in engineering science from classes taken at a local Penn State extension near Philadelphia.

The list of individuals I worked under at Franklin Institute was almost a *who's who list* for the field, including the following: Dr. Elemer Makay (centrifugal pumps), Harry Rippel (fluid-film bearings), John Rumbarger (rolling-element bearings), and Wilbur Shapiro (fluid-film bearings, seals, and rotor dynamics). I also had the privilege of working with a distinguished group of Franklin Institute's consultants from Columbia University, specifically, Professors Dudley D. Fuller, Harold G. Elrod, and Victorio "Reno" Castelli.

My Franklin Institute job gave me the opportunity to publish in my field. That bit of national recognition helped provide my next job opportunity (1971–1977) at what was then truly an internationally distinguished industrial research group, the Mechanics Department at the Westinghouse Corporate R&D Center near Pittsburgh. The main attraction for accepting that job was my new boss, Dr. Albert A. Raimondi, leader of the bearing mechanics section, whose still-famous papers on fluid-film bearings are referenced and reproduced in every undergraduate machine design book.

A bonus at Westinghouse was the presence of the person holding the department manager position, A. C. "Art" Hagg, the company's internationally recognized rotor vibration specialist. My many interactions with Art Hagg were all professionally enriching. At Westinghouse, I was given the lead role on several "cutting edge" projects, including nonlinear dynamics of flexible multi-bearing rotors for large steam turbines and nuclear reactor coolant pumps, bearing load determination for vertical multi-bearing pump rotors, seal development for refrigeration centrifugal compressors, and turning-gear slow-roll operation of journal bearings, developing both test rigs and new computer codes for these projects. I became the junior member of an elite ad hoc trio that included Al Raimondi and D. V. "Kirk" Wright (manager of dynamics section). They encouraged me to pursue my PhD part time, which I completed at the University of Pittsburgh in early 1977. I express special gratitude to my PhD thesis advisor at Pitt, Professor Andras Szeri, who considerably deepened my understanding of the overlapping topics of fluid dynamics and continuum mechanics.

Since entering academia in 1977 I have benefited from the freedom to publish widely and to apply and extend my accrued experience and knowledge through numerous consulting projects for turbomachinery manufacturers and electric utility companies. I appreciate the many years of support for my funded research provided by the Electric Power Research Institute (EPRI) and the NASA Glenn Research Laboratories.

Academic freedom has also made possible leaves to work abroad with some highly capable European technologists, specifically at the Brown Boveri Company BBC (Baden, Switzerland), Sulzer Pump Division (Winterthur, Switzerland), KSB Pump Company (Frankenthal, Germany), and the Swiss Federal Institute (ETH, Zurich). At BBC I developed a lasting friendship with my host Dr. Raimund Wohlrab. At the Sulzer Pump Division, I was fortunate to interact with Dr. Dusan Florjancic (Engineering Director), Dr. Ulrich Bolleter (Vibration Engineering), and Dr. Johan Guelich (Hydraulics Engineering). At the KSB Pump Company, I was fortunate to interact with Peter Hergt (Head of KSB's Central Hydraulic R&D, 1975–1988) and his colleagues. I particularly cherish the interactions with my host and dear friend at the Swiss Federal Institute ETH-Zurich, the late Professor Dr. Georg Gyarmathy, the ETH turbomachinery professor, 1984–1998. This book rests upon the shoulders of all whom I have here acknowledged.

Author

Maurice L. Adams, Jr. is founder and past president of Machinery Vibration Inc. and is professor emeritus of mechanical and aerospace engineering at Case Western Reserve University. The author of over 100 publications and the holder of U.S. patents, he is a life member of the American Society of Mechanical Engineers. Professor Adams received the BSME degree (1963) from Lehigh University, the MEngSc degree (1970) from the Pennsylvania State University, and the PhD degree (1977) in Mechanical Engineering from the University of Pittsburgh. Dr. Adams worked on rotating machinery engineering for 14 years in-industry prior to becoming a professor in 1977.

Since then he has been retained as a rotating machinery consultant by several machinery manufacturers and users in the United States and abroad, including GE Aircraft Engine Group, InVision Technologies, ABB Corporate Research, Rolls-Royce Power Systems, ABB Large Rotating Apparatus, United Technologies Carrier Group, Electric Power Research Institute, Eaton Corporation Manufacturing Technologies Center, Reliance Electric Motors Group, Caterpillar Engine Division, several electric power plants in the U.S. and abroad, Brown-Boveri Large Steam Turbines, Battelle Research, Sulzer Company Pump Division, Oak Ridge National Laboratories, TRW Aerospace Systems, MTorres Co., and John Deere Tractor Group.

Dr. Adams has authored three other CRC Press/Taylor & Francis books: (1) *Rotating Machinery Vibration* (2010), (2) *Power Plant Centrifugal Pumps* (2017), and (3) *Rotating Machinery Research and Development Test Rigs* (2017). He has been the MS thesis and PhD dissertation advisor to over 30 graduate students, three of whom are now endowed-chair professors. He was the recipient in 2013 of the Vibration Institute's Jack Frarey Medal for his contributions to the field of rotor dynamics.

Section I

Bearing Basic Technology

1

Sliding Bearings and Lubrication Mechanics

Sliding-surface bearings have long been around, starting in various geological phenomena and animal species anatomy. Even manufactured sliding bearings have been around for quite a while, such as in Roman chariot axel sleeve bearings lubricated with animal fat. Sliding bearings are now at the heart of the modern world. Broad categories of systems in our daily lives that employ sliding bearings include automobiles, railroads, aircraft, naval ships, space vehicles, as well as manufacturing, food processing, electric power generation, petrochemical, and home appliances. These are only some of the widely recognized uses of sliding bearings.

1.1 Dry Friction, Boundary, and Mixed Lubrication

The well-known Stribeck (2002), (or Stribeck-Hersey) curve, Figure 1.1a, is a curve-fit of many test results on sleeve bearings. It plots the sliding friction coefficient as a function of a bearing dimensionless speed, clearly delineating three different lubrication regimes. The fundamental physical phenomena at the heart of lubrication are microscopic. Whereas the sliding friction coefficient is clearly not microscopic but instead is a summation over a macroscopic sliding area of what is occurring at the microscopic level. Therefore, lubrication researchers have long used the friction coefficient as a relevant readily measureable parameter, in contrast to the various microscopic lubrication phenomena. Although modern measurement sensors have now provided researchers means to interrogate the microscopic lubrication features, the friction coefficient is still a major measurement used by researchers since it is a reliable summation of what is taking place at the microscopic level.

The *mixed lubrication* zone is considered to be an *unstable operating zone* as explained (Figure 1.1a) by postulating a small disturbance about some operating point a, as follows. Consider first a momentary temperature reduction, which produces a momentary increase in oil viscosity, which increases the dimensionless speed, which yields a reduction in friction coefficient that will further reduce oil temperature thus further increasing dimensionless speed, and so on. So, the initial temperature perturbation drives operation to point a'. That will further increase the dimensionless speed, driving operation into the beginning of the stable hydrodynamic zone. Using the same argument, a momentary temperature increase shows the operating point moving to a point a'' and on to the boundary lubrication zone. This is a classic example of *instability*, like an inverted pendulum.

Corresponding to the Stribeck curve, Figure 1.1b illustrates three journal-to-bearing relative positions corresponding to the Stribeck curve. As shown, with dry friction and boundary-lubricated sliding, the journal "climbs up" the bearing surface due to the relatively high friction coefficient. As also shown, when the hydrodynamic zone is approached, the corresponding drastic reduction in friction gives way to the hydrodynamic-pressure

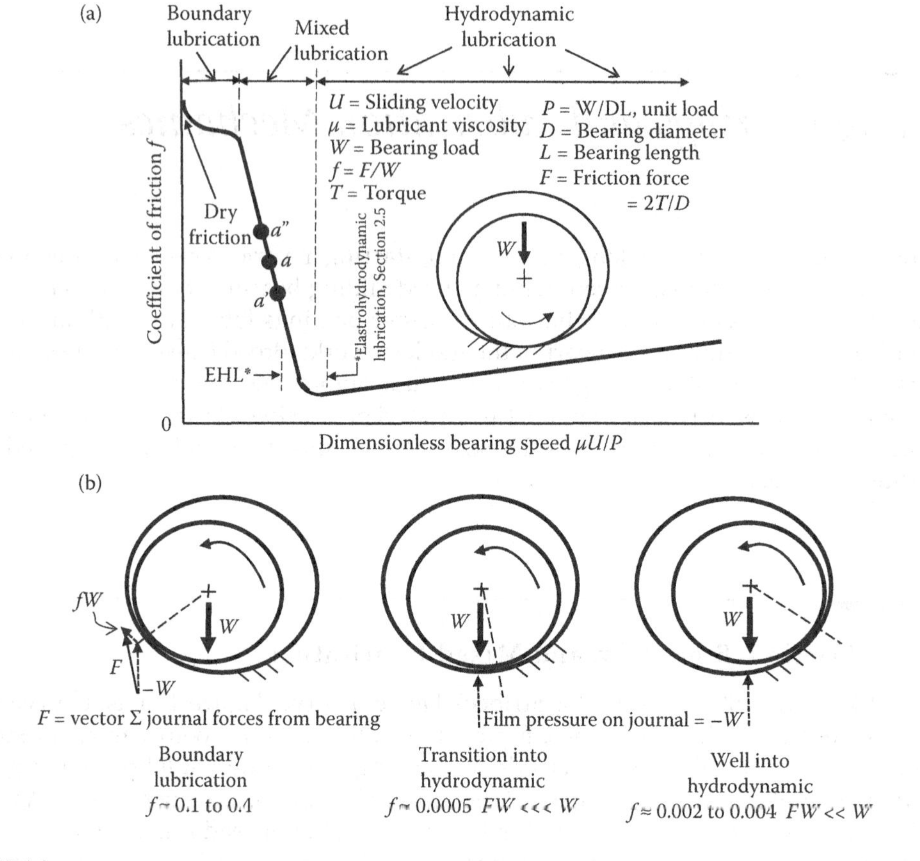

FIGURE 1.1
(a) Stribeck curve, (b) journal-to-bearing relative positions. (*EHL not part of basic Stribeck curve.). *(Continued)*

distribution as the dominant factor that determines the journal position needed to satisfy static equilibrium (see Figure 1.8).

Figure 1.1c, Raimondi et al. (1968), is by far the most complete set of viscosity-versus-temperature curves available in the open literature, covering not only oils but also virtually all other liquid types as well as gases for which viscosities were already measured at the time of its publication. Note that the well-known variation-of-temperature effect on viscosity for gases is completely opposite that for liquids. For liquids, the viscosity (measure of resistance to be sheared) is produced by the molecular mutual attraction forces which weaken considerably as temperature is increased. But for gases, the viscosity is produced by the rate of gas molecular collisions, which increase with temperature increase. Viscosity chart for commonly used oils is given in Chapter 7.

In *dry friction sliding bearings, adhesive friction* is the most probable of all unlubricated surface-to-surface rubbing phenomena. But even with unlubricated surfaces, naturally formed oxides become a *dry lubricant*. Significant adhesive friction is also possible to occur between boundary-lubricated rubbing surfaces, but at a reduced rate depending on the rubbing materials' strength properties and the *boundary lubrication* capability of the intervening lubricant. As well-known and universally used by engineers, the classical Coulomb friction coefficient is usually approximated as a "constant" independent of the

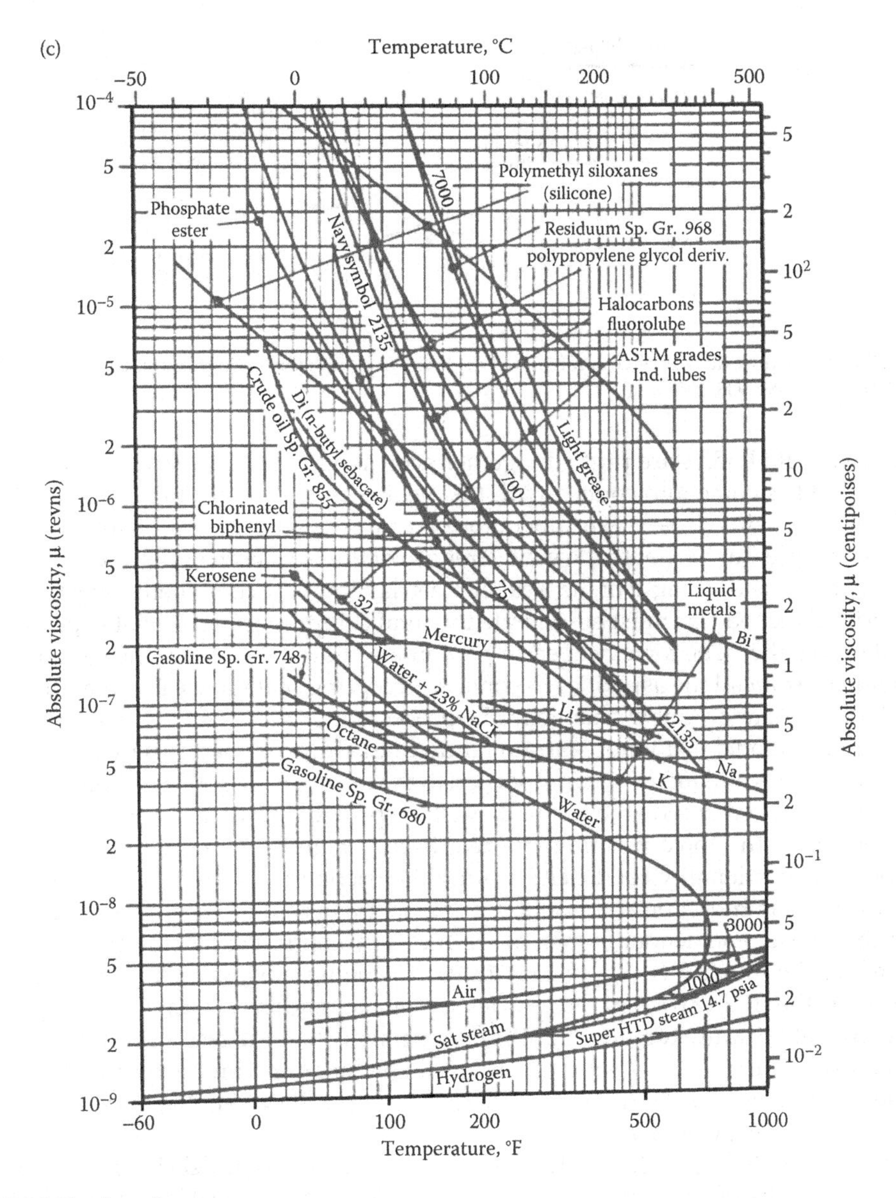

FIGURE 1.1 (Continued)
(c) Viscosity versus temperature. (Adapted from O'Connor, J. J., Boyd, J. and Avallone, E. A., *Standard Handbook of Lubrication Engineering, American Society of Lubrication Engineers*, McGraw-Hill, 1968.)

sliding speed and independent of the normal force magnitude that pushes together the two bodies in relative sliding motion. Archard (1953) hypothesized *adhesive friction and wear* as illustrated in Figure 1.2, by postulating what might occur at the microscopic level between two surface asperities in rubbing contact. The two contacting asperities are postulated to momentarily form an adhesive bond called *cold welding*. In one possibility, this bond is simply broken as the two surface asperities each go unaltered their own way as illustrated. In the other possibility shown, the adhesive bond between the two asperities overcomes the yield strength of the weaker of the two asperities, creating a wear particle.

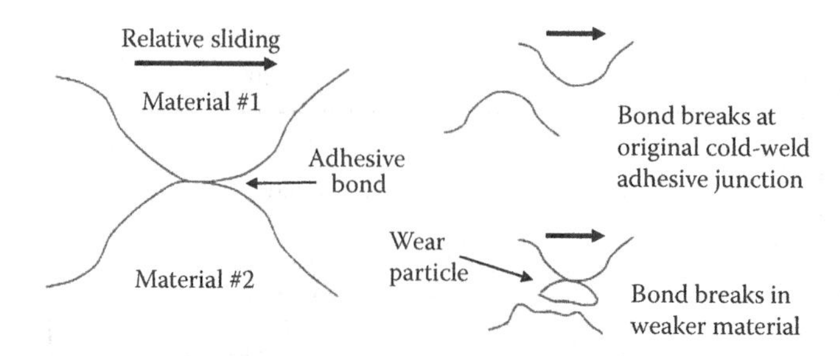

FIGURE 1.2
Two possibilities within the adhesive wear model of Archard (1953). (From Archard, J. F., *Journal of Applied Physics*, Vol. 24, pp. 981–988, 1953.)

When two solid bodies are pushed together even at light loads, relatively few asperities of the two bodies come into contact with the weaker asperity deforming plastically, that is, yielding at the tip. These plastically deformed asperity tips are reasonably assumed to have a contact pressure equal to the yield strength of the weaker material. *Static equilibrium* dictates that the yield strength of the weaker material times the combined contact area of all the asperity flattened tips must equilibrate the total compressive force that is pushing the two bodies together. The force needed to perpetuate sliding will equal the sum of sliding friction forces from all the asperities in contact. Accepting that the weaker material's yield strength does not appreciably change in the process (i.e., the perfect elastic-plastic stress-strain response model that ignores strain hardening), the combined surface area of all the flattened asperities will increase in proportion to the force pushing the two bodies together. So, the force required to slide the one body with respect to the other will be approximately in proportion to the force pushing the two bodies together. That naturally implies a constant-of-proportionality coefficient relating the friction force to the compressive force, that is, the Coulomb friction coefficient. Of course, this is a quite useful simplification of a quite complex phenomenon. In fact, modern closely controlled laboratory experiments, for example, Rabinowicz (1965), on dry friction rubbing have shown that with a fixed normal force, the sliding friction force varies somewhat randomly with the sliding velocity magnitude, but typically remains significantly smaller than the breakaway static friction force needed to start sliding. As is well known, to better approximate this for engineering purposes, two values for the friction coefficient are typically given, the *static* value μ_S and the *dynamic* value μ_D, where $\mu_S > \mu_D$.

Referring again to Figure 1.2, it illustrates the release of a wear debris particle when an asperity tip breaks off. *Archard's law* is a statistical-based approach for predicting wear volume or rate of volumetric wear derived from the concept illustrated in Figure 1.2. A detailed rendition and derivation of Archard's equation for predicting wear rate is given by Adams (2017a).

When a lubricant such as oil is entrapped between two contacting surfaces in relative sliding-motion, the lubrication regime labeled *boundary lubrication* in Figure 1.1a is likely to occur. Boundary lubrication is essentially a chemical process, the effectiveness of which depends on the propensity for *long organic* or *organic-like molecules* to form between the asperities of the two contacting surfaces, Figure 1.3. When this chemical action is sufficiently sustained, the very thin separation so produced between the two surfaces is only a few molecules thick. When fully functioning, the rate of long molecule production keeps up with the rate at which the sliding between the two solid surfaces wipes away those long molecules. The higher the sliding speed and/or the higher the

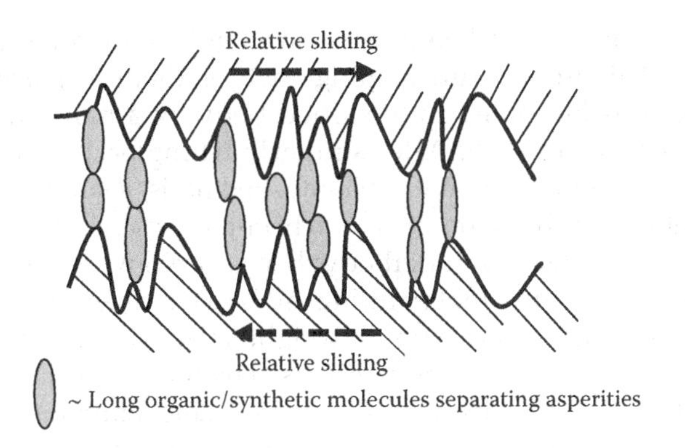

FIGURE 1.3
Heuristic illustration of boundary lubrication.

load pushing the two surfaces together, the faster the long surface-separating molecules are wiped away. Thus, it follows that when the molecules are wiped away faster than the chemical reaction can replace them, wear of one or both surfaces will occur, roughly in proportion to the sliding velocity times the bearing load. In Part 2 of this book (*Bearing Design and Application*), design methodologies for boundary-lubricated bearings are presented, including how well various oil types and oil additives enhance the formation and durability of the long organic or organic-like molecules that produce sustainable boundary lubrication.

Solid-film lubrication really does not have a welcomed place in the Stribeck curve since it is neither boundary lubrication nor hydrodynamic fluid-film lubrication. One could thus refer to it as the *black sheep* of the lubrication family, which ironically fits its physical appearance, that is, black carbon graphite. The two most common solid lubricants are *carbon graphite* and *molybdenum disulfide*. Solid-film lubrication basically entails the use of such solid materials with *extremely low grain boundary shear strength*. The Figure 1.4 sketch illustrates the low shear strength characteristic at the grain boundary between two adjacent crystals of a solid lubricant. One most common long-time example is the *writing lead pencil* with a thin cylinder of carbon graphite in its center. When the pencil tip is slid across the writing paper it slides easily by solid-lubricating itself, and leaving behind on the paper the carbon graphite wear-particle debris in what has been written, sketched, or scribbled on the paper.

In other lubrication applications, the solid lubricant is in the form of a fine powder such as readily available at the local hardware store, and long recommended for lubricating key-type locks since it will not gum up like a squirt of household oil will likely do, thus long used on car keyed door locks.

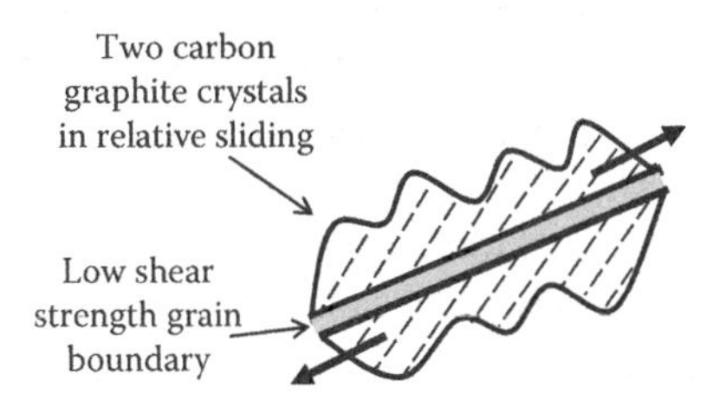

FIGURE 1.4
Two adjacent carbon graphite crystals under shear stress.

Having four sons (all mechanical engineers, one a university professor and department head) the author had first-hand fun helping each of them at *pinewood derby* time, as follows. Each wheel was journaled to the wood body by a nail that was first polished while turning in a stationary drill. After similarly truing each wheel OD with a slight crown and installing onto the body with the polished nails, a small puff of *carbon graphite power* was blown into the hub region of each wheel-nail assembly. Finally, small plugs of weights were countersunk into the underbody then covered with wood filler sandpapered flush with the underbody before final painting, all while managing to get the total weight as close as possible to the 150-gm maximum weight limit. All sons were highly successful in their respective pinewood derby competitions. The more significant uses of solid-film lubrication take full advantage of its capability for extreme temperature applications (very high or very low) such as in space-travel vehicles and earth satellites.

1.2 Reynolds Lubrication Equation for Fluid-Film Bearings

The modern starting point for modeling fluid mechanics problems is encompassed in the three-coupled *fluid-momentum partial differential equations*, Navier-Stokes Equations (NSEs), plus the single *conservation-of-mass partial differential equation*, Continuity Equation. The three scalar NSEs, which are nonlinear, are obtained by applying Newton's Second Law $\Sigma\vec{F} = d(m\vec{v})/dt$ to an inertial differential control volume of a continuum flow field. Attempting to solve these equations for 2D and 3D problems has historically been the challenge occupying the careers of mathematically inclined fluid mechanics theoreticians. That is because the three NSEs are coupled nonlinear partial differential equations. The ingenious contributions of the precomputer age fluid mechanics *"giants"* sprang from their considerable physical insight into specific problems, leading them to make justifiable simplifying assumptions that yielded mathematically solvable formulations for important applications. That is essentially equivalent to identifying and excising those terms in the NSEs of secondary importance for a specific problem. The most important example of this is in Reynolds' (1886) original paper on *hydrodynamic lubrication theory*, which is nothing short of a masterpiece, though he did not formally start with the NSEs. The *Reynolds Lubrication Equation* (RLE) applies to an *incompressible laminar* (no turbulence) strictly *viscous* (no fluid inertia) *thin fluid film* between two closely spaced surfaces in *relative motion*. Figure 1.5 illustrates a generalized thin fluid film between two solid boundaries moving with film-normal velocities (V) and film-sliding velocities (U).

Assumptions for following derivation of the Reynolds Lubrication Equation

1. The lubricant film is an incompressible uniform-density *Newtonian fluid* in purely viscous laminar thin-film flow between two smooth impermeable surfaces.
2. *Fluid inertia* is negligible, consistent with sufficiently low Reynolds number, and the flow is fully developed.

   Because of the close spacing of the two wetted surfaces, many further simplifying assumptions are invoked. These include the following.
3. Neglect of local *surface curvature* in film. That is, the film is modeled as unwrapped into a plane, the primary example being a journal bearing, Figure 1.6.

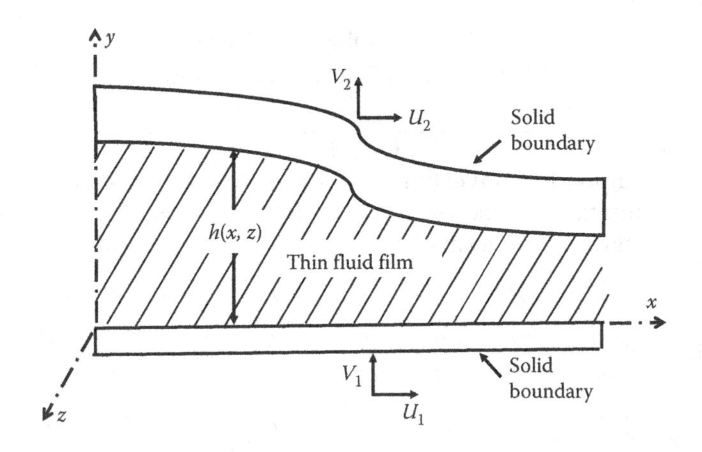

FIGURE 1.5
Generalized thin fluid film between two moving surfaces ($h \ll 1$).

4. Neglect of fluid velocity and local pressure change normal to the local plane of the film, justifying the following.

$$v << u \,\&\, w \quad \text{and} \quad \frac{\partial p}{\partial y} << \frac{\partial p}{\partial x} \,\&\, \frac{\partial p}{\partial z}$$

5. Neglect of the shear stress gradient components in the local plane of the film because they are much smaller than the shear stress gradients across the thin film.

$$\frac{\partial^2 u}{\partial x^2} \,\&\, \frac{\partial^2 u}{\partial z^2} << \frac{\partial^2 u}{\partial y^2} \quad \frac{\partial^2 v}{\partial x^2} \,\&\, \frac{\partial^2 v}{\partial z^2} << \frac{\partial^2 v}{\partial y^2} \quad \frac{\partial^2 w}{\partial x^2} \,\&\, \frac{\partial^2 w}{\partial z^2} << \frac{\partial^2 w}{\partial y^2}$$

6. Neglect of body forces, for example, gravity, electromagnetic.

Consistent with the approach in which fluid mechanics theory is now taught, the derivation here of the RLE starts with the three full NSEs and mass conservation for an *incompressible Newtonian fluid in laminar flow* for a differential control volume (CV) $dx\,dy\,dz$, as follows.

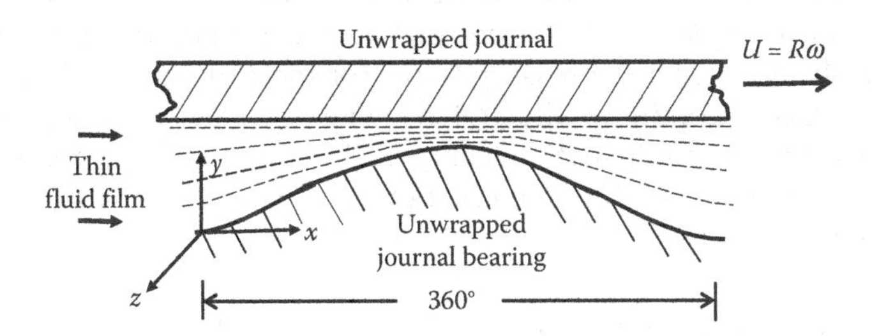

FIGURE 1.6
Unwrapped journal and bearing (film thickness exaggerated, $h \ll 1$).

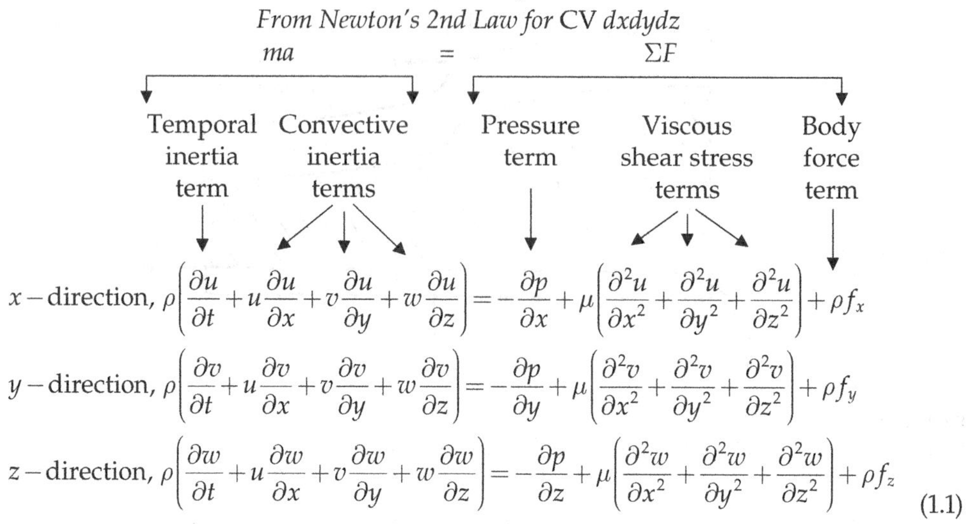

$$
\begin{aligned}
&x-\text{direction},\ \rho\left(\frac{\partial u}{\partial t}+u\frac{\partial u}{\partial x}+v\frac{\partial u}{\partial y}+w\frac{\partial u}{\partial z}\right)=-\frac{\partial p}{\partial x}+\mu\left(\frac{\partial^2 u}{\partial x^2}+\frac{\partial^2 u}{\partial y^2}+\frac{\partial^2 u}{\partial z^2}\right)+\rho f_x\\
&y-\text{direction},\ \rho\left(\frac{\partial v}{\partial t}+u\frac{\partial v}{\partial x}+v\frac{\partial v}{\partial y}+w\frac{\partial v}{\partial z}\right)=-\frac{\partial p}{\partial y}+\mu\left(\frac{\partial^2 v}{\partial x^2}+\frac{\partial^2 v}{\partial y^2}+\frac{\partial^2 v}{\partial z^2}\right)+\rho f_y\\
&z-\text{direction},\ \rho\left(\frac{\partial w}{\partial t}+u\frac{\partial w}{\partial x}+v\frac{\partial w}{\partial y}+w\frac{\partial w}{\partial z}\right)=-\frac{\partial p}{\partial z}+\mu\left(\frac{\partial^2 w}{\partial x^2}+\frac{\partial^2 w}{\partial y^2}+\frac{\partial^2 w}{\partial z^2}\right)+\rho f_z
\end{aligned}
\tag{1.1}
$$

Conservation of mass for an incompressible-fluid dx dy dz CV

$$\frac{\partial u}{\partial x}+\frac{\partial v}{\partial y}+\frac{\partial w}{\partial z}=0 \tag{1.2}$$

Here, ρ = mass density, p = pressure, μ = fluid viscosity, ρf = body force, u, v, and w are the x, y, and z velocity components, respectively, in a Cartesian x-y-z coordinate system. Equations 1.1 and 1.2 apply to an infinitesimal differential control volume of rectangular orthogonal dimensions dx, dy, and dz.

Based on the previously listed assumptions for this derivation of the RLE, deletion of the neglected NSE terms reduces Equation 1.1 to the following.

$$
\begin{aligned}
&x-\text{direction}, && 0=-\frac{\partial p}{\partial x}+\mu\frac{d^2u}{dy^2}\\
&y-\text{direction}, && 0=0\\
&w-\text{direction}, && 0=-\frac{\partial p}{\partial z}+\mu\frac{d^2w}{dy^2}
\end{aligned}
\tag{1.3}
$$

The y-direction NSE is totally eliminated while the x-direction and z-direction in-plane NSEs become *linear decoupled* second order ordinary differential equations (ordinary in y), left only with shear stress and pressure terms for their respective directions. With the surface velocity boundary conditions illustrated in Figure 1.5, the physically well-established *no-slip boundary condition* is mathematically imposed for fluid velocity boundary conditions at the two wetted solid surfaces. Integration of the x-direction and z-direction Equation 1.3 is detailed as follows.

$$\frac{1}{\mu}\frac{\partial p}{\partial x}=\frac{d^2u}{dy^2}\therefore\frac{du}{dy}=\frac{1}{\mu}\frac{\partial p}{\partial x}y+\mathrm{A}\therefore u(y)=\frac{1}{2\mu}\frac{\partial p}{\partial x}y^2+\mathrm{A}y+\mathrm{B}$$

Similarly, $w(y)=(1/2\mu)(\partial p/\partial z)y^2+\mathrm{C}y+D$, with A, B, C, D the constants of integration.

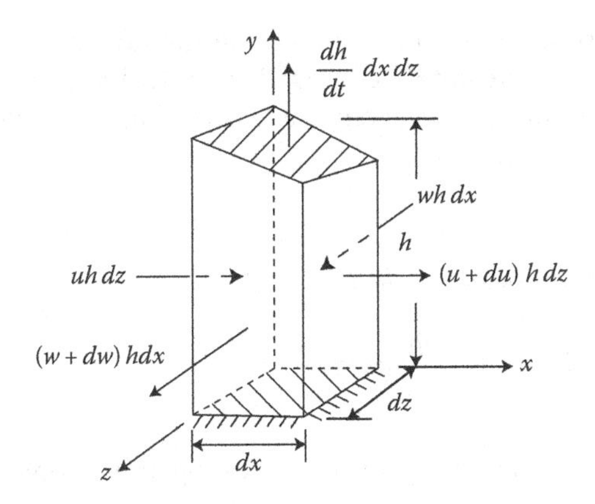

FIGURE 1.7
Flow balance components of fluid-film *differential control volume.*

The four boundary conditions that determine A, B, C, and D are $u(0) = 0$, $u(h) = U_2 + U_1 \equiv U$, and $w(0) = 0$, $w(h) = 0$, which yield the following solutions for the x-direction and z-direction fluid-film velocity distributions.

$$u(y) = \frac{1}{2\mu}\frac{\partial p}{\partial x}(y^2 - hy) + \left(\frac{Uy}{h}\right) \quad \text{and} \quad w(y) = \frac{1}{2\mu}\frac{\partial p}{\partial z}(y^2 - hy) \tag{1.4}$$

The $u(y)$ and $w(y)$ velocity distributions are integrated over their respective film differential flow areas [$h\,dz$ for $u(y)$ and $h\,dx$ for $w(y)$] to determine their corresponding flow components, Figure 1.7. The fluid is assumed to be incompressible and of uniform density, so conservation of mass flow is equivalent to *conservation of volume flow.*

x-direction flow per unit width of flow path:

$$Q_x = \int_0^h u(y)dy = \int_0^h \left[\frac{1}{2\mu}\frac{\partial p}{\partial x}(y^2 - hy) + \left(\frac{Uy}{h}\right)\right] dy = -\underbrace{\frac{h^3}{12\mu}\frac{\partial p}{\partial x}}_{\text{Hagen–Poiseuille flow}} + \underbrace{\frac{Uh}{2}}_{\text{Couette flow}} \tag{1.5}$$

z-direction flow per unit width of flow path:

$$Q_z = \int_0^h w(y)dy = \int_0^h \left[\frac{1}{2\mu}\frac{\partial p}{\partial z}(y^2 - hy) + \left(\frac{Uy}{h}\right)\right] dy = \int_0^h \frac{1}{2\mu}\frac{\partial p}{\partial z}(y^2 - hy)\, dy = -\frac{h^3}{12\mu}\frac{\partial p}{\partial z} \tag{1.6}$$

The differential control-volume h, dx, dz flow balance is additionally augmented to include a *storage term* to accommodate solid-boundary velocities normal to the local plane of the film, Figure 1.5. Reynolds' (1886) derivation did not include this. Since fluid-film velocity is neglected in the y-direction, $\partial v/\partial y = 0$, Equation 1.2 with the added storage term becomes the following.

$$\frac{\partial u}{\partial x} + \frac{\partial w}{\partial z} + \text{storage term} = 0 \tag{1.7}$$

Consistent with Equation 1.7, conservation of volume flow at a point (x, z) in a thin fluid film is imposed on the infinitesimal rectangular control volume h, dx, dz illustrated in Figure 1.7. Conservation of flow mass at a point in an incompressible uniform-density fluid film is thus satisfied by the following.

Thus,

$$\begin{aligned} &h(u+du)dz - h\,du\,dz + h(w+dw)dx - h\,dw\,dx + \frac{dh}{dt}dx\,dz = 0 \\ &h\,du\,dz + h\,dw\,dx + \frac{dh}{dt}dx\,dz = 0 \end{aligned} \tag{1.8}$$

Thus, consistent with Equation 1.7 and employing Equations 1.5 and 1.6, flow mass conservation for a thin fluid film is given by Equation 1.9. For clarity, it is relevant to mention that the differential control volume here (h, dx, dz) may initially appear to be inappropriately labeled as mathematically *differential* consistent with the mass conservation equation for incompressible flow, Equation 1.2, because the film thickness h while "small" is not infinitesimal. The explanation for this possibly perceived dichotomy is the fact that the flow velocity v normal to the local plain of the film is $v \ll u \,\&\, w$, fully justifying the earlier stated assumption that $v = 0$ in this derivation of the RLE. That is, the flow is assumed to be 2D, thus the RLE is a 2D partial differential equation (PDE).

$$\frac{\partial Qx}{\partial x} + \frac{\partial Qz}{\partial z} + \frac{dh}{dt} = \frac{\partial}{\partial x}\left(-\frac{h^3}{12\mu}\frac{\partial p}{\partial x} + \frac{Uh}{2}\right) + \frac{\partial}{\partial z}\left(-\frac{h^3}{12\mu}\frac{\partial p}{\partial z}\right) + \frac{dh}{dt} = 0 \tag{1.9}$$

Rearranging Equation 1.9 yields the RLE as typically presented, with the right-hand side containing the two driving influences and the left-hand side the pressure terms. In this form, the RLE has units of velocity.

Sliding velocity term, source of bearing static-load capacity

Squeeze-film velocity term, source of bearing damping

Reynolds Lubrication Equation

$$\frac{\partial}{\partial x}\left[\frac{h^3}{\mu}\left(\frac{\partial p}{\partial x}\right)\right] + \frac{\partial}{\partial z}\left[\frac{h^3}{\mu}\left(\frac{\partial p}{\partial z}\right)\right] = 6U\frac{dh}{dx} + 12\frac{dh}{dt} \tag{1.10}$$

Here, $p = p(x,z)$ is the film-pressure distribution and $h = h(x,z)$ is the film-thickness distribution. The RLE is still recognized as the governing theoretical basis for hydrodynamic fluid-film bearings. Furthermore, it is also the foundation for hydrostatic fluid-film bearings, which were not even developed until over half a century after Reynolds' (1986) landmark paper. Concerning the rotor-dynamic characteristics of journal bearings, it is the *sliding velocity* term that gives rise to the bearing *stiffness coefficients* and the *squeeze-film velocity* term that gives rise to the bearing *damping coefficients.*

It was Reynolds' objective to explain then recently published experimental results for rail-locomotive journal bearings which showed a capacity to generate film pressures to keep the rotating journal from contact rubbing of the bearing. Reynolds' derivation established why the sliding action of the rotating journal surface, shearing oil into the converging thin gap between an eccentric journal and bearing, produced a *hydrodynamic-pressure distribution* which could support static-radial loads across the oil film without the journal and bearing making metal-to-metal contact. This is one of the most significant discoveries in the history of engineering science. Furthermore, Reynolds' derivation clearly showed that this hydrodynamic-load capacity was in direct proportion to the sliding velocity and the lubricant viscosity. Reynolds' derivation did not include the *squeeze-film storage term*. Virtually every undergraduate text in machine design has a chapter devoted to hydrodynamic-bearing design based on the Raimondi and Boyd (1958) early 2D finite-difference computer solutions of the RLE for bearing static-load capacity, flow, and friction (see reference to Dr. Albert A. Raimondi in the Acknowledgments of this book).

In utilizing numerical solutions of the full 2D RLE to develop design charts such as those of Raimondi and Boyd, the RLE can be converted to dimensionless form so that solutions are applicable to any size bearing of the same proportions, length (L)-to-width (B) ratio. This is standard with solutions for PDEs in general when building design charts, for example, heat conduction in solid bodies. Equation 1.10 can be made dimensionless in many selected ways.

1.3 Hydrodynamic Journal Bearings

Before the existence of digital computers, Equation 1.10 was solved for journal bearing (x, z) pressure distribution conditions by neglecting either the axial-pressure flow term (*long-bearing* formulation) or neglecting the circumferential-pressure flow term (*short-bearing* formulation). As Equation 1.11 shows, with either approximation, the RLE is reduced to an ordinary differential equation (i.e., one independent spatial coordinate) and thus solvable without computerized numerical methods.

Long-Bearing RLE *Short-Bearing RLE*

$$\frac{d}{dx}\left[\frac{h^3}{\mu}\left(\frac{dp}{dx}\right)\right]=6U\frac{dh}{dx} \qquad \frac{d}{dz}\left[\frac{h^3}{\mu}\left(\frac{dp}{dz}\right)\right]=6U\frac{dh}{dx} \tag{1.11}$$

The RLE long-bearing (LB) approximation assumes that all flow sheared into the converging film moves only in the sliding velocity x-direction, that is, no flow perpendicular to the sliding velocity. Conversely, the short-bearing (SB) RLE assumes that all the flow sheared into the converging film leaves only perpendicular to the sliding velocity. It is relatively easy to visualize with the LB condition that no lubricant is allowed to flow axially out the bearing sides. Conversely, for the SB condition it is not as easily visualized that the sheared-in flow progressively exits only out the bearing sides.

For full 2D-RLE pressure solutions, with the pressure distribution directionally integrated, static-load magnitude and direction relative to the journal-to-bearing eccentricity are obtained as detailed in Chapter 7. With the full 2D-pressure solution, two flow categories are identified: (1) *carry-over flow* and (2) *side leakage flow*. As covered in Chapter 7, both two flow categories are needed in computing the net bearing through flow and operating average *film temperature* based on a heat balance involving the fluid-film viscous *friction dissipation energy* and the bearing net heat-removal *through-flow*. The importance of that in design analyses is to achieve a bearing-tolerable operating temperature which is heavily influenced by the lubricant viscosity being a very strong function of temperature. For full 2D RLE solutions, film flow in both the x and z directions is more easily visualized with the 2D-solution film-pressure distributions, as exampled for $0.5 < L/D < 1$ in Figure 1.8.

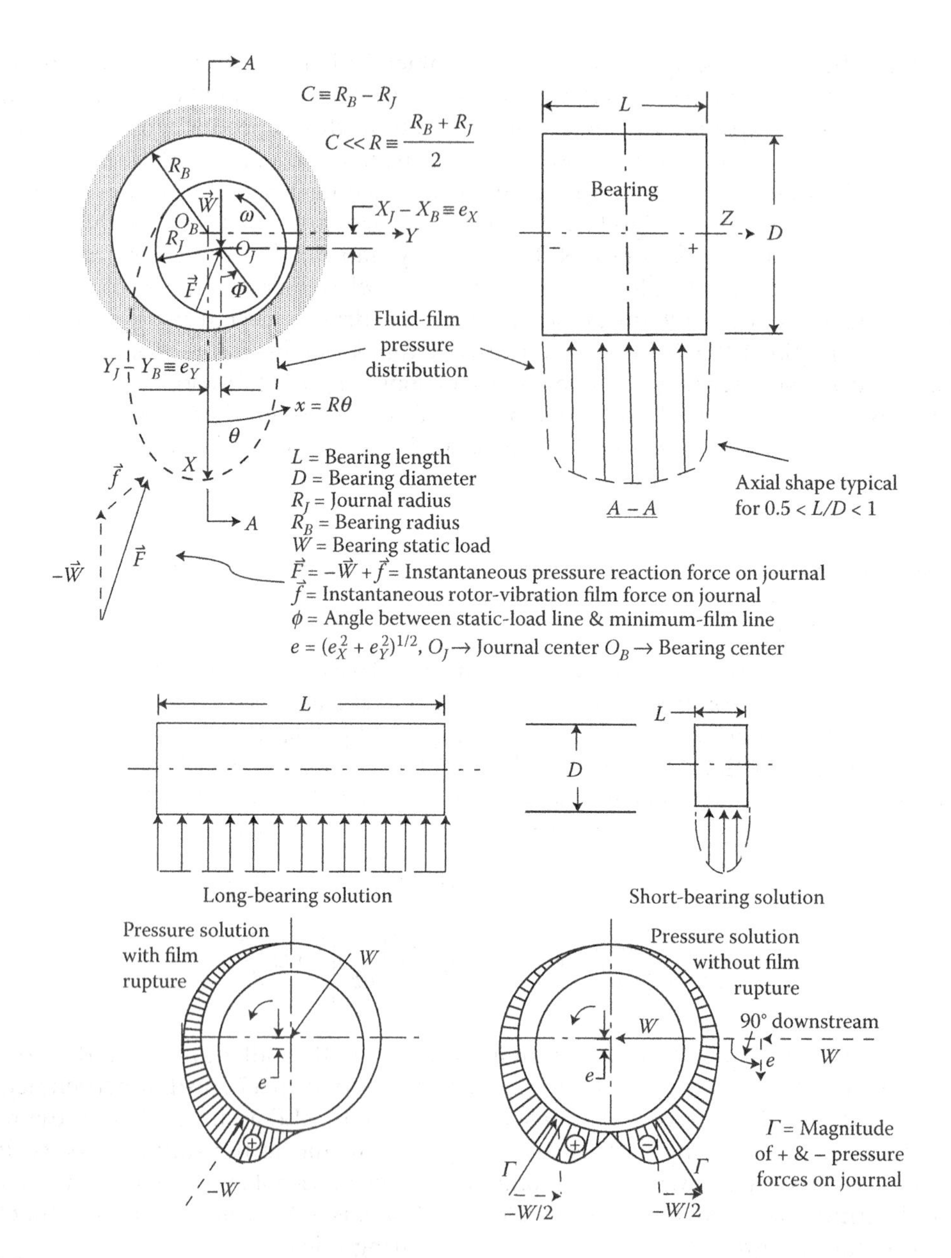

FIGURE 1.8
Journal-bearing pressure distributions and nomenclature.

For journal bearing static-load capacity, the LB and SB approximations provide an upper bound and lower bound for load capacity, respectively, for the "exact" 2D numerical (finite difference) solutions to Equation 1.10. When solving the full 2D RLE, two-axial and two-circumferential pressure boundary conditions must be specified since the RLE is second order in both x and z directions. That is, the RLE is a 2D second *order elliptic* PDE. Alternatively, the two-axial-pressure boundary conditions are replaced by two joined-boundary conditions $\partial p/\partial x$ (up stream) = $\partial p/\partial x$ (down stream), and p(up stream) = p(down stream), at some angular position when there is a continuous bearing cylindrical surface with no lubricant axial feed grove specified. More typically, there is at least one axial lubricant feed groove that provides the two-axial boundary-condition pressures. However, both approximate RLE formulations in Equation 1.11 can of course mathematically accommodate only two-pressure boundary conditions. The fluid-film pressure solutions of the LB and SB approximations clearly show why the LB solution provides the load capacity upper bound for 2D solutions of the RLE Equation 1.10 while the SB solution provides the lower bound, as Figure 1.8 clearly illustrates.

For a journal bearing in the most common operating environment, the bearing fluid-film ambient boundary pressures are near atmospheric pressure compared to the film pressure distribution. In that case, *film rupture* begins just downstream of the minimum film location, Figures 1.8 and 1.9. Film rupture occurs because unlike a solid, a fluid cannot sustain tension without *rupturing*, sometimes called *cavitation* even though in a bearing fluid-film it is quite a different phenomenon than centrifugal pump cavitation, Adams (2017a). The pressure boundary conditions $p = p_{\text{vapor}}$ and $\vec{\nabla} p = 0$ are typically imposed in full 2D RLE numerical solutions to locate the boundary between the full-film load-carrying region and the film-ruptured region, as illustrated in Figure 1.9. The $\vec{\nabla} p = 0$ condition is used to impose the physical requirement that lubricant mass flow must be conserved across the interface boundary separating the full-film and ruptured-film regions. This is clear because the *Couette flow* can't change across the boundary since there is only one value of h at each point on the boundary. Assuming the ruptured region will not support a significant-pressure gradient, only with $\vec{\nabla} p = 0$ at the boundary (between full film and ruptured film) is fluid mass flow conserved across the boundary. Otherwise there would be *Hagen–Poiseuille flow* leaving the full-film region at the boundary but not entering the rupture-film region at boundary. Obviously, imposing $\vec{\nabla} p = 0$ eliminates this physical impossibility from the RLE solution involving film rupture. Some early generation full 2D numerical RLE solutions did not correctly handle this, simply truncating the negative-pressure portion of the RLE solution without film rupture (see bottom of Figure 1.8). It is the imposed $\vec{\nabla} p = 0$ boundary condition that positions the 2D RLE solution boundary between the full-film and ruptured regions slightly downstream of the minimum film-thickness location instead of positioning it right at the minimum film-thickness location.

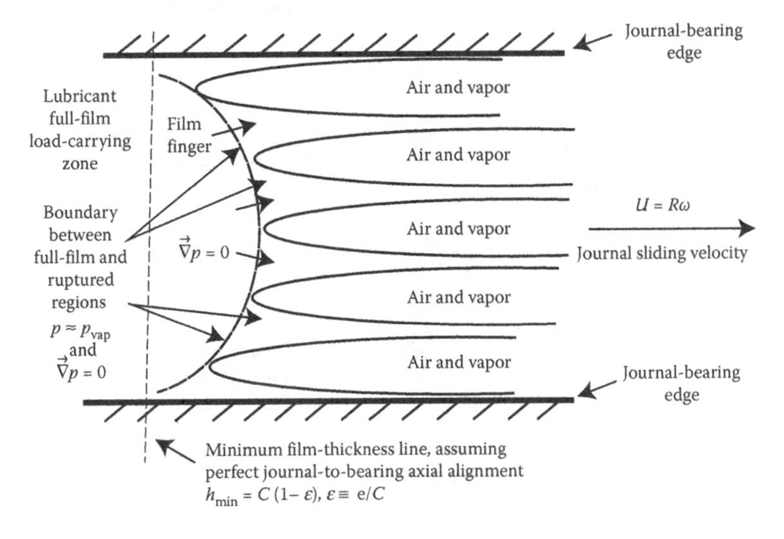

FIGURE 1.9
Unwrapped journal-bearing fluid film showing rupture striation fingers.

In much less typical operating environments, the bearing ambient pressure is sufficiently high to maintain sufficiently high positive pressures (compression not tension) to prevent most or all film rupture from occurring. One notable application where this occurs is in the *primary reactor coolant pumps* for pressurized water reactor (PWR) nuclear powered units (Adams 2017a), occurring both in commercial nuclear power plants as well as Navy nuclear powered submarines and aircraft carriers. This is because the journal bearings (water-lubricated) are submerged in the PWR-primary loop pressure, typically 160 bars. Figure 1.8 illustrates the two contrasting circumferential-pressure distributions between the ruptured and non-ruptured cases. It is clearly shown why with the non-ruptured-film case that journal-radial eccentricity displacement is 90° downstream from the applied-bearing static load (W) direction. That is, the journal moves perpendicular to the direction in which it is pushed. This is surely an unusual load-deflection phenomenon. With this unique load-deflection property, it should come as no surprise that rotor-vibration characteristics of non-rupturing hydrodynamic journal bearings are quite different than typical journal bearings with significant film rupture. Given that fact, and given that the typical PWR reactor coolant pump has a vertical rotational centerline, thus yielding considerable uncertainty in journal-bearing static loads, reactor coolant pumps have poorly predictable and randomly variable rotor-vibration characteristics (Jenkins 1993; Adams 2010).

The first journal-bearing illustration in Figure 1.8 shows the typical pressure distribution in a cylindrical 360° journal bearing with film rupture in the diverging portion of the film-thickness distribution. There are several variations of the journal bearing that are not simply of the 360° purely cylindrical configuration, such as the three common examples illustrated in Figure 1.10, and covered further in Part 2 of this book, *Bearing Design and Applications*.

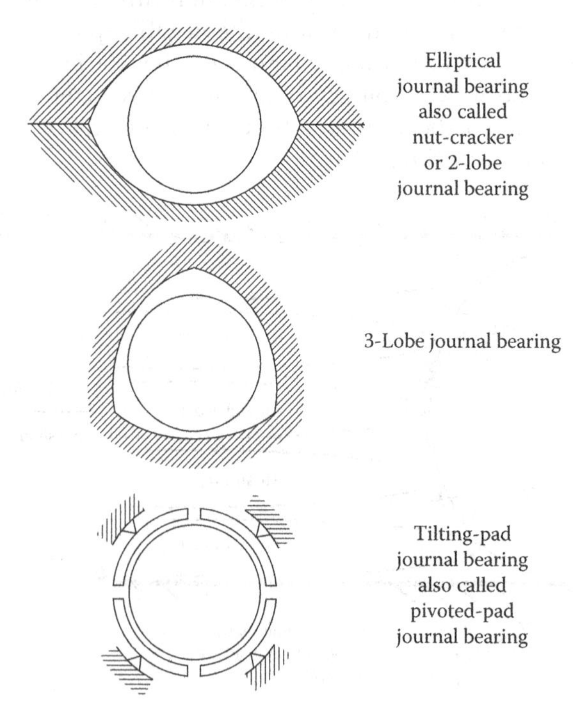

FIGURE 1.10
Other journal-bearing configurations (film thickness exaggerated).

1.4 Hydrodynamic Thrust Bearings

Bearings that support rotor axial loads are commonly referred to as *thrust bearings*. The types of rotating machinery employing fluid-film hydrodynamic thrust bearings (HTBs) abound. These applications include most types of turbomachinery such as centrifugal pumps and compressors, steam and gas turbines, gear sets, naval ship propulsion drivelines, and some large electric motors.

The fundamental theory for HTBs is the same as for journal bearings, the RLE (Equation 1.10). The simple plane slider-bearing pad is illustrated in Figure 1.11 that consolidates the fundamental underpinnings of hydrodynamic lubrication. As shown, the local fluid sliding-direction velocity profiles are each the local sum of Hagen–Poiseuille flow and Couette flow, where the Hagen–Poiseuille flow is zero at the peak of the lubricant-film pressure distribution since the pressure gradient there is zero. Film flow perpendicular to the Figure 1.11 view contains only Hagen–Poiseuille flow, as Equation 1.4 shows for the z-component of film velocity, $w(y)$. HTB pads are essentially slider bearings, albeit with the pads having cylindrical coordinate boundaries. The two major types of hydrodynamic thrust bearings are pictured in Figures 1.12 and 1.13, further illustrating the wide variety of pad configurations for fixed-profile configurations. Design information for HTBs is presented in Part 2 of this book.

As appropriate for journal bearings, the original RLE was derived in Cartesian rectangular x, y, z coordinates, yielding Equation 1.10. However, HTB pads are not of purely unwrapped

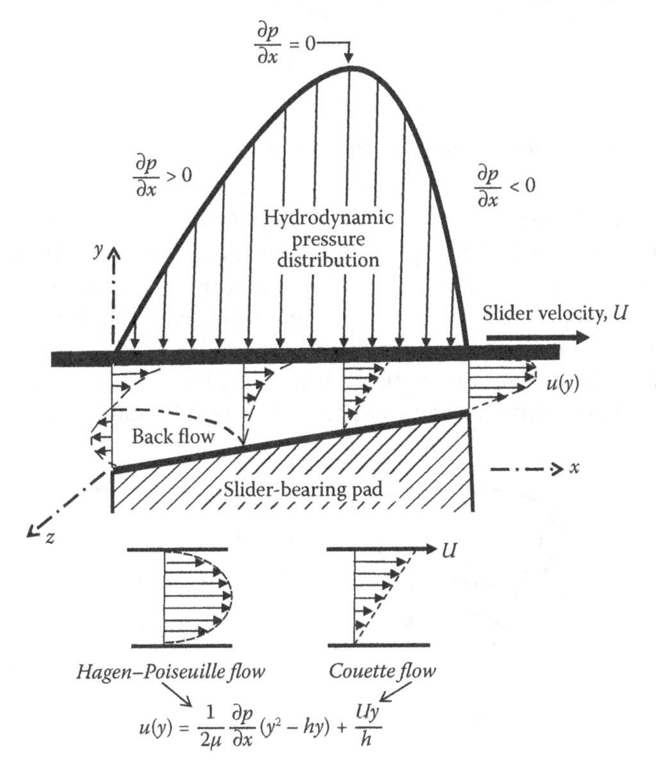

FIGURE 1.11
Slider bearing with pressure distribution and x-velocity profiles.

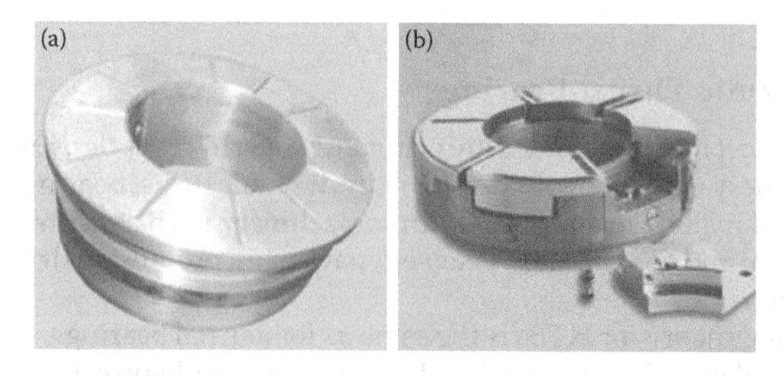

FIGURE 1.12
Hydrodynamic thrust bearings. (a) Fixed profile and (b) Self-aligning tilting-pad.

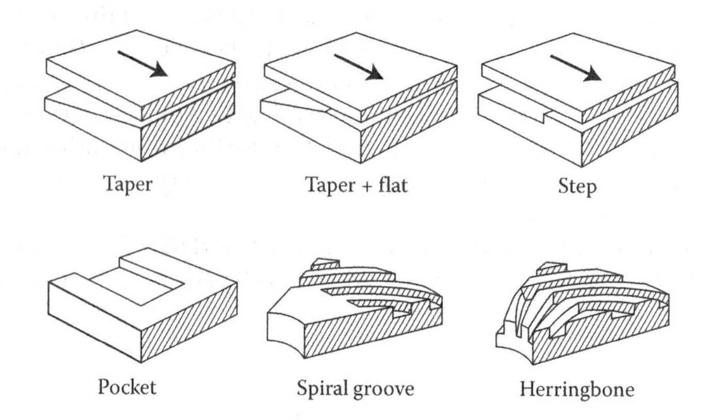

FIGURE 1.13
Pad configurations for fixed-profile hydrodynamic thrust bearings

rectangular like journal bearings, as Figure 1.12 makes obvious. Solving the RLE for an HTB pad was initially handled by approximating the actual cylindrical-sector pad as a rectangular pad, as shown in Figure 1.14. This is a simplifying assumption but one that is a good engineering approximation. Computer codes available today for general fluid-film bearing solutions have both rectangular (x, y, z) and cylindrical (r, θ, y) solver options. In the interest of space, derivation of the RLE in cylindrical coordinates is not presented here. However, using the same procedure presented here to obtain Equation 1.10, starting

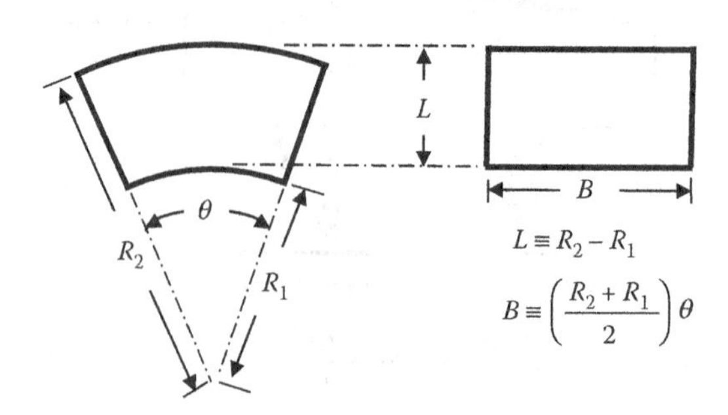

FIGURE 1.14
Equivalent HTB rectangular pad for cylindrical-sector pad.

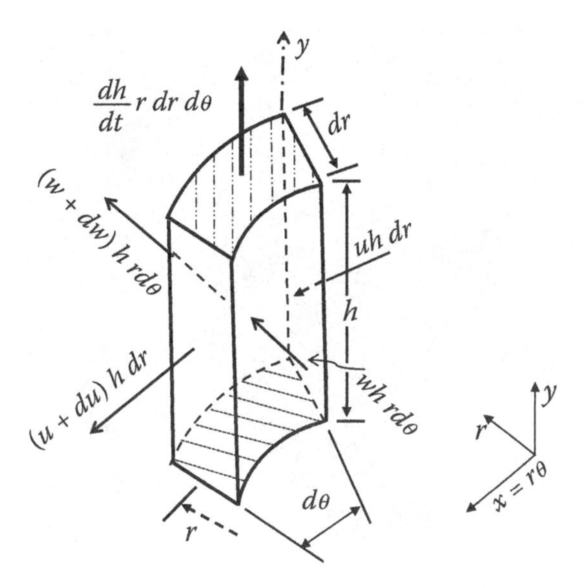

FIGURE 1.15
Fluid-film differential control volume for cylindrical coordinates.

from Equations 1.1 and 1.2 in cylindrical coordinates can of course be done. The cylindrical differential control volume equivalent to that in Figure 1.7 is shown in Figure1.15. Conveniently, the superb landmark book by Bird, Stewart, and Lightfoot (1960) has tables containing the Navier-Stokes, continuity, and other partial differential equations governing transport phenomena, all listed in *rectangular, cylindrical,* and *spherical* coordinates. The author is still as much impressed by that textbook as when he first discovered it as a graduate student. The RLE in cylindrical coordinates is already given by Hamrock (1994), Szeri (1998), and Khonsari and Booser (2001).

1.5 Hydrostatic and Hybrid Fluid-Film Bearings

Perhaps the best account of the evolution of important applications and technology for hydrostatic bearings is given by Professor Fuller (1984), with whom the author had the privilege of sharing an office at the Franklin Institute (1967–71) (see Acknowledgments). His historical account of hydrostatic bearings starts with a description of a rather odd exhibit at the famous Paris Industrial Exposition in 1878. The exhibit consisted of a very large mass supported on four flat-bottomed legs, each with a recessed pocket, resting on a planar flat horizontal plate. When high-pressure oil was pumped down into each of the four recessed leg pockets, the mass levitated on a virtually frictionless thin oil film, allowing the levitated mass to be moved horizontally by a slight hand push. Although considered an oddity at the time, it was probably the first demonstration of a hydrostatic bearing.

Fuller's historical account continues with the designer Professor Karelitz's 1938 account of the hydrostatic bearing support for the 200-inch Hale telescope on Mount Palomar, Figure 1.16. An illustration for one of the hydrostatic levitation bearing-pad sets for the Hale telescope is shown in Figure 1.17. Hydrostatic support pads were primarily used because

FIGURE 1.16
200-inch Hale telescope on Mount Palomar.

of the nearly frictionless sliding of the telescope on the hydrostatic oil films. The supported levitation force (weight) is approximately 1 million pounds (4.4 million N). The quite small force required to rotate the telescope showed that the slow-motion friction coefficient is less than 0.000004. Correspondingly, the power required to slowly rotate the Hale telescope is less than 1/10 hp, with extremely smooth slow movement of its quite heavy structure to provide precise tracking of celestial bodies.

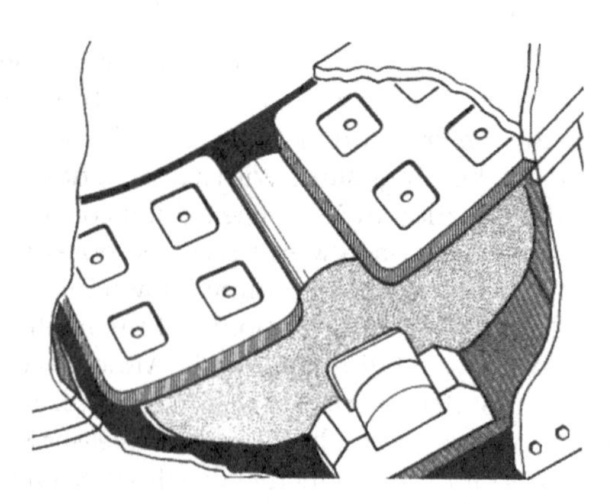

FIGURE 1.17
View of one pair of hydrostatic pads that levitate the Hale telescope.

Another highly visible modern application of hydrostatic bearings is to move major portions of sports stadium seating sections to conveniently accommodate both football and baseball. A prominent example is the original Denver Mile High Stadium 1977–78 upgrade, where the entire section of the grandstand weighing 4500 tons with 21,000 seats, was levitated for movement by water-lubricated hydrostatic rubber-bearing pads at 46 levitation points.

The most expansive manual for the design of hydrostatic bearings is given by another former Franklin Institute colleague of the author, Harry C. Rippel (1964). This design manual first appeared as a consecutive series of 10 articles in the Penton Publication *Machine Design* magazine, beginning with the August 1, 1963 issue. Topics covered by Rippel include (1) basic concepts and pad design, (2) controlling flow with restrictors, (3) influence of restrictors on performance, (4) bearing friction and film thickness, (5) bearing temperature and power, (6) practical flat-pad bearing design, (7) conical and spherical pads, (8) cylindrical-pad performance, (9) single and multiple-pad journal bearings, and (10) multi-recess journal bearings. In Part 2 of this book, the overall approach in Rippel's manual is demonstrated in the design examples presented.

To help explain to students in his undergraduate machine design course how hydrostatic bearings work, the author devised the simple DC-circuit-analogy explanation illustrated in Figure 1.18. Under levitation in the absence of any significant sliding motion of the supported load, the flow out of the bearing pad across the sills is purely Hagen–Poiseuille flow neglecting sill-entrance effects, and therefore has a flow resistance proportional to $1/h^3$ (see Equation 1.6). For example, if an increase in load W causes the film thickness to decrease by 50%, the resistance to flow over the sills is increased by a factor of $2^3 = 8$, accompanied by the corresponding increase in recess pressure to equilibrate the incremental-load increase. Therefore, by having a fixed resistance (flow restrictor) upstream of a variable resistance (film-thickness flow gap), astoundingly high fluid-film stiffness is achievable. In the analogous simple DC electrical circuit illustrated in Figure 1.18, as the downstream variable resistance is increased, the voltage (V_r) at the intermediate point between the two resistors correspondingly has to increase, because the total voltage drop across these two in-series resistances remains constant, just like the analogous-pressure drop ($p_s - p_a$) across the two hydraulic flow restrictions is constant. Of course, as the load is increased, the flow through the bearing pad correspondingly reduces, just like the current in the DC circuit correspondingly reduces when the variable resistance is increased. The reverse effect naturally occurs when the applied-bearing load W is incrementally reduced, yielding the inherent positive bearing-pad film stiffness. In application where extremely precise positioning is required, the very high stiffness potential is invaluable. Figure 1.19 schematically illustrates a hydrostatic-bearing pad with all its essential fluid supply system components.

The theory model for hydrostatic-bearing pad sills is the RLE, Equation 1.10. For a hydrostatic-bearing pad under levitation of a static load only and no sliding velocity ($U = 0$), uniform film thickness ($h =$ constant, so $\partial h/\partial x = \partial h/\partial z = 0$), and no vibration ($dh/dt = 0$), Equation 1.10 reduces to the following.

$$\frac{\partial^2 p}{\partial x^2} + \frac{\partial^2 p}{\partial z^2} = 0 \quad \text{or} \quad \nabla^2 p = 0 \tag{1.12}$$

This is the well-known Laplace equation, the same mathematical model for heat conduction and electromagnetic fields. With the addition of film-boundary velocity normal to the plane of the fluid film, the well-known Poisson equation is obtained as follows.

$$\nabla^2 p = \text{Constant} = 12\mu(dh/dt) = 12\mu V \tag{1.13}$$

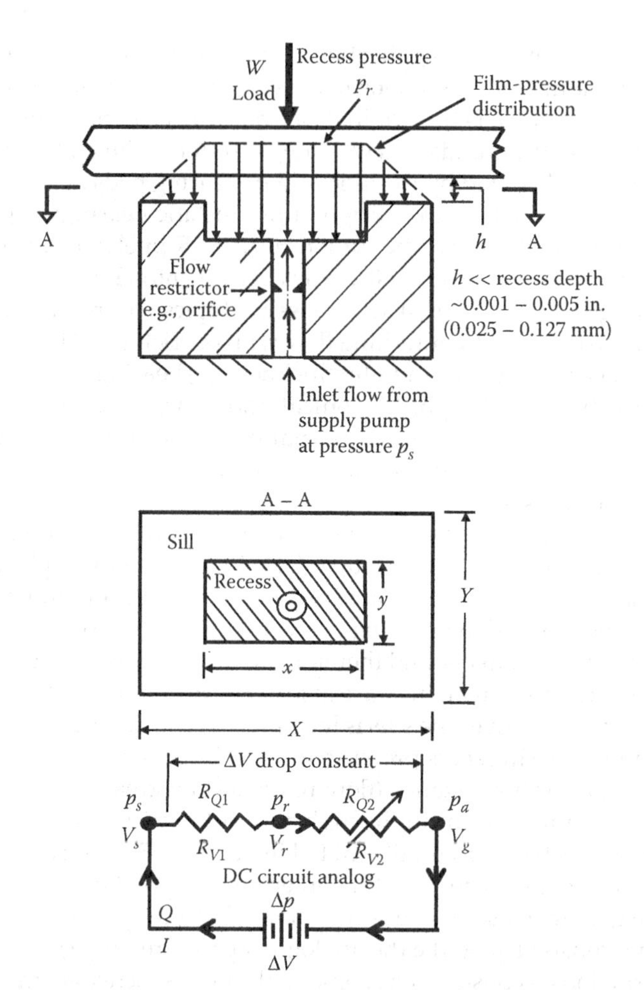

ΔV = Voltage rise produced by battery
$\Delta p = p_s$ pressure rise produced by supply pump
I = DC current
Q = Flow through hydrostatic-bearing pad, p_r = Recess pressure, p_a = Ambient pressure
R_{Q1} = Flow resistance of restrictor
R_{Q2} = Variable flow resistance of
V_g = Ground voltage

FIGURE 1.18
Functioning of a single-recess rectangular hydrostatic-bearing pad.

For a circular pad with a concentric recess such as utilized in the Figure 1.19 illustration, the RLE for static load with uniform film thickness is $\nabla^2 p(r) = 0$, thus yielding an ordinary differential equation because there is only one independent spatial variable, the radial coordinate r. That naturally yields a closed-form mathematical solution for the pad pressure distribution. However, for the more typical pad geometries, for example, the rectangular pad shape utilized in the Figure 1.18 illustration, the RLE retains two independent spatial variables, coordinates x and y, and thus does not have a mathematically closed-form solution. So, its solution requires a numerical algorithm like *finite difference*. Prior to the development by Castelli and Shapiro (1967) of their computationally efficient non-iterative finite difference numerical solution software for the RLE, film-bearing technologists used a non-numerical solution approach for hydrostatic bearings based on $\nabla^2 p(x, y) = 0$

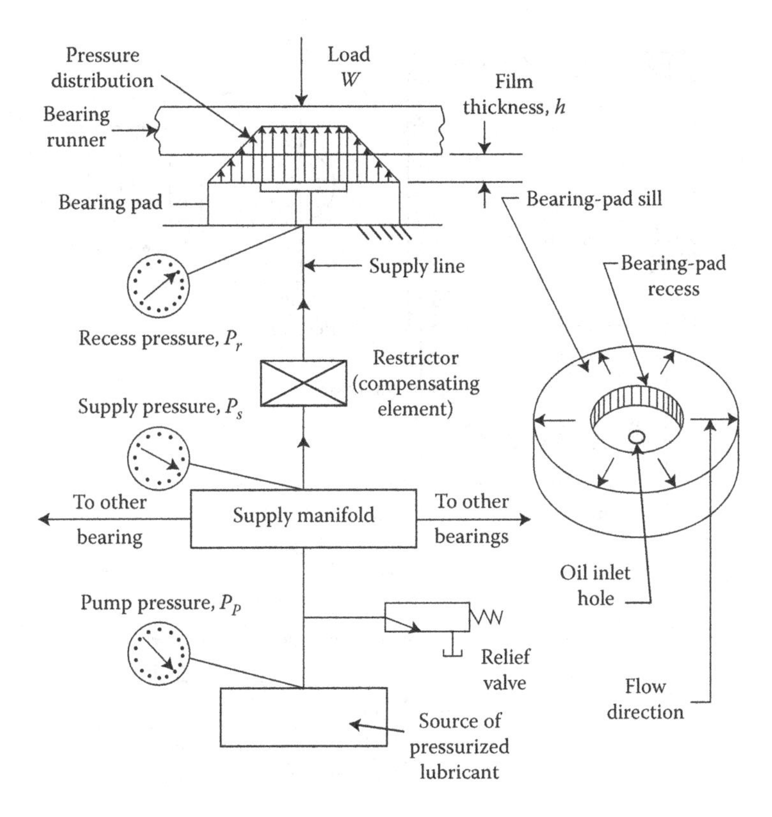

FIGURE 1.19
Typical hydrostatic-bearing system.

being the same exact PDE that models a 2D electric field. So, for a hydrostatic-bearing pad with uniform film thickness, a sheet of electrically conducting paper, with uniform and directionally isotropic properties, was used to manually determine lines of constant voltage, the analog of lines for constant pressure on hydrostatic-bearing-pad sills. Figure 1.20 illustrates the electrically conducting flat paper test apparatus used by Loeb (1958) to construct flat hydrostatic-pad design charts such as provided by Rippel (1964). The conducting paper was originally developed by Western Union, called Teledeltos paper, and has nearly isotropic electrical resistance in all in-planar directions.

Quoting Fuller (1984), "The boundaries and pressure sources of the bearing model being investigated are marked with aluminum or silver paint. An electric potential voltage difference, analogous to a hydraulic-pressure difference, is applied to these aluminum-painted boundaries and the voltage at any intermediate point is determined by means of a suitable potentiometer arrangement (as illustrated here in Figure 1.20). Once the voltage (pressure) distribution is known, the load-carrying capacity may be determined by numerical integration" (i.e., Simpson's rule for 2D array).

Hydrostatic bearings are quite useful and versatile devices in machinery test rigs. The author has designed hydrostatic bearings in many rotating machinery test rigs (Adams 2017b). The most elaborate example is the test-bearing load applicator illustrated in Figure 1.21. It was devised by the author to simulate a constant direction vertical load on a journal-bearing test rig developed to study journal bearings under very slow rotational speed operation. That is, slow speeds (3–5 rpm) as imposed on large power plant steam

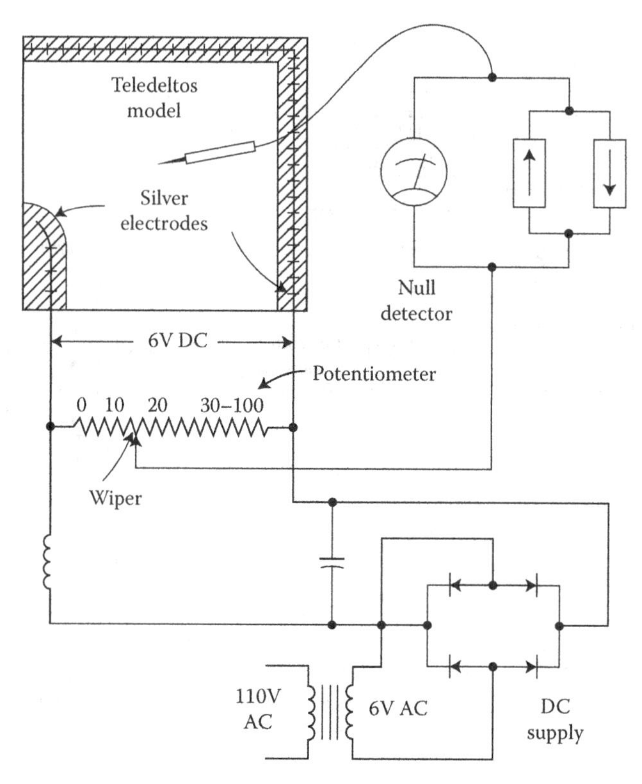

FIGURE 1.20
Electric analog to a hydrostatic-bearing pad.

turbines after a turbine coast down. This is to prevent difficult-to-undue major rotor thermal bowing that would otherwise occur as the rotor cools down to a temperature sufficiently below operating temperature. The hydrostatic floating piece (Figure 1.21a) is circular with a concentric circular recess on both the flat bottom pad and the top spherical-seat pad. The flat bottom pad frees the test bearing to move horizontally as needed to prevent any horizontal side loading on the test bearing. The spherical-seat top pad allows the test bearing to self-align with the test journal. Precision of the spherical portion of the floating piece was achieved by finish lapping together the two mating spherical surfaces, yielding near perfectly conforming mating surfaces. As illustrated, the loading piston also employs hydrostatic-bearing pockets to keep it centered without friction. The loading piston diameter was made the same as that of the concentric recesses on both floating piece hydrostatic pads. This was done for ease of controlling test rig operation by ensuring that the loading piston could not exceed the floating-piece load capacity with both conveniently pressure fed from the same source.

Figure 1.21b illustrates how the time-varying test-bearing friction force between test bearing and test journal was prevented from contributing any extraneous time-varying radial load to the test bearing, since the two torque-measuring restraint forces were equal in magnitude but opposite in direction. Test-bearing frictional torque was accurately calibrated to the pressure in the sealed Bellofram hydraulic line shown. The third sealed top hydraulic Bellofram cylinder was used to adjust the static angular position of the test bearing by a threaded adjustment on the third hydraulic cylinder piston position, as shown.

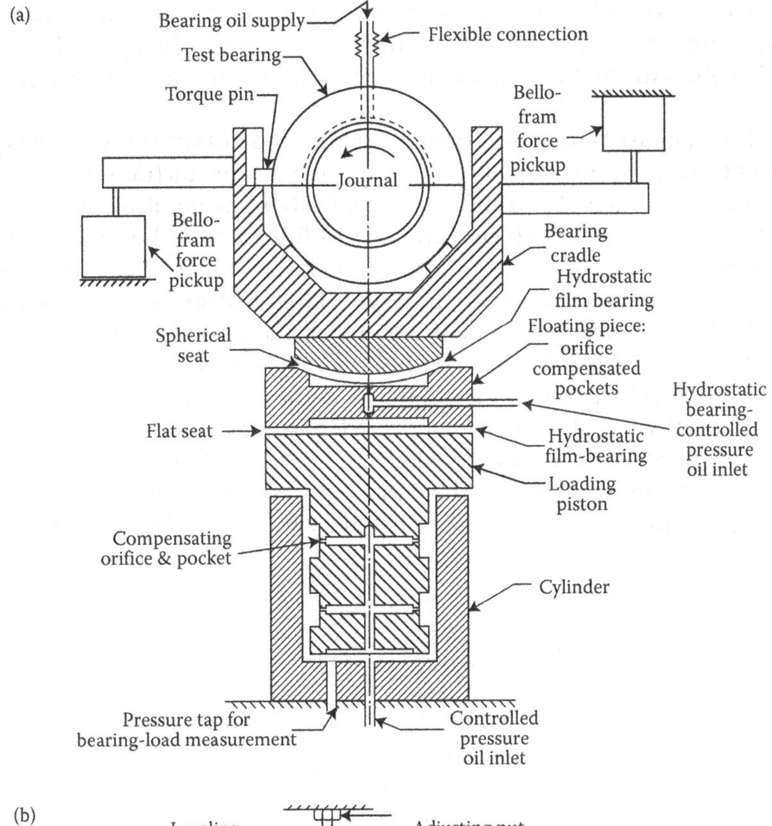

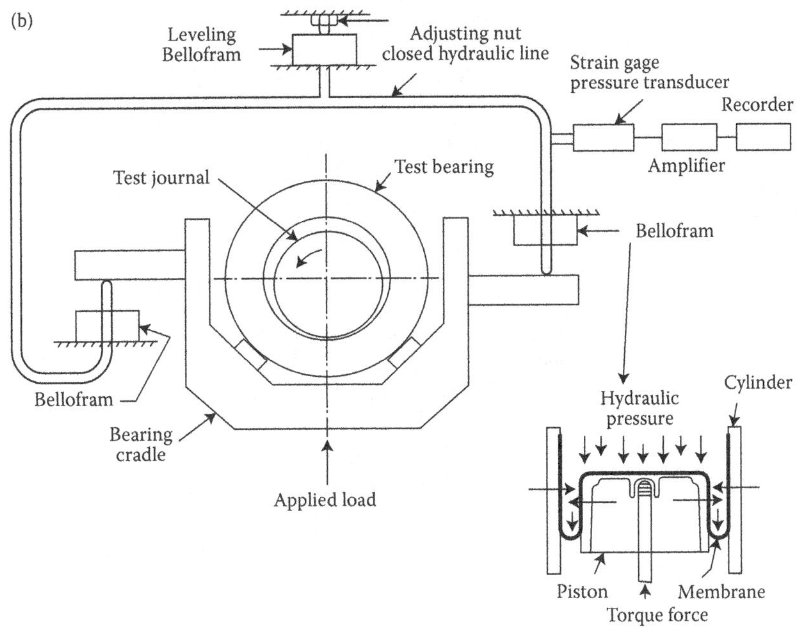

FIGURE 1.21
(a) Vertical test-bearing force applicator train employing hydrostatic-bearing pads, (b) torque measuring system utilizing opposed Bellofram sealed pistons in closed hydraulic circuit. (From Adams, M. L., *Rotating Machinery Research and Development Test Rigs*, CRC Press/Taylor & Francis, Boca Raton, FL, 2017b.)

Applications for a hydrostatic bearing can entail a significant sliding velocity like in a purely hydrodynamic bearing. That is, both hydrostatic and hydrodynamic effects occur simultaneously within the thin lubricating fluid film. Such bearings are referred to as *hybrid bearings*.

A long-time frequent use is in applications where the hydrostatic component is primarily to lift the rotor off metal-to-metal contact with the bearing for startup from zero rpm, and subsequently for coast down at shut-off. Such applications were utilized even before the computer software yet existed to fully solve the RLE for hybrid bearings. For example, many large electric power plant steam turbine-generator units in the United States have long employed small hydrostatic pads within their large hydrodynamic journal bearings to better handle *slow-roll turning-gear* operation when the unit is brought down, keeping the rotor slowly rotating (typically 3–5 rpm) to prevent rotor thermal bowing as it cools down. Another longstanding similar application is in the tilting-pad hydrodynamic-thrust bearing of the typical vertical-rotor primary *reactor coolant pump* for pressurized-water-reactor (PWR) commercial nuclear power plants. Figure 1.22 shows such a pump (Adams 2017a). With a vertical rotor centerline, the entire weight of the rotor and any downward or upward-pump hydraulic-axial loads are carried by the single double-acting tilting-pad

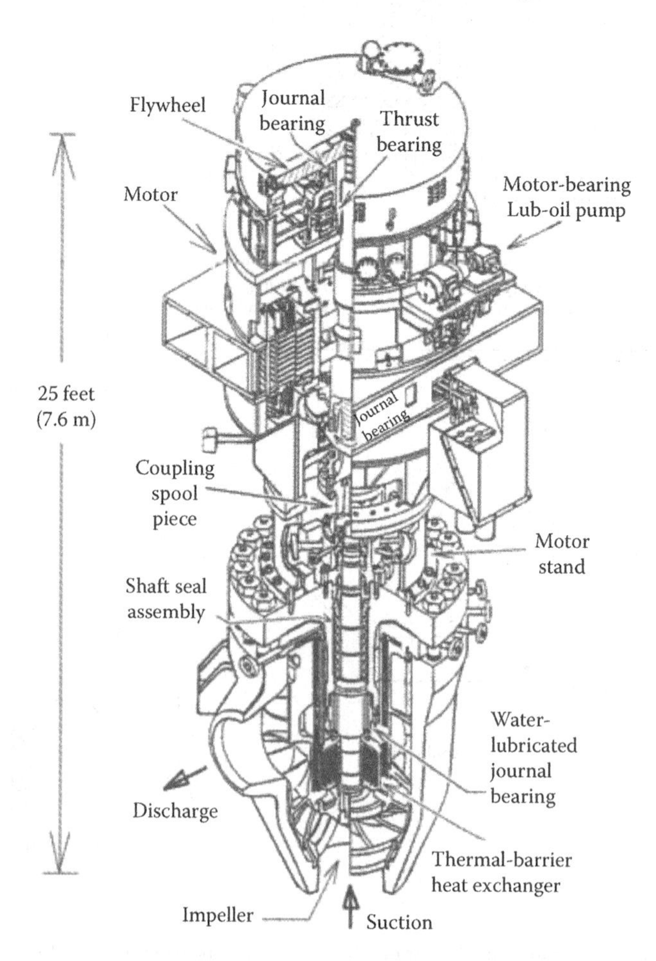

FIGURE 1.22
Pressurized water reactor primary reactor coolant pump, Adams (2017a).

thrust bearing. A significant portion of the rotor weight is the quite large flywheel at the top of the rotor. Its purpose is its yield of an *extended coast-down time* in the event of an emergency pump motor power loss. That is, to keep the reactor sufficiently cooled during the brief time period required to fully insert the reactor control rods that stop the heat produced by the nuclear reaction.

1.6 Gas Bearings

In developing the original RLE, one of the simplifying assumptions made by Reynolds and repeated in Section 1.1 in the derivation of Equation 1.10 is that the lubricant fluid is incompressible. A twentieth-century direct offshoot of the original RLE is its generalization to compressible fluids. In that case, the conservation-of-mass equation for an incompressible fluid (Equation 1.2) must be generalized as follows, where ρ is the fluid mass density.

Continuity Equation for a Compressible Fluid

$$\frac{\partial \rho}{\partial t}+\frac{\partial(\rho u)}{\partial x}+\frac{\partial(\rho v)}{\partial y}+\frac{\partial(\rho w)}{\partial z}=0 \tag{1.14}$$

When Equation 1.14 is employed instead of Equation 1.2 in deriving the RLE, the following general RLE for a compressible fluid is obtained.

Reynolds Lubrication Equation for a Compressible Fluid

$$\frac{\partial}{\partial x}\left[\frac{\rho h^3}{\mu}\left(\frac{\partial p}{\partial x}\right)\right]+\frac{\partial}{\partial z}\left[\frac{\rho h^3}{\mu}\left(\frac{\partial p}{\partial z}\right)\right]=6U\frac{\partial(\rho h)}{\partial x}+12\frac{\partial(\rho h)}{\partial t} \tag{1.15}$$

In addition to the assumptions invoked in deriving Equation 1.10 for an incompressible fluid, the following *two additional assumptions* are also needed to be invoked in order for Equation 1.15 to be valid for characterizing gas-film bearings.

1. The *ideal gas* law $p=\rho\mathcal{R}\,T$ holds for the gas lubricant fluid.
2. The gas-bearing film *Knudsen number* K_N, (ratio of the *mean free path* λ between molecular collisions to the representative physical length L of a system) is sufficiently small to justify treating a gas as a *continuum*. $K_N<0.01$ is judged necessary (Khonsari and Booser 2001).

Substitution of $p=\rho\mathcal{R}\,T$ into Equation 1.15 for ρ yields Equation 1.16.

Reynolds Lubrication Equation for a Compressible Fluid

$$\frac{\partial}{\partial x}\left[\frac{ph^3}{\mu}\left(\frac{\partial p}{\partial x}\right)\right]+\frac{\partial}{\partial z}\left[\frac{ph^3}{\mu}\left(\frac{\partial p}{\partial z}\right)\right]=6U\frac{\partial(ph)}{\partial x}+12\frac{\partial(ph)}{\partial t} \tag{1.16}$$

Note that in contrast to Equation 1.15, Equation 1.16 is nonlinear in the dependent variable p. Therefore, a 2D finite-difference numerical solution method must be combined with a time-marching computation to track in time-transient gas-film pressure solutions. Simplification of Equation 1.16 for computing time-invariant steady-state solutions is facilitated by dropping the $\partial(ph)/\partial t$ term. For rotor-vibration simulations involving self-acting gas bearings, the complete Equation 1.16 is employed in a time-marching algorithm, whether the long-term dynamic solution remains transient or converges to a periodic steady-state rotor vibration. Equation 1.16 can be made non-dimensional as follows (Gross 1980).

$$\frac{\partial}{\partial X}\left(H^3 P\frac{\partial P}{\partial X}\right)+\frac{\partial}{\partial Z}\left(H^3 P\frac{\partial P}{\partial Z}\right)=\Lambda\frac{\partial}{\partial X}(PH)+\sigma\frac{\partial}{\partial T}(PH) \tag{1.17}$$

$X = x/L_{ref}$; $Z = z/L_{ref}$; $H = h/h_{ref}$; $P = p/p_a$; $T = \omega t$
$\Lambda = ((6\mu_a UB)/p_a h_{\min}^2)$ for a slider bearing, $\Lambda = ((6\mu_a\omega)/p_a)(R/C)^2$ for a journal bearing
P_a = ambient pressure, μ_a = ambient viscosity
$\sigma = ((12\mu_a B)/(P_a[h(@t=0)]^3)(\partial h/\partial t) = ((12\mu_a\omega B^2)/P_a[h(@t=0)]^2)$ h = reference film thickness at $t = 0$

Regarding the postulated requirement (Gross 1980) that the gas-bearing film *Knudsen number* $K_N < 0.01$ is needed for adherence to the no-slip boundary condition, and that a gas-bearing film thickness is of quite small length, one is therefore justified to inquisitively ask how the film thickness compares to the *mean free path* λ between molecular collisions. If $0.01 < (K_N = \lambda/L) < 15$, the *no-slip boundary assumption* begins to break down. That would be manifest by a reduction in hydrodynamic pressure produced similar to what a viscosity reduction would cause. That is, boundary slip would correspondingly reduce shear stress gradients across the film. And if $K_N > 15$, molecular impact motion becomes fully developed and is distinctly different from continuum fluid flow mechanics (Gross 1980). This becomes an issue especially in high-speed hard-disk read/write magnetic recording devices, requiring analysis modeling of the air film based on the kinetic theory of gasses, not the RLE (Bhushan 1999). Fukui and Kaneko (1995) provide a detailed presentation of equations for large K_N lubrication flow, referred to as *molecular gas film lubrication* (MGL). A heuristic modified version of the RLE to account for boundary slip was proposed by Burgdorfer (1959) early in the development of gas bearings.

The nondimensional parameter Λ, Equation 1.17, is a relative indicator of the degree to which the gas film compressibility matters. That is, the smaller the value of Λ, the more the hydrodynamic-bearing film reacts as an incompressible fluid and vice versa. Figure 1.23 illustrates a comparison of hydrodynamic-pressure-distribution shapes between low Λ and high Λ cases. As the comparison between the shown two normalized pressure distributions implies, the larger Λ the closer the pressure-distribution peak approaches the pad trailing edge minimum film thickness h_2 point. This is the progressively increasing effect on film shear stress gradients resulting from the gas density increase corresponding to the pressure increase. As consistent with the right-hand side of Equations 1.15 and 1.17, the steady-state gas-bearing pressure distribution is not instantaneously achieved as with the solution for film pressure with an incompressible fluid. There is a physically natural time-transient characteristic of pressure distribution convergence to reach the steady-state result. Correspondingly, when numerically solving for the steady-state pressure distribution, an initial-condition pressure distribution is specified, for example, the small

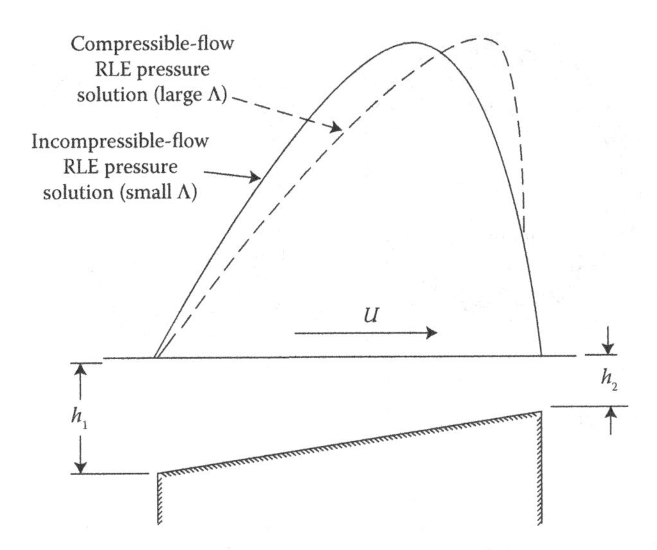

FIGURE 1.23
Slider-bearing normalized pressure distributions.

Λ incompressible solution. A forward marching time-step integration then converges faster to the steady-state solution.

1.7 Compliant Surface Foil Gas Bearings

Gas film bearings of both hydrodynamic and hydrostatic functioning were already being investigated and used in a few novel applications over 50 years ago. However, use of those bearings never achieved wide industrial use, primarily because of quite low-load capacity at modest rotational speeds and rotor dynamical instability problems at speeds sufficiently high to provide useable static-load capacities. Hydrostatic-gas bearings utilizing porous media-bearing sleeves were also shown to be feasible in laboratory testing and analysis. The foil gas-bearing concept achieved success in the predigital age high-speed tape-deck heads by manufacturers such as Ampex. The main modern application of the hydrodynamic-air bearing, initially on mainframe computer high-speed flying-head disk readers, has found its present place in PC hard drives. Within the last 10 years, a major Cleveland-based manufacturer of MRI medical scanners successfully developed and employed hydrostatic air bearings to support the main rotational positioning barrel, improving the position resolution in this scanning product.

About 30 years ago, the gas foil-bearing concept evolved into a new family of configurations, the *compliance surface foil gas bearing* (Heshmat et al. 1982). Figure 1.24 illustrates two typical compliance surface foil gas journal bearings. Similar concepts for axial-load thrust bearings have also been developed.

The main selling feature touted for the modern compliant surface foil gas bearing has been that it is *oil-free*, which can be a definite advantage for a number of applications, not just aerospace applications. Though magnetic bearings (Chapter 3) were touted by the academically inclined developers primarily for their controllable characteristics like rotor

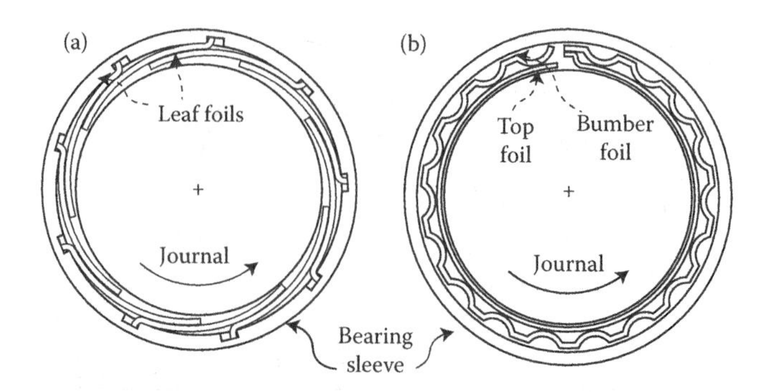

FIGURE 1.24
Two types of compliant surface foil gas journal bearings, (a) leaf-type foil bearing and (b) bumper-type foil bearing.

dynamical stiffness and damping, it is actually the *oil-free* nature of magnetic bearings that resulted in their use in a few heavy industry applications, like large-distance pipeline compressors. However, compliant surface *foil bearings* have the *added benefit* over *magnetic bearings* of use at quite elevated temperatures as high as 1100°F (593°C).

Most of the development work has been focused on the bumper-type foil bearing illustrated in Figure 1.24. These bearings are assembled with a modest amount of elastic preload, which means that at start-up the journal is in rubbing contact with the inner surface of the *top foil*. Special coatings on the interior surface of top foils are thus required to counter the potential for significant wear accumulation from many starts and stops. As the journal accelerates up to operating speed, the hydrodynamic gas film overcomes the initial preload and thereby separates the journal and top foil. As the bearing-radial load comes into play, the top foil and bumper foil elastically deform under the load to spread the load-carrying hydrodynamic separating pressure distribution more uniformly over the load-carrying area. This yields a bearing-load capacity superior to the original hydrodynamic gas bearings of rigid construction.

An interesting dichotomy between compliant surface *foil bearings* and *magnetic bearings* is the following. Foil bearings are relatively simple in their configuration, with operating properties resulting from the fundamental ambient-gas hydrodynamics of the gas film interacting with the elastically deformable foils. That is, the foil bearings *don't have to be "smart"*, they only have to *let nature take its course.*

In stark contrast, operation of the magnetic bearing is anything but simple, involving position sensors, with A-to-D, microprocessor and D-to-A power amplifiers (Adams 2010); see Figure 3.1. But the magnetic bearing has quite predictable performance, for example, *load capacity, dynamic stiffness,* and *damping* coefficients. Whereas the much simpler configured foil bearing surely presents considerable challenges in predicting its operating performance characteristics, especially rotor dynamical stiffness and damping properties. The elastic deformation of the foils presents a significantly hardening nonlinearity with load, and it is the foil hardening nonlinear deformations that dominate both the static and dynamic-bearing properties. The beneficial damping inherent in the dynamic friction rubbing between the top and bumper foils is also a nonlinear mechanism. In consequence, use of foil bearings in a specific application requires significantly more development testing than other alternatives require. The main applications thus far for these foil bearings include turbochargers and micro-turbine engines for land-based electric power generation. The (1) compact oil-free nature and (2) high-temperature capability of the compliant surface foil gas bearing is significant.

One can appreciate the unpredictability of rotor dynamical stiffness and damping coefficients from the work of Howard et al. (2001). Their large scatter of results to fit linear dynamical models leads one to the conclusion that perhaps foil bearings inherently can't be sufficiently well modeled for rotor-vibration predictions using small-perturbation linear-vibration models. This assertion is supported by the work of San Andres and Kim (2007). Their nonlinear rotor-vibration simulations compare amazingly close with their laboratory tests on a small precision 2-bearing high-speed rotor. Their results clearly demonstrate that the rotor-vibration characteristics are dominated by the structural nonlinearities of the foils, showing phenomena that could not be predicted by any linear-vibration predictive-simulation model. For example, they demonstrate that gas foil bearing-supported rotors are prone to subsynchronous whirl orbits, albeit at tolerable vibration levels. Specifically, they show subsynchronous orbit frequencies that track subharmonics of the rotor speed (i.e., 1/2, 1/3, etc. of rotor spin frequency), but more often locking into a system subsynchronous natural frequency. These subsynchronous orbital motion components may persist over a range of operating speeds, with disappearance and then subsequently reappearance with further operating changes. Their results also show that adding more rotor unbalance can trigger and worsen the severity of the subsynchronous orbital vibration components. This makes fundamental sense since the addition of rotor unbalance increases the overall vibration level and thereby increases the degree of dynamic nonlinearity, which in turn increases the propensity for the synchronous forcing function to drive energy into harmonics of itself (Adams 2010). Their results confirm that the subsynchronous-vibration components are a consequence of the structural hardening nonlinearity akin to a Duffing resonator, not a hydrodynamic rotor dynamic instability energized by the gas film.

For a conventional cylindrical journal bearing with a machined journal outside diameter (OD) and machined-bearing bore diameter (ID), the radial clearance of course has a precise definition. Conversely, the equivalent clearance of a compliant surface foil journal bearing is not as precisely definable because the bearing surface is compliant. If one attempts to progressively increase bearing load to determine at what load magnitude the compliant surface bottoms out on the bearing sleeve, the bearing is likely to fail from over-heating before that bottom-out condition is reached. Of course, for what it's worth, that test can be done with the journal not rotating.

The somewhat nebulous nature of compliant surface foil journal-bearing clearance alarms the author concerning the other non-bearing small radial rotor-to-stator radial clearances in an application, for example, radial-seal clearances, turbomachinery blade-tip clearances, and balance drum clearance. An important role of any radial bearing is to prevent these other small radial-clearance components from becoming inadvertent bearings during operation.

1.8 Elastrohydrodynamic Lubrication

Strictly speaking, this section is not in keeping with the title of this chapter, but it is a fluid-film-type of lubrication. This section relates to lubrication of primary importance to rolling element bearings. It is the author's choice to place this summary of elastrohydrodynamic lubrication (EHL, also abbreviated earlier as EHD) in this chapter since its treatment draws on the Reynolds lubrication equation. But treating the EHL phenomenon is not as straightforward as treating the relatively thicker film hydrodynamic lubrication with the assumptions of incompressible constant-viscosity Newtonian fluids between rigid solid surfaces, as derived in Section 1.2. It is more complex. A comprehensive treatment of EHL is Hamrock (1994), which covers the 10 1974–1981 research papers coauthored by Hamrock and his PhD thesis advisor, Professor Duncan Dowson of Great Britain.

A more recent thesis by Goodyear (2001) demonstrates how more recent computational methods have considerably facilitated computational predictions for EHL. The author's mission for this section is not to provide all the analysis intricacies of the developments by Hamrock, Dawson, and others, but to provide an insightful summary of EHL. The primary points are all illustrated in Figure 1.25, devised by the author to explain EHL in his undergraduate machine design class, employing the most important EHL application example, the *ball bearing*. Figure 1.25 illustrates a rolling ball at its maximum-load point, that is, lined up with the bearing load W, interacting with its outer raceway. The same type of contact stress also occurs simultaneously on the inner raceway.

The contact footprint between the ball and raceways is categorized as a Hertzian contact. The ball-on-raceway loaded-contact footprint is much smaller than the rolling element since the ball radius-of-curvature is significantly smaller than both the inner and outer

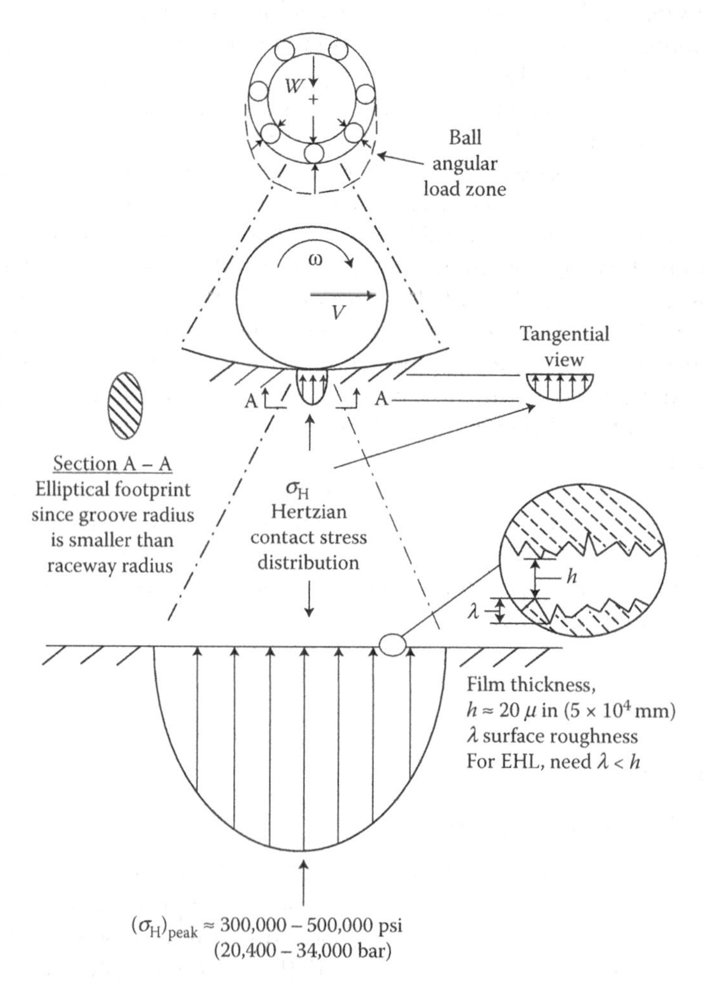

FIGURE 1.25
Ball bearing under radial load, zooming-in on the EHL phenomenon.

raceway diameters. The very high maximum contact stress has a large effect on increasing lubricating oil viscosity within the contact zone. Szeri (1980) provides a historical account of how the influence of pressure on oil viscosity was first correctly characterized by Grubin (1949) after over 30 years of erroneous efforts by a number of other investigators. Equation 1.18 is the formula typically used for the pressure effect on lubricating oil viscosity, based on a semi-log straight-line curve fit approximation to viscosity measurements made under controlled hydrostatic pressure. The pressure effect on viscosity is reversible, existing momentarily while the oil is under the elevated pressure.

$$\mu = \mu_0 e^{\alpha p} \tag{1.18}$$

μ = viscosity at pressure p
μ_0 = viscosity at atmospheric pressure
α = pressure-viscosity coefficient for specific oil

The very thin EHL oil films at lubricated rolling element contacts combined with the large viscosity increase from the high contact stress pressure, momentarily traps the oil within the contact footprint. Thus, if the rolling elements and raceways are manufactured with extremely smooth surface finishes, hard metal-to-metal direct contact is prevented by the entrapped EHL oil film. Of course, any accrued surface finish degradation in service (e.g., corrosion, dirt ingestion, subsurface initiated fatigue) will degrade the EHL effectiveness.

1.9 Squeeze-Film Dampers

Vibration-damping capacity of a rolling element bearing (REB) is extremely small and therefore to measure it is virtually impossible since any test rig for this purpose would have its own damping that would mask that of a tested REB. As is well known and shown by Figure 1.26 (Adams 2010), the benefit of damping is in preventing excessively high-vibration amplitudes at resonance conditions. Thus, for the many machines running on REBs that have the maximum running speed well below the lowest critical speed, the absence of any significant REB damping presents no problem.

Since an REB can usually operate for sustained periods of time after the normal lubrication supply fails, REBs are usually a safer choice over fluid-film bearings in aerospace applications such as modern aircraft gas turbine jet engines. In such applications, however, the inability of the REBs to provide adequate vibration-damping capacity to maintain tolerable unbalance vibration levels through critical speeds frequently necessitates the use of SFD. Typically, an SFD is formed by a cylindrical annular oil film within a small radial clearance between the OD cylindrical surface of an REB's outer raceway and the precision cylindrical bore into which it is fitted in a machine. The radial clearance of the SFD is similar to that for a comparable diameter journal bearing, possibly a bit larger as optimized for a specific application. Figure 1.27 illustrates a configuration that employs *centering springs*.

An SFD is like a journal bearing without journal rotation. Referring to Equation 1.10, the "sliding velocity term" in the RLE is then zero, leaving only the *squeeze-film term* to generate hydrodynamic pressure within the small annular clearance. The factors of *film rupture* and *dissolution of air* in the SFD oil film produce considerably more complication

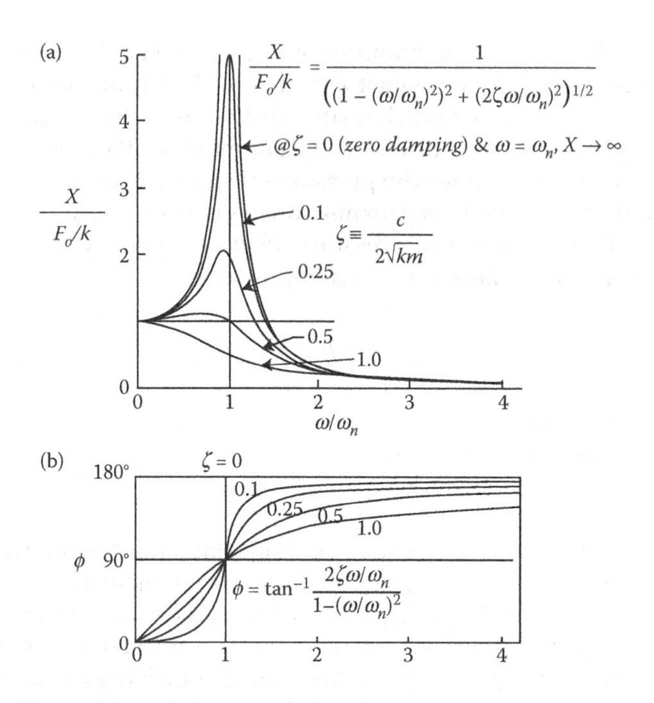

FIGURE 1.26
One-degree-of-freedom steady-state response to a sinusoidal force. (a) $X/(F_0/k)$ vs ω/ω_n (b) ϕ vs ω/ω_n, $\varsigma = 1$ @ critically damped.

and uncertainty in computational predictions of damping coefficients than these factors do in journal bearings as discussed by Adams (2010). Also, the neglect of fluid inertia effects implicit in the RLE is not as good an assumption for SFDs as it is for journal bearings.

The SFD configuration shown in Figure 1.27 employs *centering springs* since there is no active *sliding velocity term* to generate static-load-carrying capacity in the hydrodynamic oil film. To create a static equilibrium position about which the vibration occurs and is damped by the SFD, *centering springs* are used to negate the bearing static load and maintain damper approximate concentricity. The radial stiffness of the *centering springs* is far less than the radial stiffness of the REB. Thus, the radial stiffness is essentially the isotropic-radial stiffness of the *centering springs*.

Eliminating the centering springs makes the SFD more compact. And obviously the possibility of centering spring fatigue failure does not need to be addressed if there are no

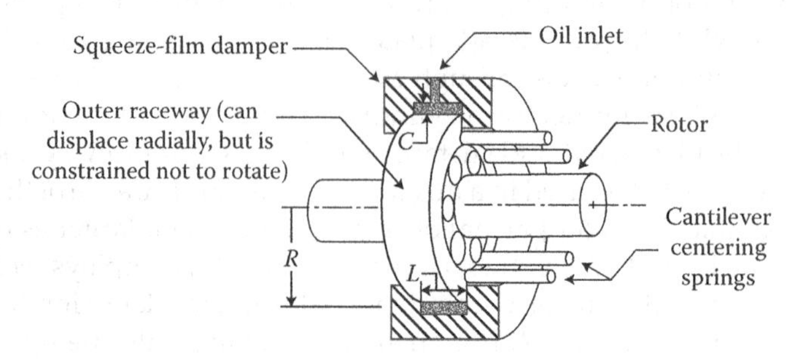

FIGURE 1.27
Squeeze-film damper with centering springs.

centering springs. However, from a rotor-vibration point-of-view, eliminating the centering springs makes the system considerably less simple to analyze. The damper now tends to "sit" at the bottom of the clearance gap and it requires some vibration to "lift" it off the bottom. That is a very nonlinear dynamics problem as experimentally studied, for example, Adams (2017b).

Some modern aircraft engines are fitted with springless SFDs while some have spring-centered SFDs. Under NASA sponsorship, Adams et al. (1982) devised methods and software to retrofit algorithms for both types of dampers into the general purpose nonlinear time-transient rotor response computer codes used by the two major U.S. aircraft engine manufactures. Adams et al. (1982) show a family of nonlinear rotor-vibration orbits that develop in springless SFDs as a rotating unbalance force magnitude is progressively increased. With a static decentering force effect (e.g., rotor weight) and small unbalance magnitudes, the orbit barely lifts off the "bottom" of the SFD, forming a small orbital trajectory that has been likened to a *crescent moon.* If vibration magnitude increases, the SFD overcomes the static decentering force with the orbit progressing from the small crescent moon trajectory to a distorted ellipse to a nearly concentric circular orbit as the vibration overpowers the static decentering force effect.

1.10 Lubrication Supply

There are various arrangements employed in supplying lubricant to slider bearings. Hydrodynamic fluid-film bearings function best on a forced-fed continuous supply of lubricant flow, for example, with an oil lube pump supplying axial feed grooves for journal bearings and radial-feed groove for thrust bearings, with feed grooves located near the beginning of the converging section of the film-thickness distribution. Other feed groove configurations are also employed. Figure 1.28 illustrates various journal-bearing feed groove arrangements. Regarding the two pressure distributions illustrated in Figure 1.28d, the discussion in Section 7.1.3 concerning Figure 7.18 explains how improper angular placement of axial oil-feed groves relative to the radial-load direction can seriously reduce the operating minimum film thickness.

Pressurizing the feed grooves above bearing ambient drain pressure is typically accomplished with an oil pump. Ideally, the feed pump supplies a flow rate *equal to* (or *larger than*) what the hydrodynamic bearing requires to satisfy the fluid-film inlet-flow boundary condition (called *classical supply flow*). Figure 1.11 illustrates film-flow velocity profiles and hydrodynamic-pressure formation with classical supply flow.

A hydrodynamic bearing is in fact a viscous pump that must be continuously primed by the lubricant supply source, for example, lube pump. However, if the flow rate supplied is less than the *classical supply flow*, the bearing runs *partially starved* and then the hydrodynamic fluid-film pressure distribution does not start at the beginning of the converging section of the fluid-film-thickness distribution, but starts further downstream where the amount of continuously supplied lubricant just begins to fill up the gap between the bearing and slider surface. If oil supply is totally interrupted during operation, hydrodynamic bearings may become *totally starved* of lubricant and are then likely to incur some surface damage, quite serious damage unless the lubrication supply interruption is immediately detected (e.g., monitored bearing surface temperature; see Figure 5.21.) and the machine is quickly shut down.

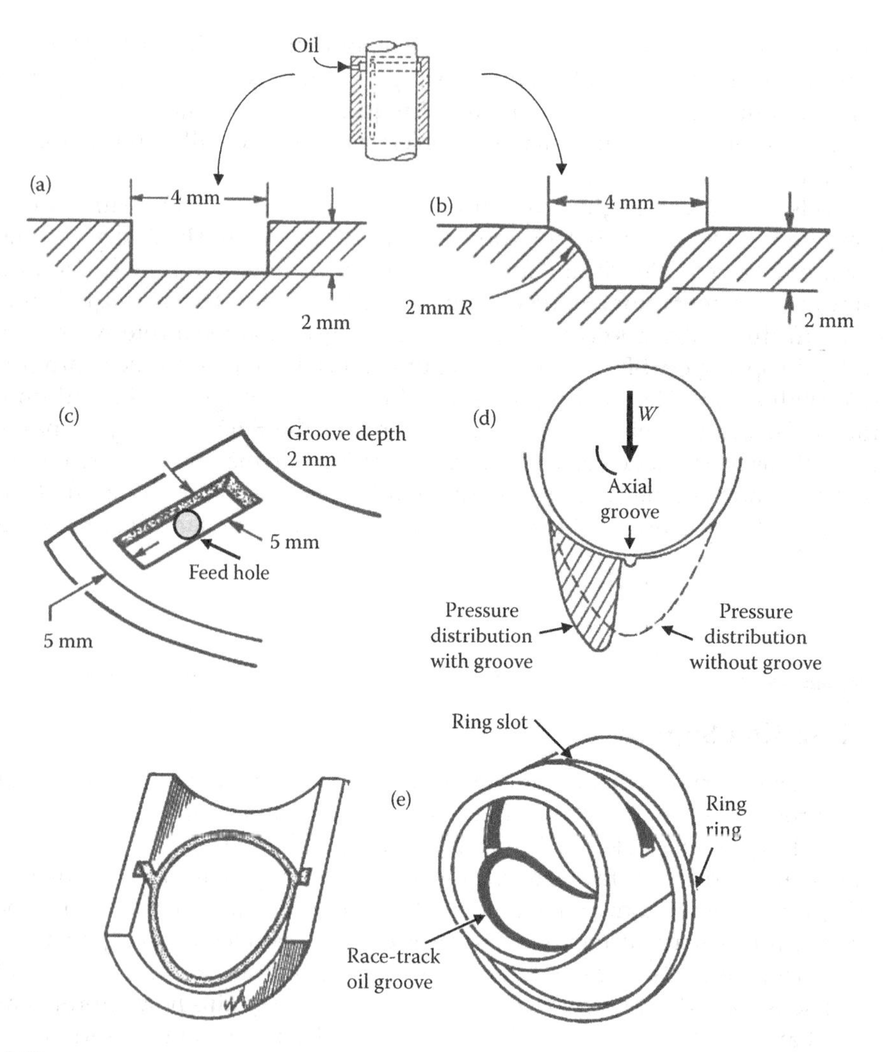

FIGURE 1.28
Typical journal-bearing oil feed groove configurations; (a) and (b) auto engine con rod-bearing circumferential grooves, (c) and (d) general industrial machinery journal-bearing grooves, (e) for oil-ring-lubricated journal bearings with load direction varying widely in operation from vertical to horizontal.

Hydrostatic bearings also only function with a continuous supply of lubricant, albeit at significantly higher supply pressures than required for hydrodynamic bearings. Other more marginal lubricant supply methods suffice in lower-duty applications, including (1) oil-ring lubricant supply, (2) oil-saturated bearing wicks, (3) oil-impregnated bearings, and (4) porous bearings.

Supply of lubricant to a hydrodynamic journal bearing with a *journal riding oil ring*, illustrated in Figures 1.28(e) and 1.29, eliminates the need for an oil pump. The oil ring rotates by riding freely on the bearing journal. The oil ring thereby picks up oil from a trapped pool of oil in the bottom of the journal-bearing housing and continuously deposits some of that oil onto the axial center of the journal and feed groove there, possibly with the aid of an oil scraper. Oil rings are used in rotating machinery of a few hp up to thousands of hp, limited only by the factor of maximum journal surface speed of 3000 ft/min (900 meters/min).

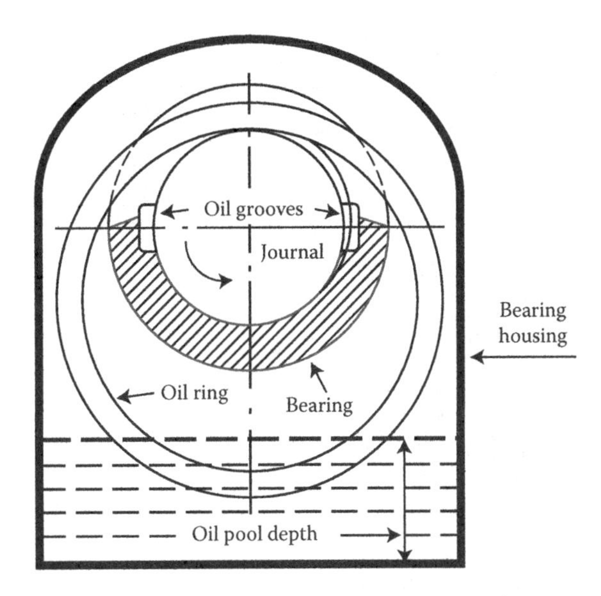

FIGURE 1.29
Journal-bearing oil ring lubrication.

Fuller (1984) describes in depth several wick and felt-pad-lubricating devices, and their performance data, that supply oil to bearings at low rates adequate for many sliding bearing applications. The bearings so lubricated are typically not operating in a complete hydrodynamic regime, instead operating in *mixed* and/or *boundary lubrication* regimes (see Figure 1.1a). Figure 1.30 shows a collection of these low-flow lubricating devices.

1.11 Hydrodynamic Stability and Turbulence in Fluid Films

The RLE, Equation 1.10, is derived with the assumption of laminar flow. This is a valid assumption for many hydrodynamic fluid-film bearings, and covered virtually all pre-WW II applications. However, the evolutional increasing of size and rotational speeds and the use of low-viscosity liquids as the lubricant in many modern machines have resulted in many hydrodynamic-bearing films running in the turbulence regime.

Hydrodynamic bearings can be evaluated for the occurrence and degree of turbulence through an assessment based on a proper bearing dimensionless Reynolds number that is the quotient of *inertia forces/viscous forces*, similar to pipe flow. In the various types of fluid flow problems, the transition into turbulence falls into one of the categories of (1) a direct transition from laminar flow into fully developed turbulent flow, or (2) an intervening zone where the lower Reynolds number primary laminar flow pattern first transitions into a more complex *secondary laminar flow* that then with continuing further increase in the Reynolds number transitions into a fully developed turbulent flow. Hydrodynamic bearings fall into the second transition category, encountering first a *secondary laminar flow* through the primary laminar flow becoming unstable.

Taylor (1923) determined the conditions started with the nearly intractable Orr–Sommerfeld equation (OSE) governing small flow disturbances *laminar flow instability* determination, Equation 1.19. By limiting the flow field to that between two concentric

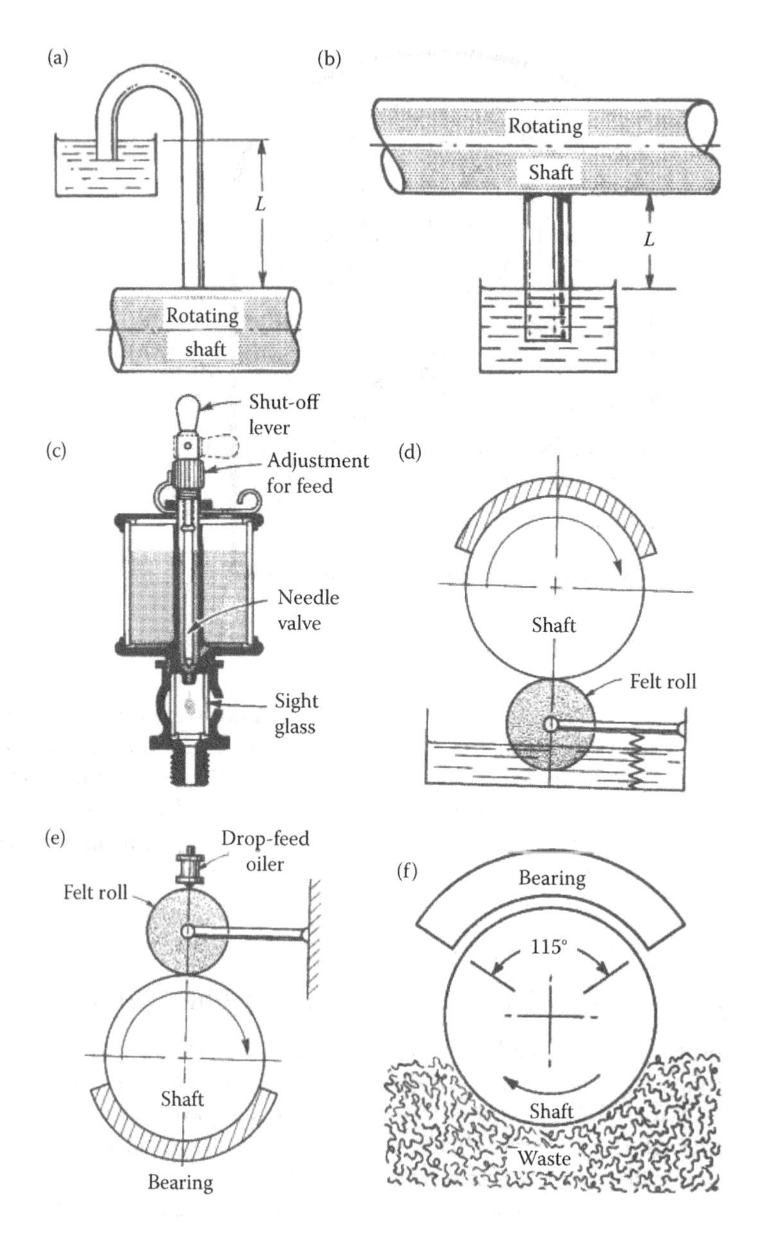

FIGURE 1.30
Low-flow lubricating devices: (a) siphon-wick oiler, (b) bottom-wick oiler, (c) drop-feed cup with sight glass, (d) and (e) felt rolls, and (f) oil-saturated waste pack (typical railroad bearing).

long cylinders, the inner cylinder (radius, r_1) rotating with the outer cylinder (radius, r_2) not rotating, Taylor developed a Reynolds-type number T which is the ratio fluid *centrifugal force* to the *viscous force*. The value of T where the primary laminar flow transitions to a more complicated secondary laminar flow was then derived from the OSE.

Taylor (1923) analyzed the stability of the mean primary laminar flow between *two concentric long cylinders*, the inner cylinder (radius, r_1) rotating with the outer cylinder (radius, r_2) not rotating, For this flow problem there is only one non-zero steady-state mean primary flow solution obtained from the Navier-Stokes equations (in cylindrical coordinates). That solution is a rotationally symmetric circumferential flow pattern, being a function of only radius r as follows (Szeri 1998).

$$v(r) = \frac{r_1^2 \omega}{r_2^2 - r_1^2}\left(\frac{r_2^2}{r} - r\right) \quad (1.19)$$

Since Equation 1.19 is based on the cylinders' boundary velocities satisfying the no-slip boundary condition, the circumferential velocity v is zero at the out-cylinder radius r_2 and $v(r_1) = r_1\omega$ on the inner cylinder. Consequently, the centrifugal acceleration of the flow is largest on the inner cylinder and zero on the outer cylinder. Taylor formulated a dimensionless number T which incorporating the Reynolds number Re is the ratio of the fluid *centrifugal force* to the *viscous force*, expressed as follows.

$$T = \frac{2(1-n)}{(1-n)}\mathrm{Re}^2 \quad \mathrm{Re} = \frac{r_1\omega(r_2 - r_1)\rho}{\mu} \quad n = \frac{r_1}{r_2} \quad \mu = \text{viscosity} \quad \rho = \text{mass density} \quad (1.20)$$

If the boundary between stable and unstable flow is stationary, and $(r_2 - r_1) \ll r_1$, the lowest value of T that marks hydrodynamic instability is expressible as follows.

$$\mathrm{Re} = 41.2\left(\frac{r_1}{r_2 - r_1}\right) \quad (1.21)$$

When this critical Reynolds number value is exceeded, a Taylor vortices small disturbance secondary laminar flow does not die out but instead grows, replacing the rotationally symmetric primary laminar flow field given by Equation 1.19. This secondary laminar flow pattern is illustrated in Figure 1.31 (Szeri 1998). It is composed of an integer number

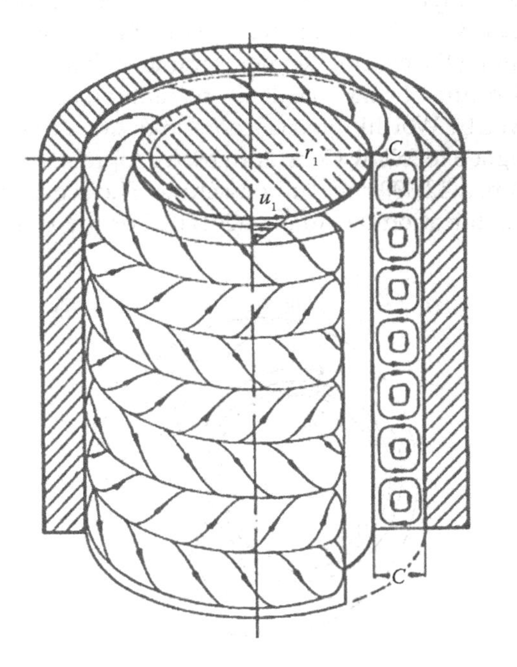

FIGURE 1.31
Taylor vortices between concentric cylinders; inner cylinder rotating, outer cylinder stationary.

of vortices, each counter spinning with respect to its two adjacent vortices. However, the Re number must exceed considerably the Equation 1.21 vortex-initiation before the flow transitions into fully developed turbulence. In extensively quoting Koshmieder (1993), Szeri (1998) provides a detailed insightful description of the quite complex and non-unique path in which the vortex cells become progressively more distorted and unsteady as the Re number is progressively increased in the transition to reach turbulence.

As stated, the Taylor vortex secondary flow pattern illustrated in Figure 1.31 is for concentric cylinders. However, when a journal-bearing supports its intended function to support radial loads, the radial eccentricity of the journal with respect to the bearing is not zero. So, extending Taylor's result to journal bearings of course was important to journal-bearing developers. Investigators determined that Couette flow such as embodied in Equation 1.19 is maintained if the eccentricity ratio between the two cylinders is $\epsilon < 0.3$. On increasing the eccentricity above 0.3, the mean primary flow departs from Couette flow with the growth of a stable recirculation cell (Kamal 1966), illustrated in Figure 1.32 (Szeri 1998).

The stable primary laminar does not lessen the Re number at which the flow pattern transitions into that illustrated in Figure 1.31. In fact, as DiPrima (1963) showed, the lowest Re number for transition to Taylor vortices is for concentric cylinders. DePrima and subsequently other investigators determined that as eccentricity is increased ($0.5 < \epsilon < 0.8$) consistent with the typical-load range of journal bearings, the critical Re number progressively increases significantly beyond that predicted by Equation 1.21.

Although full turbulence is one of nature's most complex phenomena, its treatment regarding hydrodynamic-bearing load capacity is not as complex as treating the route to turbulence through the hydrodynamic instability phenomenon covered earlier in this section. The earliest significant attention to the effects of turbulence on fluid-film-bearing performance was undertaken by Constantinescu (1959, 1962), who employed the then well-accepted Prandtl mixing-length approach to predict effects of turbulence in general. That led to the scaling of turbulence on bearing-load capacity based on a single-bearing Reynolds number.

The primary uncertainty of that approach was in knowing the best bearing-length parameter to use in defining the bearing Reynolds number, that is, the best comparison with laboratory bearing-test results. In the mid-1960s the author, as a then-recent engineering school graduate employed by Worthington Corporation's Advanced Products Division near New York City, was assigned to follow Constantinescu's approach using tests on a prototype rotor for a newly developed 42,000 rpm turbomachine. Those experiments and results are detailed in the recent publication by Adams (2017b). For the water-lubricated fixed-pad

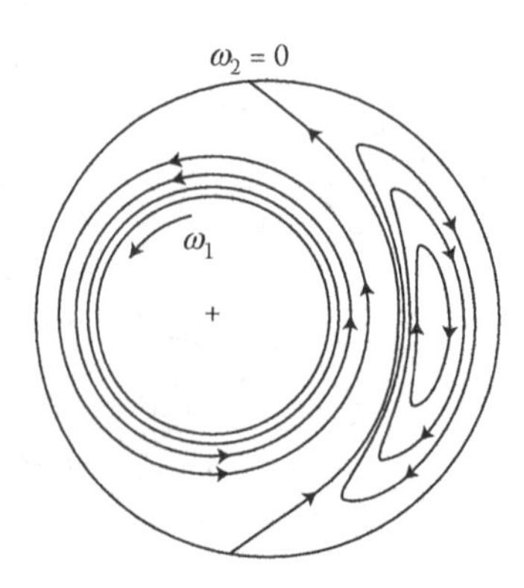

FIGURE 1.32
Stable primary laminar flow between two cylinders when $\epsilon > 0.3$.

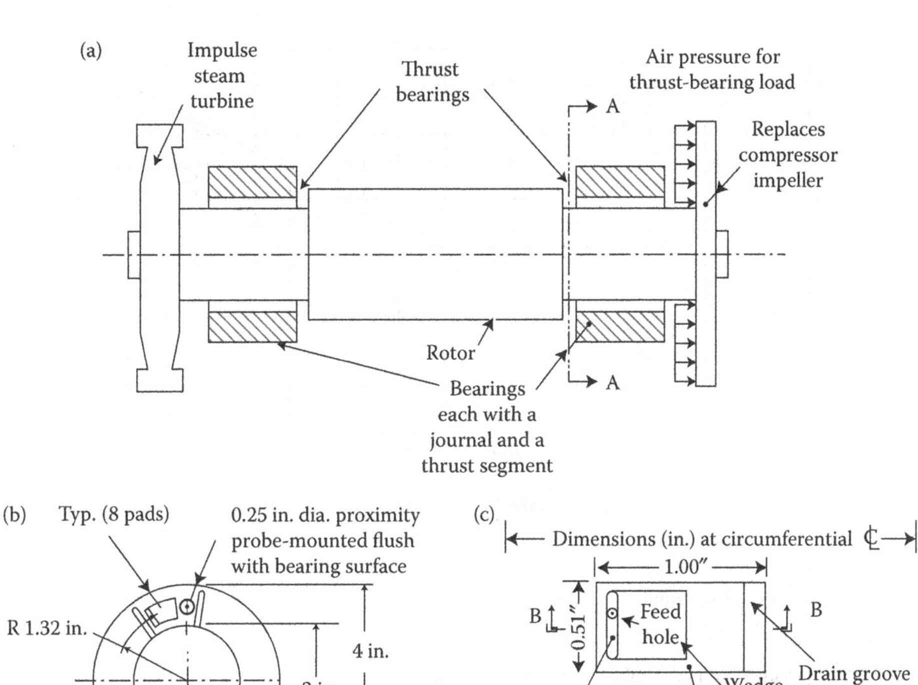

FIGURE 1.33
Test set up for water-lubricated turbulent flow thrust bearing: (a) 50,000 rpm max-speed rotor-bearing unit, (b) thrust bearing, and (c) bearing details.

thrust bearing of that machine, tests were run up to 50,000 rpm. The test apparatus and tested thrust bearing are illustrated in Figure 1.33.

A very good correlation was found to scale laminar theory-predicted load capacity to measured turbulent-load capacity employing a Reynolds-number-dependent coefficient based on a single-bearing Reynolds number. It was found that the best scaling of laminar-theory load capacity to measured turbulent-load capacity performance was obtained using the thrust-bearing entrance wedge film thickness (h_1) as the length parameter in defining the bearing Reynolds number. Figure 1.34 shows the results of the full range of tested loads, yielding a quite good straight-line on a log-log plot, thus expressible by the following equation.

$$\frac{W_T}{W_L} = A\,\mathrm{Re}^n \tag{1.22}$$

where

W_T = measured load capacity (with turbulence)
W_L = load capacity predicted from laminar theory à la Reynolds equation
Re $= \rho U h_1/\mu$, ρ = mass density, U = sliding velocity, μ = viscosity
h_1 = pad-wedge inlet film thickness $= h_2 + t$, t = pad-taper depth (Figure 1.32)

A and n are obtained from the equation for a straight line on the log-log plot.

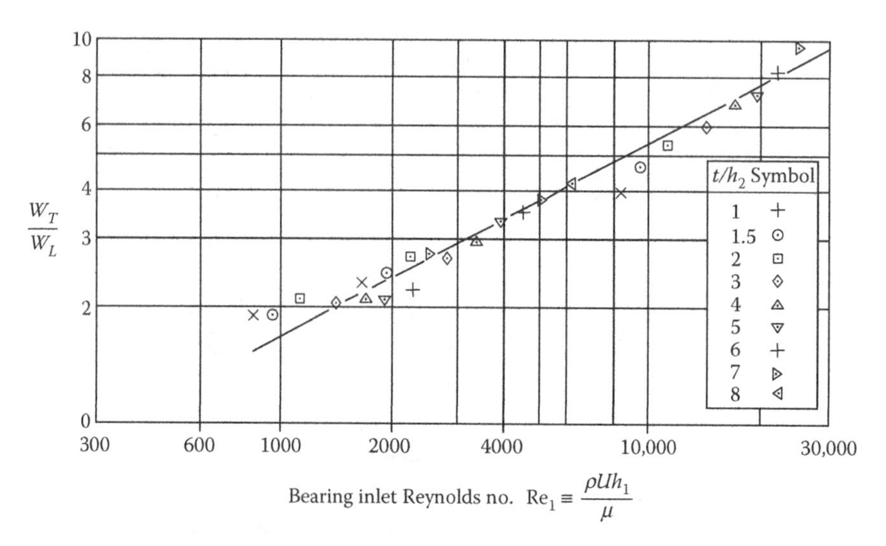

FIGURE 1.34
Load amplification versus inlet Reynolds number.

The same approach was employed by the author to correlate *turbulent-bearing stiffness* with *laminar theory prediction*. The experimental stiffness was obtained simply by numerically differentiating the experimental load versus displacement data. Figure 1.35 shows the comparison between turbulent-to-laminar load capacity and stiffness ratios. As shown, the stiffness ratio so determined is less than one for Re < 10,000 and less than the load ratio at all Reynolds numbers tested, but asymptotically approaching the load ratio as the Reynolds number approaches 100,000. That all made sense to the author at the time, since when load is reduced the film thickness increases which also increases the Reynolds number. And conversely, when the load is increased the film thickness decreases which also decreases the Reynolds number. The author got subsequent verification of this stiffness

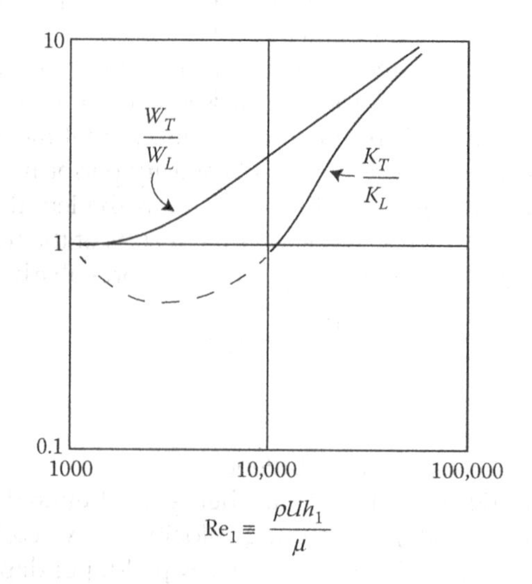

FIGURE 1.35
Comparison between load and stiffness ratios.

characteristic with turbulence during subsequent employment at the Franklin Institute utilizing advanced software for predicting turbulent-bearing performance. That is, while turbulence in general increases hydrodynamic-bearing load capacity, turbulence can reduce the bearing-film stiffness.

In the landmark paper by Elrod and Ng (1967), a procedure was presented that could be incorporated directly into newly developed software for laminar-flow bearings. That is, software employing the numerical finite-difference method to solve the Reynolds lubrication equation for the film-pressure distribution, Equation 1.10. The Elrod–Ng approach was to adjust upward the lubricant viscosity value locally at each finite-difference grid point based on the film flow Reynolds number at that location. Not only was the correlation with measured load capacity improved, but all the other bearing performance parameters like bearing friction coefficient and inlet and side flow were computationally computed. Their approach was equally well suited to hybrid bearings entailing simultaneous hydrodynamic and hydrostatic action. Available software now for predicting fluid-film-bearing performance routinely incorporates essentially the Elrod–Ng approach. So, the capability of predicting fluid-film-bearing performance factors with films in the turbulence regimes is now routine.

1.12 Water-Lubricated Rubber Liner Bearings

The marine industry, naval and commercial, uses some type of water-lubricated bearings to support both the load and dynamic characteristics associated with vessel propulsion systems. The first types of water-lubricated bearings used for marine propulsion were on steam driven engines in the early to mid-1800s. Those bearings were typically manufactured out of brass or soft metal bushings which utilized sea water as the lubricating fluid. However, the combination of bearing material and use of a low viscosity lubricant such as sea water proved futile when the iron sailing ship Great Eastern experienced both accelerated and uneven wear of the shaft bearings on its first transatlantic voyage. The first proposed solution to this issue was presented by the English engineer John Penn when he proposed to use lignum vitae, a hard and dense wood as a marine shaft bearing in 1858. Lignum vitae has also been used extensively in the twentieth century for impulse hydroelectric impulse turbines.

Since its initial use in early steam vessels, lignum vitae had been the dominant material of choice until the notion of using a flexible, elastomeric (rubber) lining for the housing of a bearing was inadvertently discovered by an engineer Charles Sherwood while working on water pumps for shaft mines back in the 1920s (Orndorff 1985). Since then the concept of using elastomeric materials in combination with low viscosity fluids as an integrated bearing has been widely accepted. Studies have been performed to better understand the solid-fluid interaction between a rotating journal, lubricant, and elastomeric material as it relates to shaft/propulsion system dynamics, friction, wear, and load-carry capacity.

In comparison, research into this class of hydrodynamic fluid-film bearings has been one of the least investigated when compared to metallic bearings operating with medium to high viscosity fluids. Annis (1927) provided an early publication describing rubber as a suitable material for fluid-film bearings, parametrically testing a variety of spiral shapes within a rubber-lined bearing. Annis presented various performance characteristics comparing each of the spiral groove geometries. Busse and Denton (1932) experimentally investigated the effects of surface roughness and journal speed on the coefficient of friction of both rubber and metal bearings. Their observations concluded that a thin fluid film was needed to separate rubber and metal, while a thicker fluid film was needed to separate two similarly rough metal surfaces. Shannikov (1939) was the first to investigate the development of fluid-film pressure on an elastic-bearing surface, concluding that there was no theoretically founded conclusive publications to explain satisfactorily the phenomena taking place in elastic bearings. Brasier and Sowyer (1937) determined that increased loading on elastic bearings lowers the coefficient of friction similar to hydrodynamic bearings.

Those early papers were primarily test driven given the lack of modern computers for solving nonlinear equations for the fluid-film pressure distribution. Current research and modeling of this special class of bearings have focused on coupling the fluid film and elastic liner deformation, to provide both static-load capacity and the rotor dynamic properties of elastomeric polymeric bearings. One of the first current researches to investigate the solid-fluid boundary by use of computation fluid dynamics (CFD) was Braun and Dougherty (1988), who modeled the coupled interaction between fluid-film pressure and compliant-bearing liner. DeKraker (2006) describe a method to generate Stribeck curves for water-lubricated bearings, coupling surface effects and material deflection. They used the Reynolds lubrication equation coupled with an average roughness model for film height variation between the journal and elastic liner. Overall surface pressures generated within the fluid film as originally proposed by Shannikov (1939) have also been studied since the development of modern data acquisition systems (DAQ) and recent advances in pressure-sensor technology.

Cabrera et al. (2005) experimentally investigated the developed fluid-film pressures on the surface of an operating-elastic bearing. They used a straight-flute rubber bearing and embedded a miniature pressure transducer in the surface of a rotating shaft. The high-frequency pressure transducer allowed for a full 360° sweep of the pressures developed on the surface of each bearing surface. Bearing rotor dynamic characteristics have also been studied, for example, Lahmar et al. (2009). There are also many additional works that are experimentally driven, for example, Daugherty and Sides (1981), and Litwin (2011), who studied the performance of water-lubricated bearings.

Use of rubber bearings typically utilize two core geometries: (1) removable staves and (2) fully molded profiles, as illustrated in Figure 1.36 These contoured geometries also utilize a

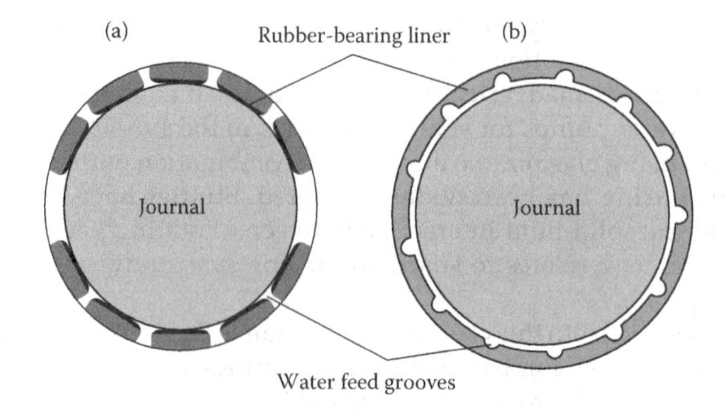

FIGURE 1.36
(a) Flat land stave and (b) contour water-lubricated elastomeric bearing.

series of axially placed water grooves for fluid flow. Without the inclusion of water grooves this type of bearing would be similar to the classical, smooth bore hydrodynamic fluid-film bearings. The second type of profile makes use of a stave, which is a long homogeneous/heterogeneous base material bonded to a layer of rubber which runs the entire length of a bearing. These staves can then be inserted into a metal/composite housing as shown in Figure 1.1a.

Stave-type-bearing configurations, elastomeric, wood, or polymer, configurations have been widely used by many marine and industrial applications due to the ease of removal of bearing liners. To understand the mechanics of fluid-film production, it is necessary to understand some of the basic physics associated with fluid-film generation. The film pressures and load-carry capacity generated within a flat stave land is of course different from a smooth continuous, walled-journal bearing. The main difference is driven by the bearing compliance and that each stave is essentially an individual partial-arc bearing, since each stave is separated by a feed groove, justifying pressure distribution from a Reynolds equation solution as illustrated in Figure 1.17. But the converging to a valid solution for a compliant surface rubber-bearing liner requires iteration between the Reynolds solution and the structural deformation of the bearing liner, until convergence to the steady-state film-thickness distribution is obtained. It is the compressive deformation of rubber staves that results in appreciable bearing-load capacity that would be unachievable with a rigid stave.

The need to properly explain the phenomena taking place in elastic bearings was initially observed from experimentation and documented by Orndorff (1985) and Orndorff and Holzheimer (1992), who described the elastic deformation of rubber and described the subsequent fluid-film generation as a phenomenon called plasto-elastohydrodynamics lubrication (PEHL) or soft EHL (Hamrock 1994). This PEHL process is unique to rubber and other highly elastic large deformation materials. When a journal is loaded on a rubber surface the material response is transient in nature and will undergo an initial deflection over a short finite period of time. As time continues, the material will slowly and continuously deflect over a much longer period than that of the initial deflection. This process is known as creep and is the material's ability to slowly deflect over time when exposed to stresses below the materials yield stress. The magnitude and rate of material creep is dependent on the material properties, stress exposure, and temperature. It should be noted from experiments that the surface friction and fluid-film development is continuously being developed during this time and is only complete when the mating material is no longer deforming. The deformation of the surface will allow for the rubber to become compliant to the shaft and thus allow for a thin fluid film to develop once the bearing reaches a steady state.

The fluid-film generation of elastic bearings can be characterized in several regimes, boundary, mixed, EHL, and hydrodynamic lubrication (see Figures 6.3 and 6.4 and accompanying text). These various regimes are dependent on the thickness of the fluid film which separates the journal and housing of the bearing. The determination and location of each respective regime is characterized by the Stribeck curve dimensionless speed, Figure 1.1a.

The ability for an elastic material to have large deflections and creep over a finite time period also influences the break-in period of a rubber bearing, as illustrated in Figure 1.37 (supplied from Duramax Marine in-house laboratory). The generic rubber formulation used for the Figure 1.37 tests show the variation of friction as a function of trial time and how the fluid-film development progresses to different lubrication regimes over the break-in process. The corresponding graphs trace the performance of a rubber bearing initially tested (test 1: initial deflection) and subsequent tests to show performance at

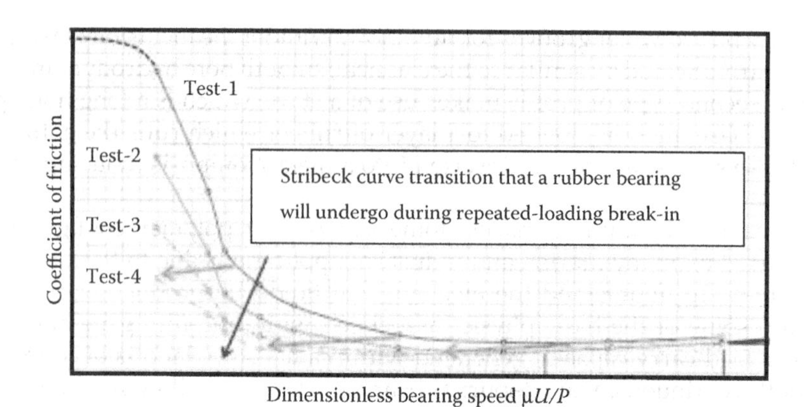

FIGURE 1.37
Stribeck curves illustrating the change in friction coefficient during break-in period of a rubber bearing.

various creeped positions over extended time (tests 2 and 3) and a steady state broken-in bearing (test 4).

It can be seen that the initial test 1 yielded the classical Stribeck curve where the fluid film progressed through the *mixed*, *soft EHL* and *hydrodynamic* fluid-film regimes as sliding speed is increased. However, as the testing progressed and the material deflected over time, two characteristic changes in bearing performance can also be seen. First, the overall coefficient of friction decreases over time from test 1 to test 4. Second, the overall Stribeck curve moves leftward. This shifting shows the transition and development of the fluid film as the material deflection compliance of the bearing reaches a steady state value. At that point, the bearing when operating in low to high speed applications will develop a fluid film faster than the initial (undeformed) test. It should be noted that the rubber formulation can highly affect fluid-film development. Therefore, it is critical to have a thorough understanding of how all compounds formulated for elastic materials affect the overall shape of the Stribeck curve and the generation of a fluid film.

References

Adams, M. L., *Rotating Machinery Vibration—From Analysis to Troubleshooting*, 2nd ed., CRC Press/Taylor & Francis, Boca Raton, FL, 2010.

Adams, M. L., *Power Plant Centrifugal Pumps—Problem Analysis and Troubleshooting*, CRC Press/Taylor & Francis, Boca Raton, FL, 2017a.

Adams, M. L., *Rotating Machinery Research and Development Test Rigs*, CRC Press/Taylor & Francis, Boca Raton, FL, 2017b.

Adams, M. L., Padovan, J. and Fertis, D., Engine Dynamic Analysis with General Nonlinear Finite-Element Codes, Part 1: Overall Approach and Development of Bearing Damper Element, *ASME, Journal of Engineering for Power*, Vol. 104(3), pp. 81–87, 1982.

Annis, B. B., Cutless Rubber Bearings, *Mar. Eng. Ship. Age*, Vol. 32(5), pp. 275–277, 1927.

Archard, J. F., Contact and Rubbing of Flat Surfaces, *Journal of Applied Physics*, Vol. 24, pp. 981–988, 1953.

Bhushan, B., *Principles and Applications of Tribology*, Wiley, 1999.

Bird, R. B., Stewart, W. E. and Lightfoot, E. N., *Transport Phenomena*, Wiley, 1960.

Brasier and Sowyer, *Rubber Bearings*, 1937.

Braun, M. J. and Dougherty, J. D. Hydrodynamic Analysis and Fluid-Solid Interaction Effects on the Behavior of a Compliant Wall (Thick) Journal Bearing, Part I and II, 1988.

Burgdorfer, A. The Influence of the Molecular Free Path on the Performance of Gas Lubricated Hydrodynamic Bearings, *ASME Journal of Basic Engineering*, Vol. 81, pp. 41–100, 1959.

Busse, W. F. and Denton, W. H. *Water-Lubricated Soft-Rubber Bearings*, Trans. ASME, 1932.

Cabrera, D. L., Woolley, N. H., Allanson, D. R. and Tridimas, Y. D., Film Pressure Distribution in Water-Lubricated Rubber Journal Bearings, *Engineering Tribology*, Vol. 219(Part J), 2005.

Campbell, W. E., *Boundary Lubrication: An Appraisal of World Literature*, ASME, New York, pp. 87–117, 1969.

Castelli, V. and Shapiro, W., Improved Method for Numerical Solutions of the General Incompressible Fluid Film Lubrication Problem, *ASME Journal of Lubrication Technology*, Vol. 89(2), pp. 211–218, 1967.

Constantinescu, V. N., On turbulent lubrication, *Proceedings, Institution of Mechanical Engineers (IMechE)*, Vol. 173, pp. 881–889, 1959.

Constantinescu, V. N., Analysis of Bearings Operating in Turbulent Regime, *ASME Transactions*, Vol. 82, pp. 139–151, 1962.

Daugherty, T. L. and Sides, N. T., Frictional Characteristics of Water-Lubricated Compliant-Surface Stave Bearings, *Tribology Transactions*, Vol. 24, pp. 293–301, 1981.

De Kraker, A., Calculation of Stribeck Curves for (Water) Lubricated Journal Bearings, 2006.

DiPrima, R. C., A Note on the Stability of Flow in Loaded Journal Bearings, *ASLE Trans.*, Vol. 6, pp. 249–253, 1963.

Elrod, H. G. and Ng, C. W., A Theory for Turbulent Films and its Application to Bearings, *ASME Transactions*, Vol. 89, pp. 347–362, 1967.

Fuller, D. D., *Theory and Practice of Lubrication for Engineers*, 2nd ed., Wiley, 1984.

Fukui, S. and Kaneko, R., Molecular Gas-film Lubrication. In: Bhushan, B. (editor) *Handbook of Micro/Nanotribology*, CRC Press, Boca Raton, pp. 559–604, 1995.

Goodyear, C. E., Adaptive numerical methods for elastrohydrodynamic lubrication, PhD thesis, University of Leeds, UK, 2001.

Gross, W., *Fluid Film Lubrication*, Wiley, 1980.

Grubin, A. N., Fundamentals of the Hydrodynamic Theory of Lubrication of Heavily Loaded Cylindrical Surfaces. In: Ketova, K. F. (editor) *Investigation of the Contact Machine Components*. Translation of Russian Book No. 30, Central Scientific Institute for Technology and Mechanical Engineering, Moscow, 1949, Chapter 2, (Available from Department of Scientific and Industrial Research, Great Britain, Trans. CTS-235 and Special Libraries Association, Trans. R-3554).

Hamrock, B. J., *Fundamentals of Fluid Film Lubrication*, McGraw-Hill, 1994.

Heshmat, H., Shapiro, W. and Gray, S., Development of Foil Journal Bearings for High Load Capacity and High Speed Whirl Stability, *ASME Journal of Lubrication Technology*, Vol. 104(2), pp. 149–156, 1982.

Howard, S. A., Valco, M. J., Prahl, J. M., and Heshmat, H. Dynamic Stiffness and Damping Characteristics of a High-Temperature Air Foil Journal Bearing, *Trans. STLE*, Vol. 44(4), pp. 657–663, 2001.

Jenkins, L. S., Troubleshooting Westinghouse Reactor Coolant Pump Vibration, *EPRI Symposium on Trouble-Shooting Power Plant Rotating Machinery*, San Diego, CA, May 19–21, 1993.

Kamal, M. M., Separation in the Flow Between Eccentric Rotating Cylinders, *ASME Journal of Basic Engineering*, Vol. 88, pp. 717–724, 1966.

Khonsari, M. K. and Booser, E. R., *Applied Tribology—Bearing Design and Lubrication*. Wiley, p. 496, 2001.

Koshmieder, E. L., *Benard Cells and Taylor Vortices*, Cambridge University Press, Cambridge, 1993.

Lahmar, M., Ellagoune, S. and Bou-Said, B., Elastohydrodynamic Lubrication Analysis of a Compliant Journal Bearing Considering Static and Dynamic Deformations of the Bearing Liner, 2009.

Litwin, W., Influence of Local Bush Wear on Properties of Water Lubricated Marine Stern Tube Bearings, *Polish Maritime Research*, Vol. 18, pp. 32–36, 2011.
Loeb, A. M., Determination of Flow, Film Thickness and Load-Carrying Capacity of Hydrostatic Bearings Through the Use of the Electric Field Plotter, *Trans. ASLE*, Vol. 1, pp. 217–224, 1958.
O'Connor, J. J., Boyd, J. and Avallone, E. A., *Standard Handbook of Lubrication Engineering, American Society of Lubrication Engineers*, McGraw-Hill, 1968.
Orndorff, R. L., Water-Lubricated Rubber Bearings, History and New Developments, *Naval Eng. Journal*, 1985.
Orndorff Jr, R. L. and Holzheimer, C., Thin Film Tribology and Rubber Bearings, *Proceeding 19th Symposium of Tribology*, Akron, Ohio, pp. 611–620, 1992.
Rabinowicz, E., *Friction and Wear of Materials*, Wiley, 1965.
Raimondi, A. A. and Boyd, J., A Solution for the Finite Journal Bearing and its Application to Analysis and Design, Parts I, II and III, *Trans. ASLE*, Vol. 1(1), pp. 159–209, 1958.
Raimondi, A. A., Boyd, J. and Kaufman, H. N., Analysis and Design of Sliding Bearings, Chapter 5, *Standard Handbook of Lubrication Engineering*, American Society of Lubrication Engineering (ASLE), Handbook Editors: O'Connor, J. J., Boyd, J. and Avallone, E. A., McGraw-Hill, 1968.
Reynolds, O., On the Theory of Lubrication and its Application to Mr. Tower's Experiments, *Philosophical Transactions of the Royal Society, London, England*, Vol. 177(Part 1), 1886.
Ripple, H. C., *Cast Bronze Hydrostatic Bearing Design Manual*, Cast Bronze Bearing Institute, Cleveland, Ohio, 1964.
San Andres, L. and Kim, T.-H., Issues on Instability and Forced Nonlinearity in Gas Foil Bearing Supported Rotors, *AIAA Joint Propulsion Conference, Paper No. 5094*, 8–11 July 2007, Cincinnati, OH.
Shannikov, V. M., *Rubber Bearings*, 1939.
Stribeck, R., *Die wesentlichen Eigenschaften der Gleit und Rollenlager* (The basic properties of sliding and rolling bearings), developed in the first half of the 20th century in Berlin at the Royal Prussian Technical Testing Institute, available in Zeitschrift des Vereins Deutscher Ingenieure, Nr. 36, Band 46, pp. 1341–1348, 1432–1438 and 1463–1470, 2002.
Szeri, A. Z., *Tribology—Friction, Lubrication and Wear*, Hemisphere, 1980.
Szeri, A. Z., *Fluid Film Lubrication—Theory and Design*, Cambridge University Press, p. 414, 1998.
Taylor, G. I., Stability of a Viscous Liquid Contained Between Two Rotating Cylinders, *Philosophical Transactions, Royal Society, A 223*, pp. 289–343, 1923.

2

Rolling Element-Bearing Mechanics

Rolling element bearings (REBs) are applied today for many consumer products like home appliances, power tools, bicycle wheel bearings, automobile wheel bearings, roller skates, as well as very high-tech products like modern aircraft gas-turbine jet engines, just to name a few. The use of intervening rolling elements to provide relatively low-friction load-bearing movement dates at least back to the Assyrians (1100 BC), to transport massive stones as depicted in Figure 2.1 (Harris and Kotzalas 2007). Leonardo da Vinci's major engineering invention was the ball bearing, upon which the success of most of his other inventions depended. Most of da Vinci's engineering creations sprung from his engineering repertoire of live on-stage theater scenery spectaculars. Leonardo da Vinci's ball-bearing invention was ahead of its time in relation to other yet-to-be invented machinery, and thus was not then widely publicized, leaving it to be reinvented at a future point in time.

At first glance to the average person, modern REBs deceptively do not appear to be all that complicated. The engineering technology of modern REBs is extensive in its complexity, stemming from (1) modern manufacturing developments, (2) many millions of hours of laboratory testing, (3) application of advanced modern computerized modeling tools, (4) bearing material research, and more. Since many of our modern devices utilize REBs, they are widely taken for granted. REBs are often referred to as *anti-friction bearings*. That characterization *could not be farther from the truth*. REBs are low-friction devices but not zero-friction devices. Otherwise, rolling element bearings would enable the thermodynamically impossible *perpetual motion machine*, which every mechanical engineering professor has seen proposed by some "bicycle-shop" mechanic at one time or another. *Anti-friction* infers an even more fallacious idea, namely that these bearings have *negative friction*, which of course would provide limitless sources of free energy, but might make walking, even crawling, unachievable.

The feasibility of wide REB application was seriously hampered prior to the twentieth century by lack of the needed precision manufacturing methods not then yet perfected. Early in the twentieth century the emerging manufacturing methods finally started to make rolling element bearings feasible for numerous machinery applications. However, up until the end of WW II, the field of rolling element bearings was considered to be more *engineering art* rather than *engineering science*. One of the earliest and widely recognized publications that shifted rolling element bearings into the realm of engineering science was that of A. B. "Bert" Jones (1946). As a young budding research engineer at The Franklin Institute Research Laboratories (FIRL) 1967–1971, the author was privileged to work under FIRL's then widely recognized bearing specialists (see Acknowledgments). They were totally in agreement that Bert Jones was the "father" of modern rolling element-bearing engineering science, and a dedicated meticulous engineer as well as a conservative gentleman. Rolling element bearings have some significant *advantages* and *disadvantages* compared to fluid-film bearings, as follows.

FIGURE 2.1
Rollers used by the Assyrians to move massive stones in 1100 BC.

Advantages of Rolling Element Bearings Compared to Fluid-Film Bearings

1. High-load capacity regardless of rotational speed
2. Low starting friction
3. Lower friction at operating speeds, thus more energy efficient
4. Single bearing can simultaneously support both radial and axial loads
5. Can run with no radial clearance (preloaded in precision spindles)
6. Machine configurations with easy-to-replace bearings
7. Can operate for a significant time with lubrication interrupted (aircraft engines)
8. Usually give warnings prior to gross failure (noise, metal debris detected)
9. Wide availability in load capacity, precision, and cost

Disadvantages of Rolling Element Bearings Compared to Fluid-Film Bearings

1. Life-rated because of wear sources such as sub-surface initiated fatigue
2. Less tolerant to corrosion and dirt ingestion damage
3. Lower maximum speed capability
4. Virtually no rotor-vibration-damping capacity
5. Less compact
6. Available only from a bearing supplier (a manufacturing specialty)

The two major REB categories are the following.

1. Ball
2. Roller

While each of these REB types has its own unique features, each one depends on the following fundamental governing technological specialties.

1. Loads and stresses
2. Component motion kinematics

3. Computer-aided design analysis
4. Lubrication chemistry and methods
5. Materials selection and processing
6. Failure modes, life, and reliability
7. Development and life testing

2.1 Ball and Roller-Bearing Descriptions

2.1.1 Ball Bearings

There is a wide variety of ball-bearing configurations, as follows: (1) deep-groove, (2) self-aligning, (3) angular-contact, (4) duplex pairs, (5) ceramic balls, (6) ball thrust, and (7) linear.

Deep groove (Conrad, 1903, 1906) ball bearings have high radial-load capacity and moderate axial-thrust-load capacity. A typical configuration is pictured in Figure 2.2. Two types of angularly *self-aligning* single-row radial-ball bearings are pictured in Figure 2.3. The self-aligning configuration in Figure 2.3a has the outer-raceway groove ground to a spherical shape, not deep groove, which gives it a relatively low-load capacity compared to the deep groove configuration, but has reliable self-aligning capability since the balls are free to roll on the spherical surface as needed to accommodate angular misalignment. The self-aligning

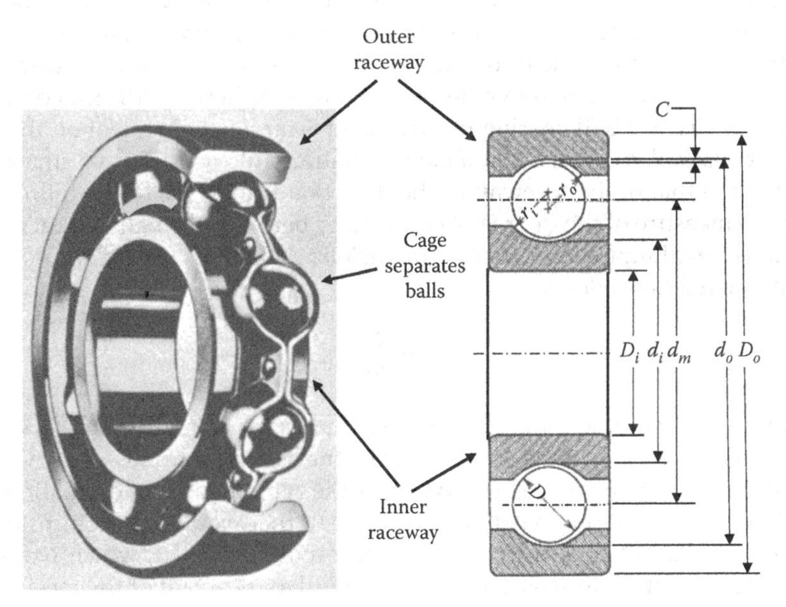

FIGURE 2.2
Single-row deep-groove ball bearing.

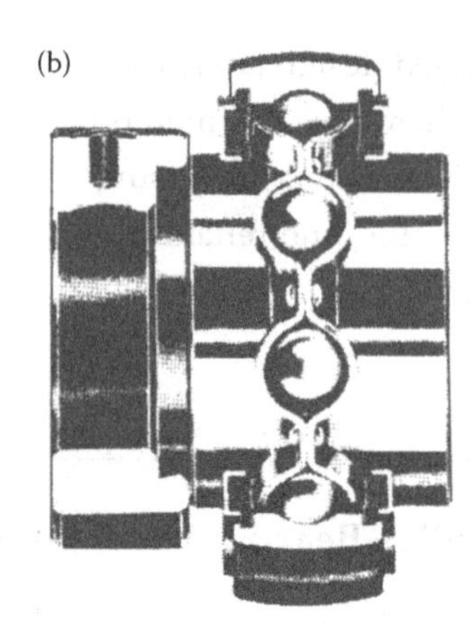

FIGURE 2.3
Self-aligning single-row ball-bearing configurations.

configuration in Figure 2.3b has the outer raceway ground with an outer spherical surface that conforms within a mating spherically ground housing fit. Since it has a deep groove outer raceway, it has a load capacity comparable to the deep-groove ball bearing. But it should be assembled *without squeeze* plus care taken, possibly using selected matched parts to maintain freedom of angular movement between the outer raceway and its conforming spherical housing. While the Figure 2.3b configuration should readily accommodate static misalignment, the Figure 2.3a configuration should more easily accommodate not only static misalignment but also dynamic misalignment such as from a shaft-whipping motion.

The relatively low load capacity of the Figure 2.3a self-aligning configuration results from the relatively low *osculation* value of the ball-on-spherical surface contact. That is because the capacity of a ball bearing to carry load is largely dependent on the osculation between the balls and raceways. *Osculation* is the ratio of the radius of curvature of the rolling element to that of the raceway in the direction transverse to the rolling direction. Osculation is a measure of the degree of conformity between the ball and raceway groove radius-of-curvature. Employing generic dimensions defined in Figure 2.2, osculation ϕ is given by Equation 2.1 as follows.

$$\phi = \frac{D}{2r} \tag{2.1}$$

Osculation is not the same for both raceways, that is, $r_1 \neq r_2$. For example, scaling from the Figure 2.3a self-aligning configuration gives the spherical radius-of-curvature $\approx 10r_o$ which yields an osculation value of 0.1. Whereas the typical value of osculation for a deep groove ball bearing is approximately $\phi = 0.5$. A ball that would conform to the raceway groove radius ($D = 2r$) of course would yield an osculation $\phi = 1$, which would not be a good design because the ball would rub over the full axial extent of the raceway yielding unacceptably high interfacial sliding friction. On the other hand, the smaller the osculation, the smaller is the elastic contact footprint and thus the smaller is the maximum contact-load capacity within the contact elastic limit. Summarizing, the two extremes of osculation are (1) osculation = zero and (2) osculation = 1. In case (1) any load will cause plastic contact deformation, that is, as $D \rightarrow 0$ contact stress $\rightarrow$ material strength, while condition (2) has a footprint that conforms with the raceway groove radius r, resulting in high interfacial slipping and friction between ball and raceway.

Obviously, the osculation value now typically used ($\phi \approx 0.5$) is basically the best compromise between the two competing objectives: (1) maximizing bearing-load capacity and (2) minimizing ball-raceway interfacial sliding friction. To visualize this, Figure 2.4 illustrates ball-raceway interfacial sliding for a typical osculation value near 0.5 and the condition of only radial-bearing load (no axial-thrust load). Exact osculation nominal values are generally part of bearing-company-specific proprietary non-published information. The Figure 2.4b interfacial sliding picture is an approximate illustration that has often been used in elementary explanation of ball-bearing interfacial slipping under only ball rolling motion. Figure 2.4c illustrates a more informative picture, showing a sliding velocity distribution (V_s) based on the rolling rotational speed (V_θ) of the ball, as follows.

$$V_\theta = \omega_{\text{ball}}\, r_\theta = \omega_{\text{ball}}\, r_{\text{ball}} \cos\theta \tag{2.2}$$

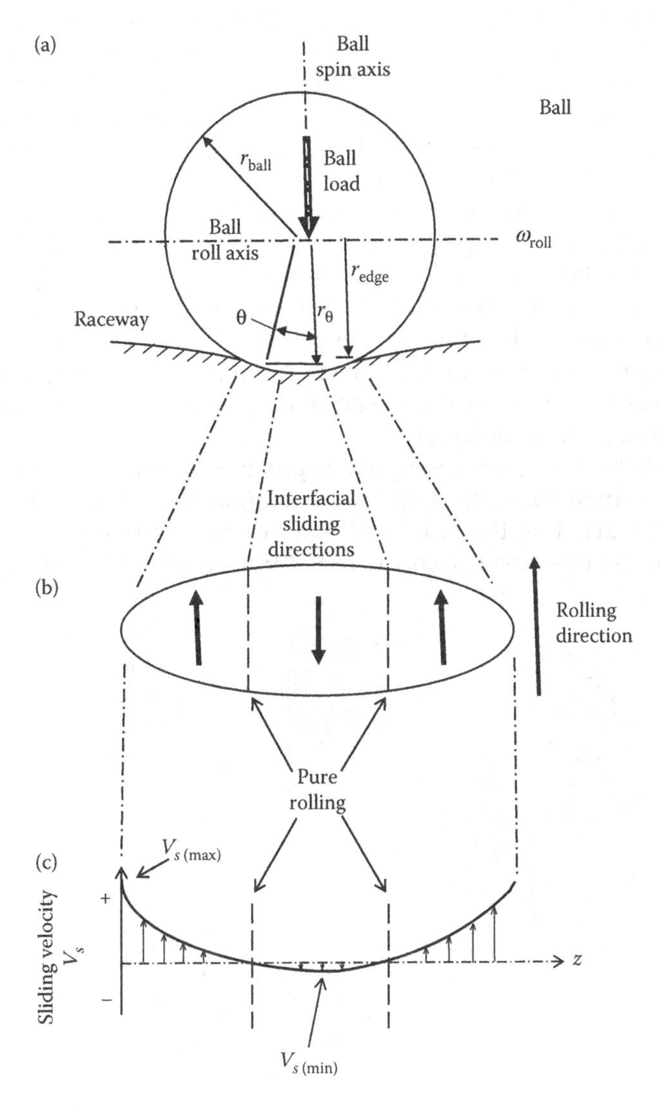

FIGURE 2.4
Interfacial slipping on contact footprint between ball and raceway.

Using Equation 2.2 and scaling proportionally from Figure 2.4c yields the following result.

$$\frac{V_\theta\ @\ \theta \cong 1^\circ\ (\text{pure rolling}) - V_\theta\ @\ \theta \cong\ 2.5^\circ\ (\text{edge})}{V_\theta\ @\ \theta = 0\ (\text{center}) -\ V_\theta\ @\ \theta \cong 1^\circ\ (\text{pure rolling})} \equiv \frac{\Delta V_{\text{edge}}}{\Delta V_{\text{center}}} \cong 5.3 \tag{2.3}$$

Elementary trigonometry (Equation 2.2) demonstrates why, as illustrated in Figure 2.4c, the magnitude of slip velocity (V_s) at the two footprint axial edges is significantly larger than the magnitude of slip velocity at the center of the footprint. Clearly, even under this simplest operating condition of radial-bearing load only, a *ball bearing is not anti-friction*.

Actual ball-raceway interfacial sliding is typically not quite as simple as Figure 2.4c depicts. Because in actual operating ball bearings there is usually some axial thrust imposed on the bearing, and that imposes additionally on the ball what is commonly referred to as *spin rotational velocity*. Figure 2.4a shows the ball spin axis, which is a radial line, but Figure 2.4c does not illustrate any ball superimposed spin interfacial sliding on rolling sliding. However, it is relatively easy for one to picture what a ball spin velocity superimposes on that shown in Figure 2.4 for rolling only. Harris and Kotzalas (2007) show illustrations similar to Figure 2.4c for a variety of interfacial sliding simultaneous combinations of rolling, spinning, and gyroscopic affects, for assumed equilibrium of moments on a ball and conditions where the ball is skidding.

As a ball approaches the load zone (see Figure 1.25), the raceway compressive forces on it are changing rapidly at a rate proportional to the shaft speed. And the corresponding contact traction forces between the ball and raceways are rapidly changing. Furthermore, the higher the shaft speed, the more the ball's inertia can dominate ball-raceway interactive forces. An appreciation of the extreme complexity of ball-raceway interactions surely provides an understanding of why development testing of ball bearings has entailed many millions of hours of testing to arrive at the present state of the art for ball as well as for the other types of rolling contact bearings.

Angular-contact ball bearings are configured to provide an axial-thrust load-capacity that is substantially higher than that of the deep groove configuration illustrated in Figure 2.2. This is accomplished by grinding the inner and outer-raceway grooves with their respective radii-of-curvature centers offset from the radial line, as illustrated in Figure 2.5. But in

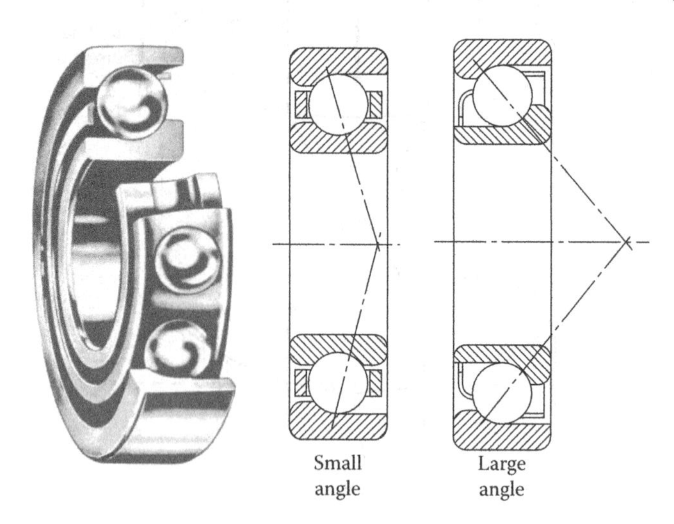

FIGURE 2.5
Angular-contact ball bearings.

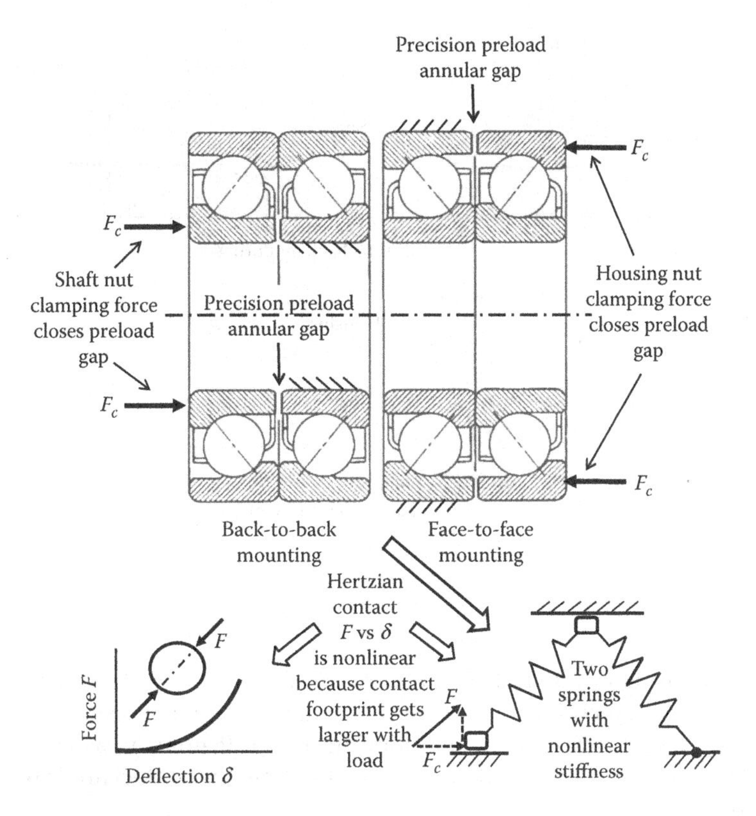

FIGURE 2.6
Mechanics of the duplex angular-contact ball-bearing pair.

contrast to the deep-groove configuration illustrated in Figure 2.2, a single angular-contact ball bearing can only support axial load in one axial direction, which is obvious from the illustration. Enabling the angular-contact ball bearing to carry axial-thrust loads in both axial directions is accomplished with the duplex pair configuration, the fundamentals of which are illustrated in Figure 2.6. Duplex sets of two angular-contact ball bearings are offered with the options for *small*, *medium*, or *large* preload, resulting from the preload gap, for example, 0.002, 0.004, or 0.006 in. (0.05 mm, 0.10 mm, or 0.15 mm). The preload gap surfaces of duplex angular-contact pairs are precision ground and matched at the bearing factory to ensure that when the preload gap is compressed closed at installation, the desired bearing preload is precisely achieved. Because of the fundamental nonlinear stiffening property of a Hertzian elastic contact (illustrated in Figure 2.6), the higher the duplex preload, the higher the bearing stiffness, both radially and axially. But the higher the duplex preload, the shorter is its life rating. The axial clamping force to close the gap is typically applied with an axial thrust nut as illustrated in Figure 2.7 for the two possible duplex configurations.

In high-speed ball-bearing applications, to better visualize the contact mechanics of both radial deep-groove and angular-contact ball bearings, it is helpful to view a ball in its raceways with an illustration highlighting the differences between the ball radius and the two-raceway radii of curvature, as illustrated in Figure 2.8. This illustration clearly shows why ball-raceway force interaction is a Hertzian contact, not a conforming contact. The elastic contact deflection from the ball-on-raceway contacts is well modeled as a uni-axial

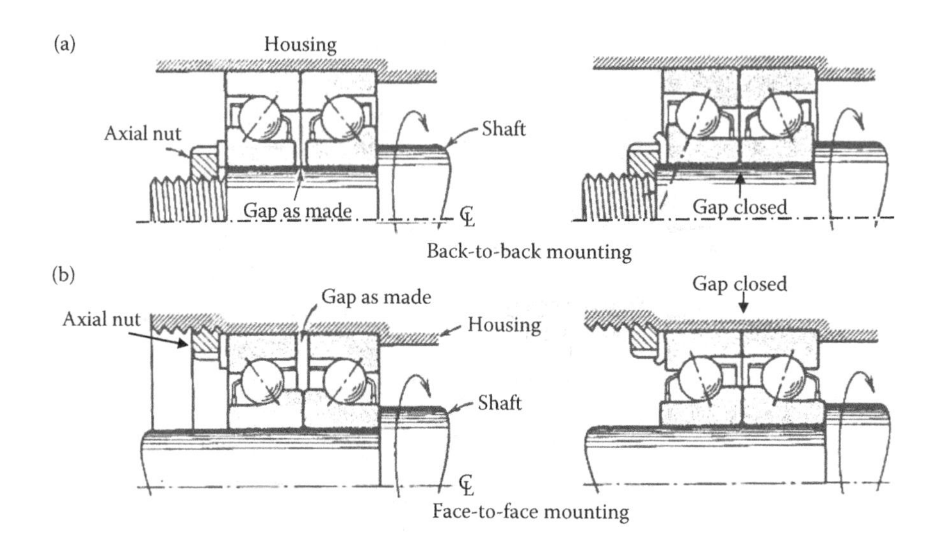

FIGURE 2.7
Preload gap closing with axial nut.

spring with a stiffening as illustrated in Figure 2.6. The *contact angles* illustrated in Figure 2.8, for deep-grove radial-contact ($\alpha = 0$), and angular-contact ball bearings, ($\alpha \neq 0$), *change* somewhat when axial-bearing loads are superimposed. Primarily, when an axial-thrust load is added to a deep-groove radial-contact ball bearing, the contact angle will not remain precisely zero.

In high-speed ball-bearing applications there is a major dynamic effect that the balls incur when the rolling axis is not parallel with the shaft rotational axis. That is a gyroscopic moment effect on the balls. For the radially loaded deep-groove radial bearing (Figure 2.8a), the ball-rolling axis is parallel to the shaft axis of rotation. For that case, the ball axis-of-rolling does not have an angular precession velocity as the balls orbit with the cage. In contrast, as clear from Figure 2.8b, when the ball rolling axis is not parallel to the shaft axis-of-rotation, the ball rolling axis traces out a conical surface. Preventing the associated gyroscopic effect from causing ball skidding requires a sufficient restraining moment from the contact friction force distribution expressed as follows.

$$\vec{M} = \vec{\omega}_{\text{cage}} \, \text{X} (I_{\text{P}} \vec{\omega}_{\text{ball}}) \tag{2.4}$$

where

Precession velocity is $\vec{\omega}_{\text{cage}}$, $\vec{\omega}_{\text{ball}} = \vec{\omega}_{\text{cage}} + \vec{\omega}_{\text{roll}}$, I_P = ball moment-of-inertia.

As application rotational speed is progressively increased, a point is reached where the ball-on-raceway contact friction force distribution will not be able to prevent skidding caused by the gyroscopic effect.

Rolling element clearance is a basic property that warrants description at this point. The ball-bearing cross-section illustration in Figure 2.2 happens to show a non-zero radial clearance, C. Some applications can tolerate a small amount of clearance. More often, applications need zero clearance, and some may need preload (i.e., negative clearance) for higher stiffness under zero-bearing-load operation, for example, to achieve high precision centering or increased rotor dynamic stiffness to elevate a rotor critical speed. Zero clearance provides

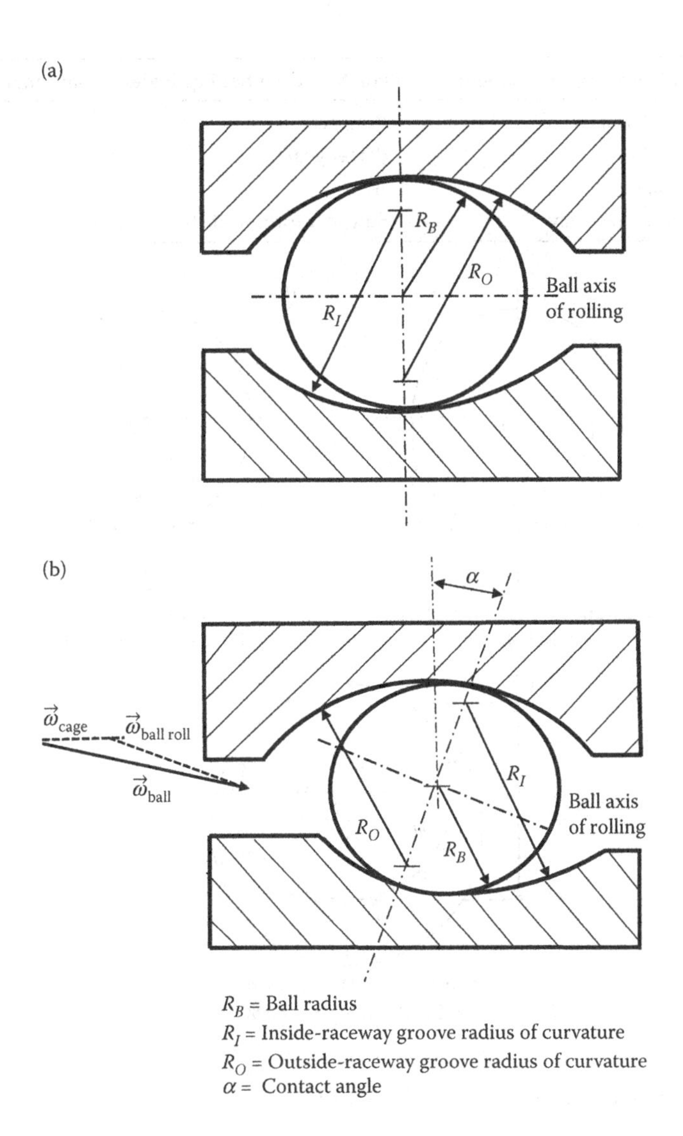

FIGURE 2.8
Ball-bearing raceway and ball radii-of-curvature. (a) Radial-contact ball bearing, (b) angular-contact ball bearing.

some degree of centering under zero-bearing load without the additional bearing ball-raceway-contact stress imposed by preload. But that is without substantial radial stiffness in an unloaded condition (see F vs. δ plot in Figure 2.6). To achieve substantial radial stiffness under all loading conditions including zero-bearing load (e.g., for machine tool spindles), a significant preload may be needed. That will expand the contact-load zone as illustrated in Figure 2.9, which shows the three categories of clearance condition. Referring to Figure 1.25 on elastohydrodynamic lubrication, the contact-load zone illustrated in that illustration is for the zero-clearance case.

In the absence of significant ball skidding, the ball-cage assembly orbital velocity can be kinematically viewed as a simple *planetary gearing set* (also called epicyclic gearing), illustrated in Figure 2.10. The bearing components and equivalent gear components are tabulated as follows.

Ball-Bearing Component		Planetary Gear Set Equivalent Component
1. Outer raceway	→	Ring gear
2. Balls	→	Planet gears
3. Cage	→	Arm
4. Inner raceway	→	Sun gear turning at shaft rotational speed

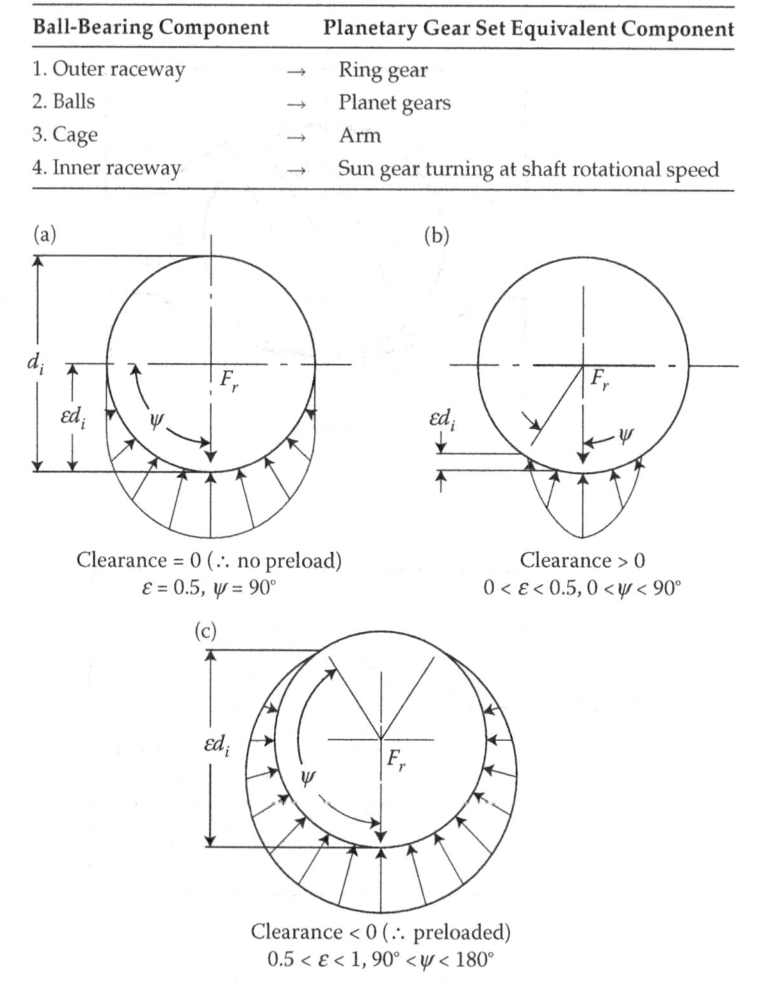

FIGURE 2.9
Ball-raceway contact-load zone as a function of clearance, *C*.

Unlike a tandem gear set that needs one speed input to obtain one speed output, the basic planetary gear sets (also called epicyclic gears) as illustrated in Figure 2.10 require two inputs to achieve one output, thus their well-known speed ratio versatility (e.g., automotive automatic transmissions). The sun, planet, and ring gear tooth numbers uniquely set the speed ratio options. The number of planet gears does not alter that, only letting the gear loads share between the planet gears. For the purpose here of determining the ball-cage orbital velocity, the single-planet configuration in Figure 2.10c suffices for this purpose. The following approach follows that found in any typical undergraduate kinematics textbook.

The rotational speeds are *referenced to the arm*. This means they hold the arm from rotating, which turns it into a tandem gear set, that is, one input and one output. Referring to Figure 2.10c, the equivalent tandem train value e is as follows.

$$e = \left(\frac{N_2}{N_4}\right)\left(\frac{N_4}{N_5}\right) \tag{2.5}$$

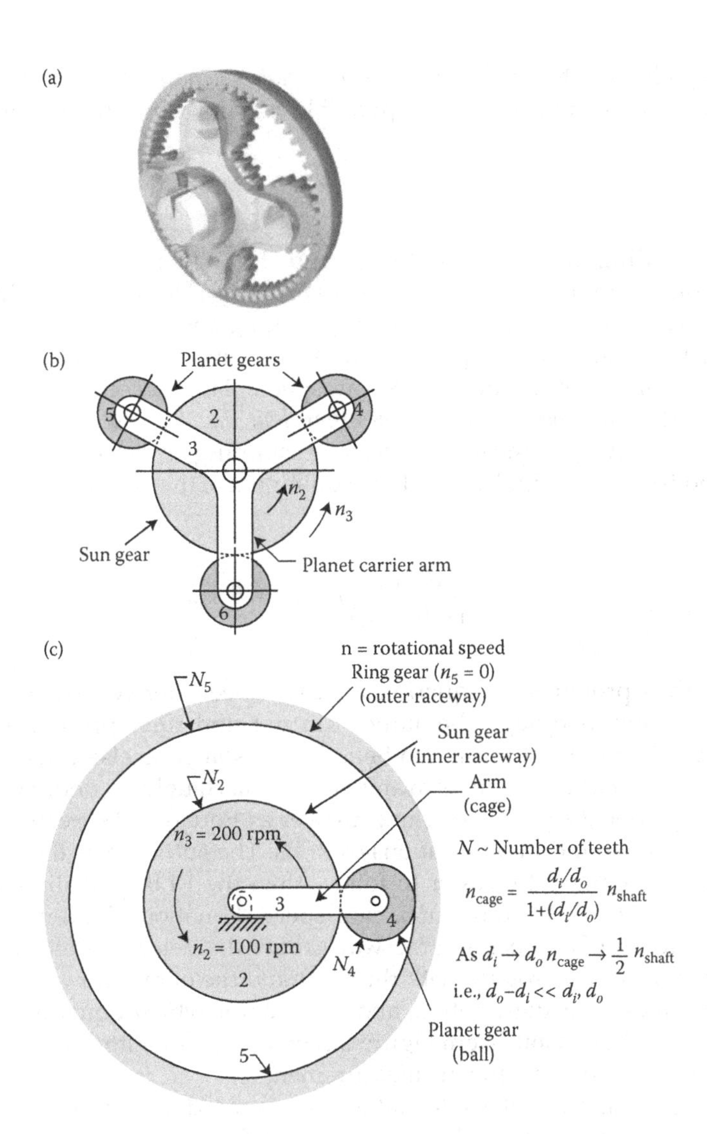

FIGURE 2.10
Planetary gear set equivalent to ball bearing. (a) Four-planet gears, (b) three-planet gears, (c) one-planet example with non-rotating ring gear ($n_5 = 0$).

N_i = number of teeth on gear number i. The train value $e > 0$ when the output rotation is in the same rotational direction of the input rotation, and $e < 0$ when the output rotation is in the opposite rotational direction of the input rotation. For the equivalent ball-bearing model, the N's are replaced as follows by the equivalent ball-bearing component diameters, as follows.

Gear Tooth Number		Ball-Bearing Component Diameter (see Figure 2.2)
N_2	→	d_i, inner-raceway bottom-of-groove diameter
N_4	→	D, ball diameter
N_5	→	d_o, outer-raceway bottom-of-groove diameter

By referencing all the rotational speeds to *arm rotational speed*, the gear set appears as a tandem gear set, with its train value (e) expressible as follows (n is rotational speed).

$$e = \frac{n_5 - n_3}{n_2 - n_3} \tag{2.6}$$

For example, scaling d_i (analogous to N_2) and d_o (analogous to N_5) from the ball-bearing illustration in Figure 2.2, and substituting into Equation 2.6 yields the following: $e = d_i/d_o = -0.58 = (n_5 - n_3)/(n_2 - n_3)$. Picking any speed for the inner-raceway rotational speed (i.e., shaft speed, plus for CCW, negative for CW) yields a cage orbital velocity of $n_3 = 0.37$ times the inner-raceway speed. This is consistent with common knowledge that the *cage speed rotates at less than half the shaft speed,* unless the cage is stuck to the inner raceway or the inner raceway is slipping on the shaft (i.e., two-bearing malfunction modes). Accordingly, based on Equation 2.6, the equation for cage speed is Equation 2.7.

$$n_{\text{cage}} = \frac{d_i / d_o}{1 + (d_i / d_o)} n_{\text{shaft}} \leq 0.5\, n_{\text{shaft}} \tag{2.7}$$

Equation 2.7 thus provides an equation for cage speed derived from the analogous planetary-gear system. Although the author did not find this approach to cage speed calculation in the literature, it is easy to believe that someone else might have already published it, since it should be obvious to anyone who has taught an undergraduate course in kinematics. Equation 2.7 yields the same answer as Equation 2.9, Section 2.3, which is derived as commonly found in publications of REBs. The author feels the derivation here for Equation 2.7 is significantly easier to follow than the REB specialists' derivation of Equation 2.9 in Section 2.3. To summarize cage speed, Equation 2.7 provides the formula for cage speed in terms of two-bearing raceway groove diameters as defined in Figure 2.2, assuming the inner raceway rotates with the shaft and the outer raceway does not rotate. For straight-line linear bearings, both d_i and $d_o \to \infty$, for which Equation 2.7 then gives $n_{\text{cage}} = 0.5\, n_{\text{shaft}}$, visually obvious from Figure 2.1. In service, monitoring of cage speed can provide a valuable real-time diagnostic measurement.

When applying Equation 2.7 it is relevant to keep in mind that unlike gears, the rolling elements and raceways are *not toothed*. So, any tendency for *net rolling slippage* (not interfacial footprint slippage a la Figure 2.4) of the rolling elements would produce a secondary effect yielding cage speed somewhat less than as predicted by Equation 2.7. Referring to Figure 2.9, it is reasonable to expect Equation 2.7 to be accurate if a substantial portion of the rolling elements are within the load-zone angle. An exception to this would be the rare instance of an unloaded, or very lightly loaded, REB assembled with clearance. In the modern era of elaborate machinery health *monitoring and diagnostics*, Chapter 5, bearing cage angular velocity can be a significant monitored signal for tracking symptoms of bearing health and other emerging machine deterioration rates and root causes.

In less typical applications such as the double-spool shaft gas turbine jet engine, the radial bearings intermediate between the inner and outer spool shafts of course experience both inner and outer raceways rotating potentially in opposite directions. For such multispool shaft applications at different speeds, the same planetary-gear-set fundamental speed ratio formulation here employed to obtain Equation 2.7 will also provide the formulation for cage speed. Specifically, the ring gear (outer raceway) illustrated in Figure 2.11 will

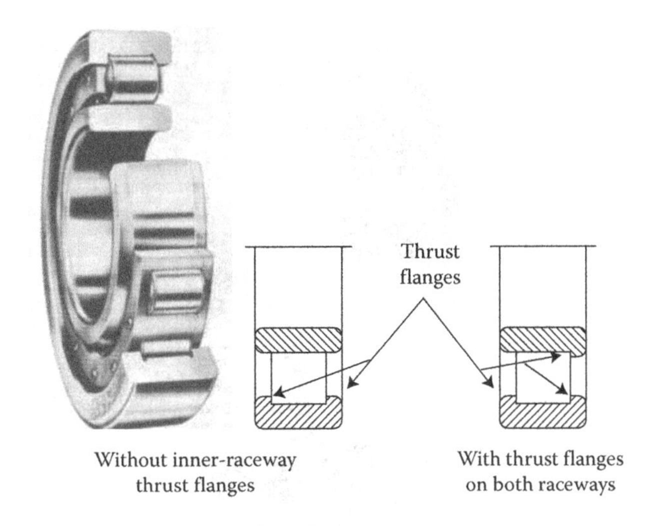

FIGURE 2.11
Single-row cylindrical bearing with and without thrust flanges.

then not have zero speed, and likewise for atypical applications where only the outer raceway rotates.

Utilizing the afore-presented introductions on (1) *contact footprint slipping* (Figure 2.4), (2) bearing *clearance* (Figure 2.9), and (3) *cage speed* (Figure 2.10 and Equation 2.6), the associated friction losses in ball bearings are addressed later in this chapter. That topic is of particular importance in the prediction of bearing operating temperatures, a quite significant limiting design factor for high-speed applications and applications with very high-bearing ambient temperatures (e.g., gas turbine jet engines). In addition, the specific lubrication-type employed in any application also is an important factor in assessing bearing operating temperatures; see Section 2.6. The following four are the major lubrication categories.

1. Dry with no lubrication
2. Grease-packed
3. Oil mist lubrication
4. Jet through-flow lubrication

Currently, most REB manufacturers have in-house proprietary computer codes, and/or commercially available computer codes that provide accurate predictions for many of the operating parameters of their products. Since about the late 1960s, codes have been available to solve for all the rolling element-raceway *statically indeterminate* interactive forces for any given combination of bearing radial and axial loading. Computer codes are now also available to predict the steady-state operating temperature of a bearing under given loading, ambient temperature, type of lubrication, and bearing manufacturing precision classification. This is a quite non-trivial heat transfer problem. Modeling approaches and analyses need precision test results to benchmark and calibrate such computer models, employing various empirically obtained computer model inputs (Adams 2017).

FIGURE 2.12
Double-row cylindrical roller bearing (precision machine tool spindle).

2.1.2 Roller Bearings

The following is a list of the available types/configurations of roller bearings. Each of these has many configuration versions as illustrated in Figures 2.11–2.22 (Harris and Kotzalas 2007).

1. Cylindrical
2. Spherical
3. Tapered
4. Needle

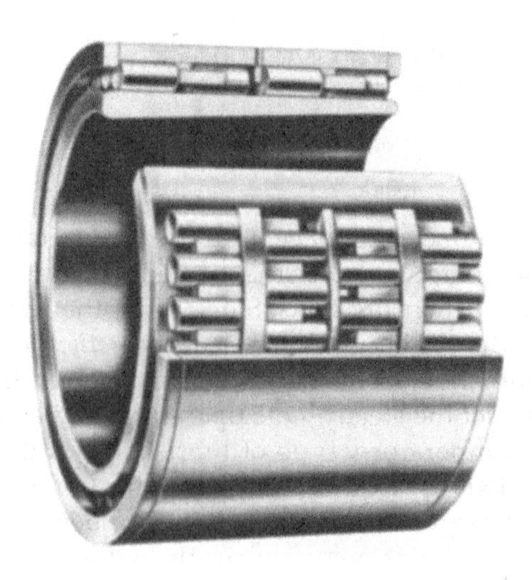

FIGURE 2.13
Multi-row roller bearing (rolling mill application).

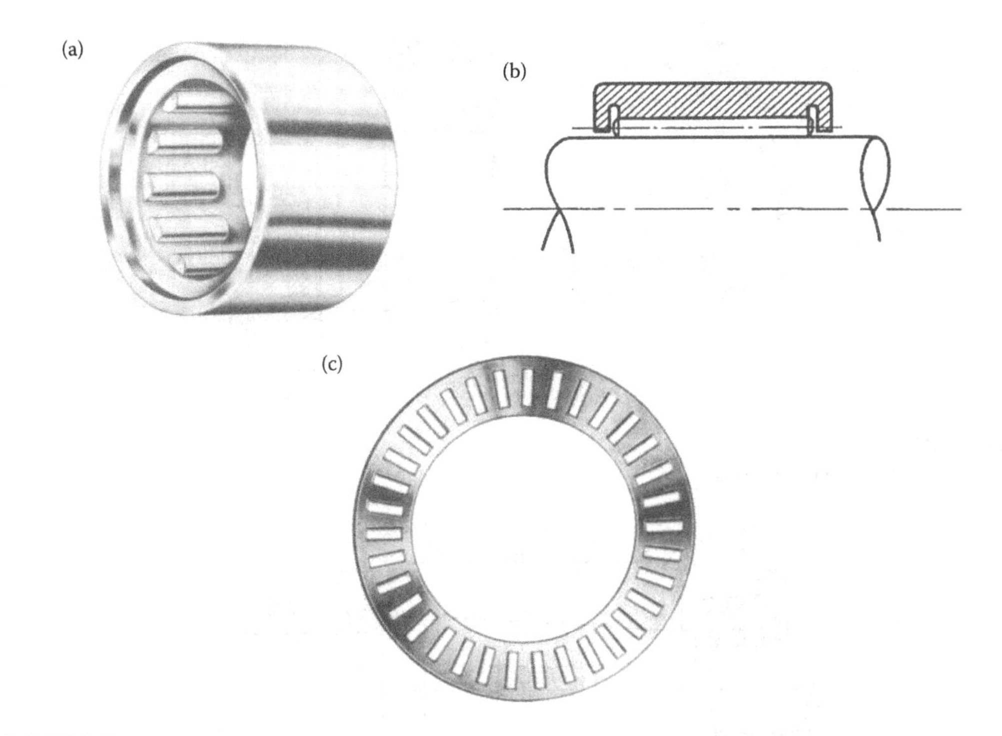

FIGURE 2.14
(a) Needle-radial bearing, (b) needle radial bearing with shaft as inner raceway, (c) needle thrust bearing cage assembly.

The *most fundamental difference* between all *roller bearings and ball bearings* is that rollers have only one natural axis about which to roll, the polar axis. Whereas any axis through the center of a spherical ball is an axis about which the ball can naturally roll. Therefore, a bearing ball's total natural instantaneous angular velocity is automatically accommodated. This is not the case for rollers.

FIGURE 2.15
Full complement (no cage) needle roller bearing.

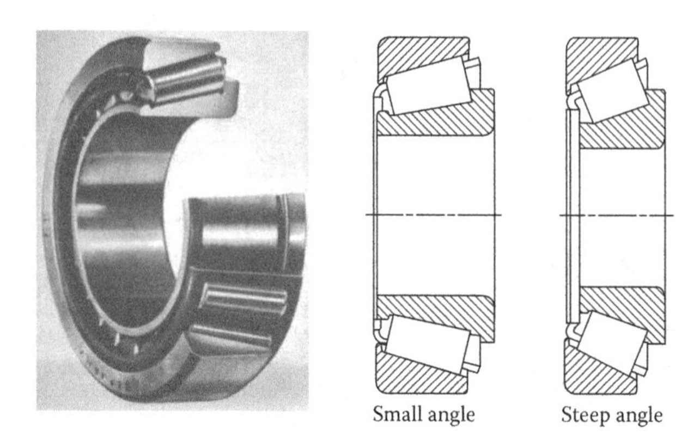

FIGURE 2.16
Single-row tapered roller bearing.

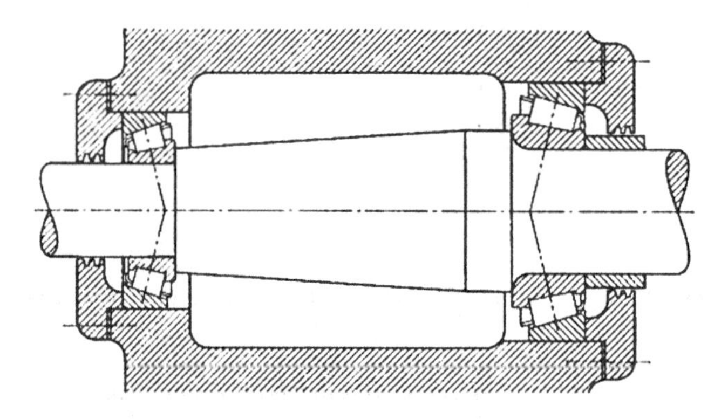

FIGURE 2.17
Typical application mounting of tapered roller bearings.

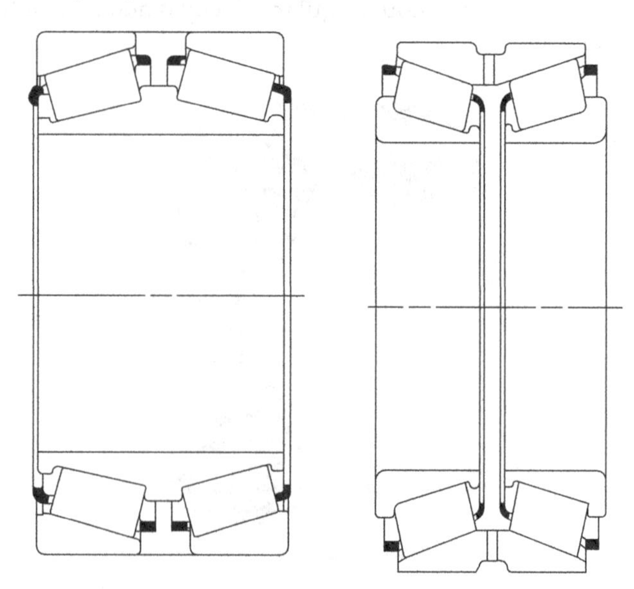

FIGURE 2.18
Double-row, double-cone tapered roller bearings.

FIGURE 2.19
Double-row spherical roller-bearing assemblies.

The *next important practical difference* between ball and roller bearings is that every axis through a spherical ball center is a *principal mass moment-of-inertia* axis. So, a spherical ball has no inertial coupling between its principle axes, that is, no products of inertia in any Cartesian coordinate system orientation. This is analogous and mathematically identical to the state-of-stress at a point when all three principal stresses are equal, that is, the 3D Mohr's circles become a single point, reflecting no shear stress. This is not the case for rollers, unless the length-to-diameter ratio is specifically so proportioned. This inherent property combined with the aforementioned fact that a roller has only one axis for rolling results in roller bearings having a significantly lower maximum speed capability than ball bearings. As rotational speed of a ball bearing is increased, the maximum useable operating speed will be limited by

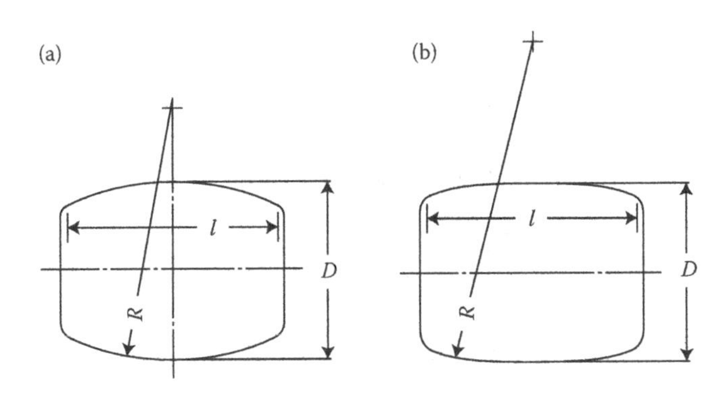

FIGURE 2.20
(a) Spherical roller (fully crowned), (b) partially crowned cylindrical roller (crown radius exaggerated for clarity).

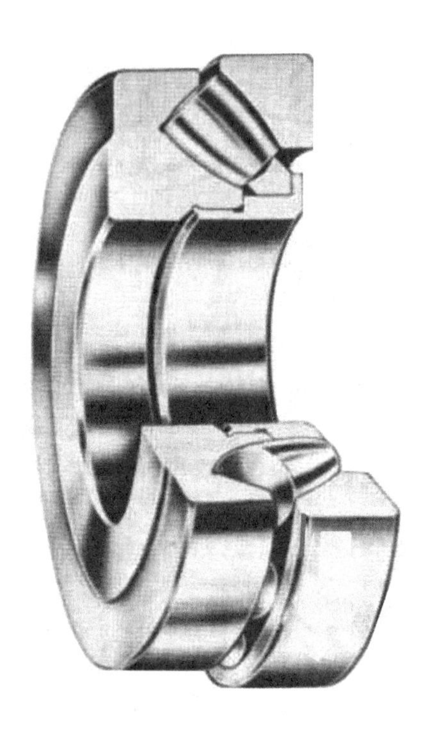

FIGURE 2.21
Spherical roller thrust bearing.

a combination of centrifugal-force ball loading on the outer raceway and/or bearing-elevated temperature. However, for a roller bearing, the maximum operating speed is usually limited by roller dynamics, that is, the onset of roller tilting motions, referred to as *skewing*. That is, as speed is progressively increased, a point is reached where the various interactive forces on the rollers, for example, from raceway thrust flanges, become strong enough to produce roller skewing. In physical terms, the aforementioned fundamental roller mass moment-of-inertia and geometry essentially initiate the roller to "attempt unsuccessfully" to emulate a ball's natural ability to simultaneously roll, spin, and skid.

A third important distinguishing difference between ball and roller bearings is the more limited capacity of a roller to handle loads co-linear with its roll axis. As a comparison, the radial-contact deep-groove ball bearing has a natural load-carrying capacity in the

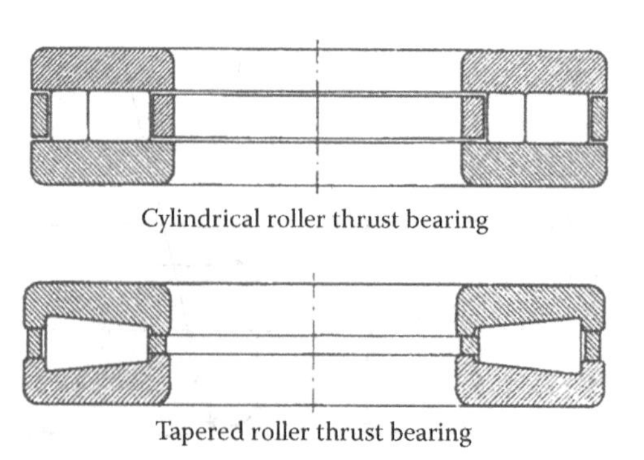

FIGURE 2.22
Cylindrical and taper roller thrust bearings.

axial thrust direction, albeit at less capacity than the radial-load capacity. In the case of a roller, its load-carrying contact footprint does not extend to the axial direction of the roller. Axial-load capacity requires the presence of raceway *thrust flanges*, Figure 2.11. But the roller contact footprint extends over the full axial length of the roller. That fact provides the *major advantage of roller bearings* over ball bearings, *much higher-load capacity*. In the next section, this is covered in more detail.

The *single-row cylindrical* roller bearing is the most frequently employed member of the roller-bearing family, used primarily to carry strictly large radial loads. By employing two or more rows of cylindrical rollers, the relative radial-load capacity is accordingly increased without instead employing a single more-prone-to-skidding equivalent-length single roller, Figures 2.12 and 2.13. When the cylindrical roller bearing employs "long" rollers (large length-to-diameter ratio) suitable for lower speed applications, it is called a *needle bearing*, Figures 2.14 and 2.15.

Tapered roller bearings combine both radial and axial high-load capacity and are available in many configurations, Figures 2.16, 2.17, and 2.18.

Double-row *spherical roller* bearings allow the roller axes to be angled with respect to the shaft axis as illustrated in Figure 2.19, to inherently provide both radial and axial-thrust-load capacity in contrast to a cylindrical roller bearing. Cylindrical and spherical roller-bearing rollers are compared in Figure 2.20. Cylindrical rollers are typically crowned at the axial ends to eliminate stress concentration there, as detailed in the next section. A single-row spherical roller bearing is also configurable to carry primarily axial-thrust capacity as pictured in Figure 2.21.

Both cylindrical and tapered roller bearings are specifically configurable for axial-thrust-load capacity, as illustrated in Figure 2.22. But the cylindrical-roller configuration clearly has significant interfacial sliding, which limits it to relatively lower speed applications. In contrast, the tapered-roller configuration can have the roller cone angle specified to avoid the gross interfacial slipping inherent in the cylindrical-roller configuration. Thus, the tapered-roller configuration has a significantly higher speed capability than the cylindrical-roller configuration.

2.2 Loads, Stress, and Deformation

The inherent characteristic of a rolling contact is that the load-carrying elastic contact footprint area is small in comparison to the area sizes of the elastic bodies in contact. Rolling contacts are illustrated in Figures 2.23 and 2.24. In the shown case of a wheel

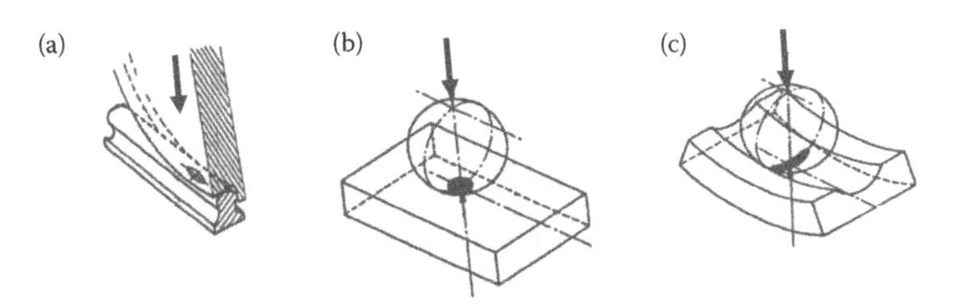

FIGURE 2.23
Variety of Hertzian elastically loaded contact footprints. (a) Wheel on rail, (b) ball on plane, (c) ball on outer raceway.

loaded against a rail, Figure 2.23a, both contact surfaces have a convex radius of curvature in planes mutually perpendicular to each other, thus yielding a small load-carrying *elliptic* contact footprint deformation. For a ball loaded against a plane, the ball radius provides a convex contact radius of curvature in all planes perpendicular to the contacted plane, thus the load-carrying contact footprint deformation is *circular* as shown, Figure 2.23b. A ball loaded against a ball-bearing raceway also has a convex radius of curvature, but the raceway has a concave radius of curvature, with the raceway groove radius of curvature approximately twice that of the ball, that is, *osculation* ≈ 0.5, resulting in an *elliptic* elastic contact footprint as illustrated, Figure 2.23c. Figure 2.24a illustrates a ball in loaded contact between its inner and outer raceways, showing both its *elliptic* elastic contact footprints and the corresponding *surface static compressive stress distribution*. Figure 2.24b illustrates a loaded cylindrical-roller elastic-contact footprint with its outer raceway and the corresponding *surface static compressive stress distribution*. Note that the footprint areas are small in comparison to overall rolling element surface areas. In the roller-bearing case, the contact footprint is the product of a small number, $2b$, and a relatively large number (L). For the ball-bearing case, the contact footprint area is from two relatively small dimensions, a and b. This characteristic comparison is the fundamental reason why a roller bearing can carry significantly more load than a comparably sized ball bearing. That is, the footprint-load-carrying area of a roller within the elastic limit is much larger than that of a ball on its raceways. All these elastic contacts are referred to as Hertzian contacts (Hertz 1896).

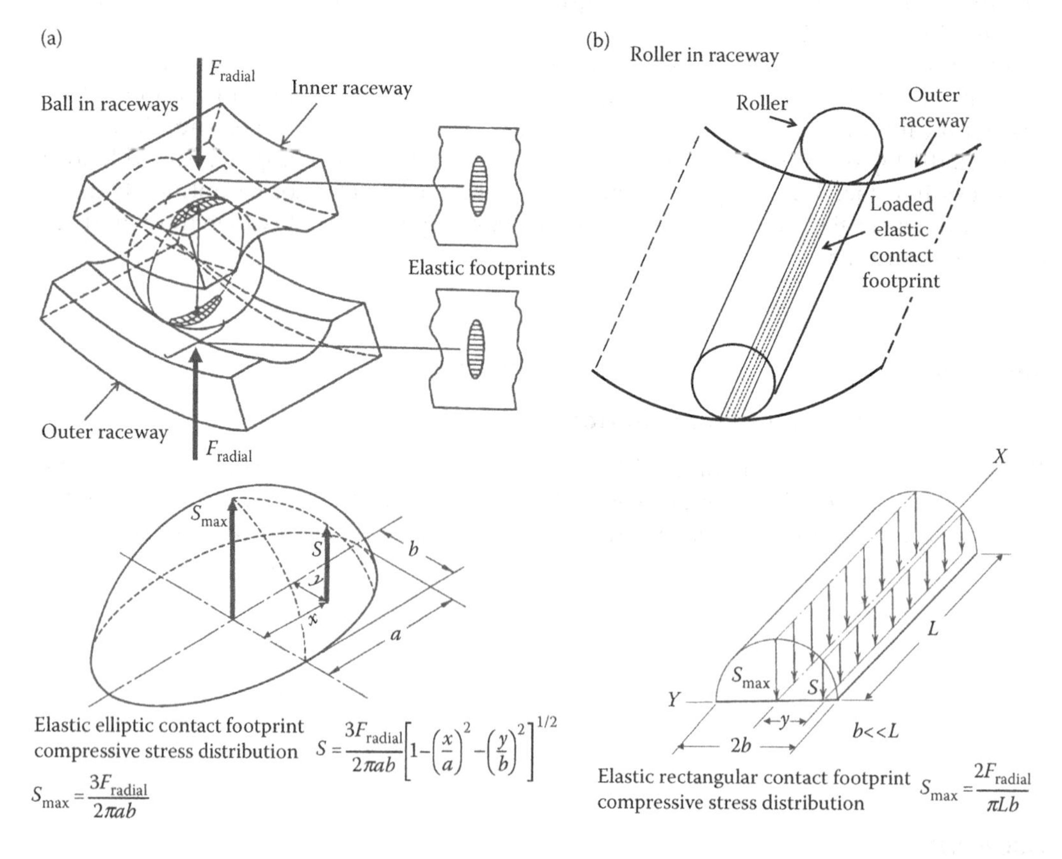

FIGURE 2.24
Loaded elastic contact footprints; (a) ball bearing, (b) roller bearing.

REB-*static load* is often the dominant factor dictating bearing selection. From an applied-bearing static load, determination of each individual rolling element-on-raceway instantaneous interactive static force for the rotationally changing element positions is *not a simple static equilibrium problem*. That is because the multiplicity of rolling elements simultaneously supporting the applied-bearing load makes the individual rolling-element loads *statically indeterminate*. This inherent complex property of REBs is deserving of its own special treatment, Sections 2.4 and 2.5. Further complicating REB internal interactive forces, as a rolling element moves through the element load zone the element-on-raceway interactive force changes in time as a function of the changing angular position of the rolling element as well as the initial assembled bearing clearance (see Figure 2.9).

With the aid of Figure 2.25 (Harris and Kotzalas 2007), the mission here is to focus only on the interactive forces between rolling element and raceways in terms of the individual element loading. Figure 2.25a illustrates the simplest example, a ball supporting only a radial-static load between raceways. The ball-on-raceway interactive static load is directly obtained from *elementary static equilibrium*, yielding the following ball-on-raceway interactive force.

$$Q = \frac{Q_r}{\cos\alpha} \tag{2.8}$$

Similarly, as illustrated in Figure 2.25b, roller-on-raceway interactive-static load on a symmetrically angled pair of spherical rollers is likewise given by Equation 2.7. The roller-on-raceway loading of a tapered roller is a bit more complicated as Figure 2.25c illustrates.

Formulating ball rotations that simultaneously contribute to body forces is complicated by the fact that as the balls orbit with the cage, the imposed loads between the raceways vary considerably as they transition into and out of the contact-load zone (see Figure 2.9). The formulation for cage rotational speed provided by Equation 2.7 based on the planetary

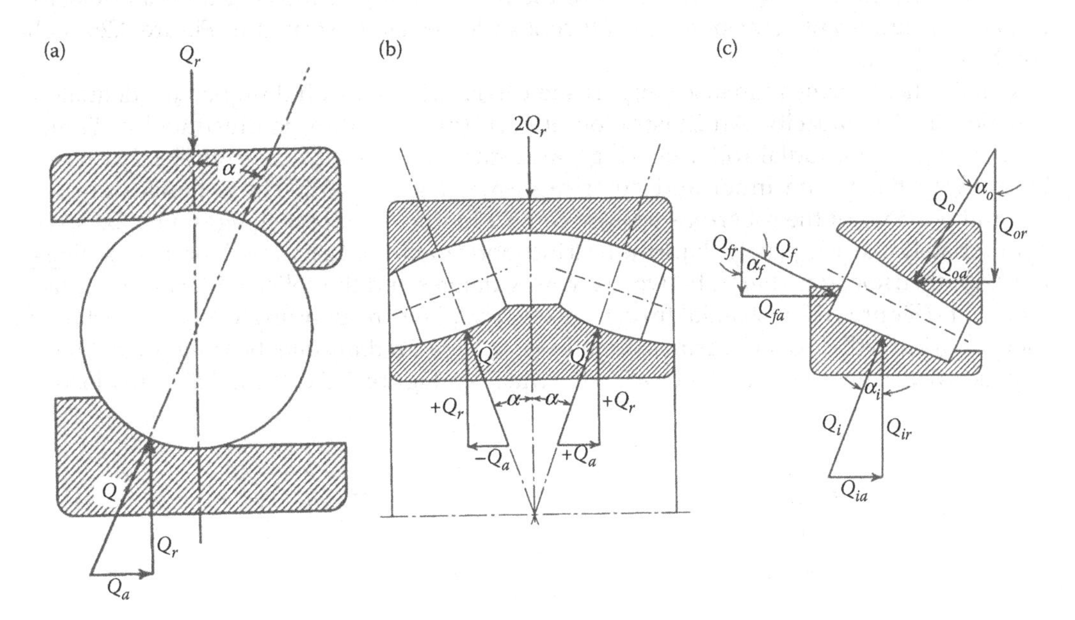

FIGURE 2.25
Radially loaded raceways with (a) ball, (b) symmetric spherical roller pair, (c) tapered roller.

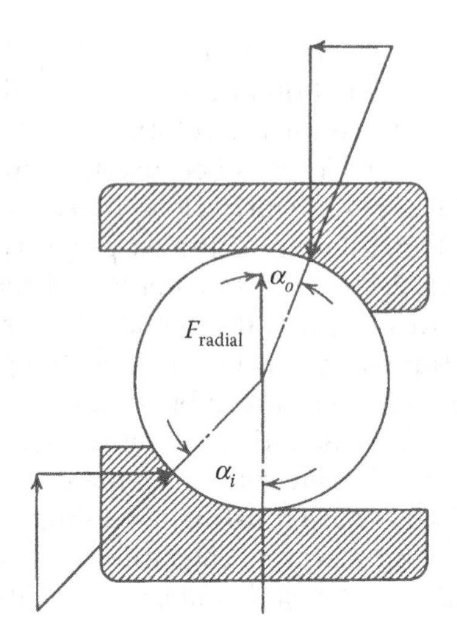

FIGURE 2.26
Ball under thrust plus centrifugal load.

gear model does not account for the potential effects of the cyclic-loading transition that rolling elements can experience once per cage orbit, although these most likely average out in favor of Equation 2.7 accuracy. The cage-rotational-frequency phenomenon produces significant accelerations and decelerations in ball angular velocities, that is, roll, spin, and skidding. The uncertainties in that phenomenon are just one factor among many that have made extensive laboratory development testing of REBs an essential component in the evolution of modern REBs. Also, advanced computer codes have been developed in modern times to model such dynamic phenomena with REBs (Section 2.7). In high-speed ball bearings, the *centrifugal acceleration* of the balls orbiting at the cage angular velocity becomes a significant contributor to ball contact forces as illustrated in Figure 2.26 with combined thrust.

Radial-roller bearings can also support some axial-thrust load if designed to adequately function in that capacity. An illustration of such thrust loading is illustrated in Figure 2.27. For *cylindrical* radial-roller bearings, axial-thrust loads must be completely carried by raceway flanges on inner and outer raceways, Figure 2.27a. Those flanges are not integral portions of the roller-on-raceway contact footprints, so the flanges must be sized appropriately and adequately share the bearing lubrication. For *tapered* radial-roller bearings, the axial-thrust load is shared between raceway flanges and the roller-on-raceway contact footprints (Figure 2.27b), so axial thrust load capacity is correspondingly more robust than for cylindrical radial-roller bearings. In contrast, *spherical* radial-roller bearings do not need any raceway flanges because as clearly illustrated in Figure 2.27c, the axial-thrust load is

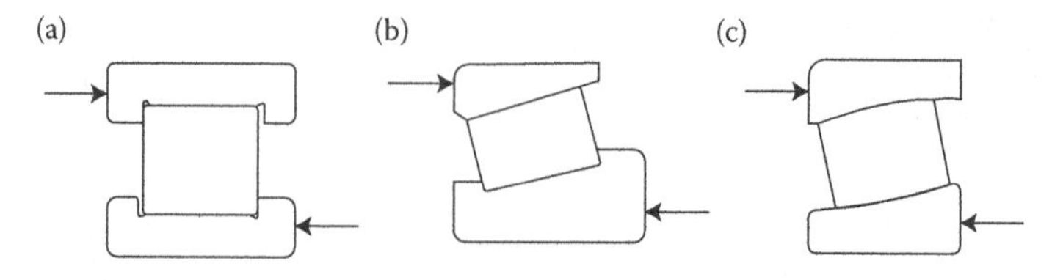

FIGURE 2.27
(a) Cylindrical, (b) tapered, and (c) spherical rollers thrust-loaded.

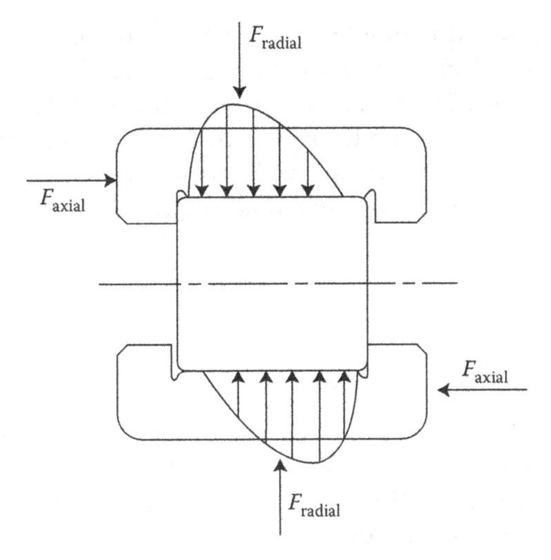

FIGURE 2.28
Cylindrical radial roller bearing contact-load distribution under combined radial and axial loads.

carried entirely by the roller-on-raceway contact footprints. The illustration in Figure 2.28 shows the asymmetric-load distributions across the roller-on-raceway contact footprints (see also Figure 2.24b) in a radial cylindrical roller bearing with combined radial and axial load. Similar asymmetric roller-on-raceway load distributions of course occur on tapered roller and spherical roller-radial bearings under combined radial and axial loading.

In high-speed applications, a combination of (1) dynamic effects of roller mass unbalance, (2) impact loading between roller and flanges, and (3) impact loading between roller and cage can produce roller skewing. Roller skewing in roller bearings is defined by Harris and Kotzalas (2007) *"as an angular rotation of the roller axis (in a plane tangent to its orbital direction) with respect to the axis of the contacting ring,"* at the slewing angle, Figure 2.29.

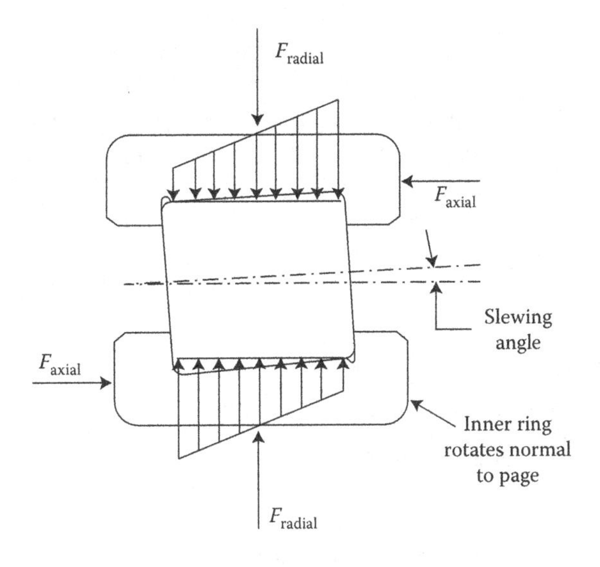

FIGURE 2.29
Snapshot of roller in skewing rotation.

To determine the maximum acceptable operating speed for a particular roller bearing design and application, such roller dynamic skewing can be modeled and simulated with present-day software, Section 2.7. As stated earlier in this chapter, such dynamic motions of a roller are the primary factor that limits the maximum practical operating speed of a roller bearing. This is in stark contrast to ball bearings since the balls are spheroids, thus yielding significantly higher maximum speed ability than roller bearings.

2.3 Component Motion Kinematics

Two kinematically equivalent rolling contact systems are illustrated in Figure 2.30. For the typical application case, the outer raceway is not rotating while the inner raceway rotates as shaft speed ω_I. That is the condition illustrated in Figure 2.30a. To best understandably visualize the pure rolling case, Figure 2.30b, just view the Figure 2.30a case from a coordinate system rotating at the cage angular velocity ω_C. That confirms that the raceway angular velocities defined in Figure 3.30(b) would in fact have the rolling element rotating without translating, that is, pure rolling.

Utilizing the *pure rolling* case (Figure 2.30b) yields the following derivation.

$$V_A = V_{A'} \therefore 1/2(E-d)(\omega_I - \omega_C) = 1/2\omega_C d \quad \text{and}$$

$$V_B = V_{B'} \therefore 1/2\omega_r d = 1/2(E+d)\omega_C \text{ and } V_{A'} = V_{B'} \text{ yields the following.}$$

Cage speed ratio:

$$\omega_C/\omega_I = 1/2[1 - d/E] \tag{2.9}$$

Equation 2.9 (Zaretsky and Poplawski 2009) provides the same result as Equation 2.7 based on the author's application of the analogous planetary gear set model (Figure 2.10). Also resulting from the derivation of Equation 2.9 is the following.

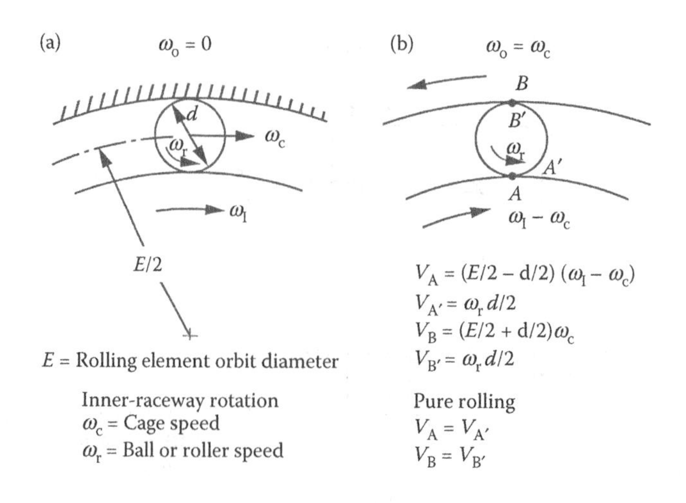

FIGURE 2.30
Ball rolling kinematics (a) inner raceway rotation, (b) pure rolling. (From Zaretsky, E. V. and Poplawski, J. V., *Rolling Element Bearing Technology*, Handout for 3-day short course, Case Western Reserve University, Cleveland, Ohio, p. 973, June 23–25, 2009.)

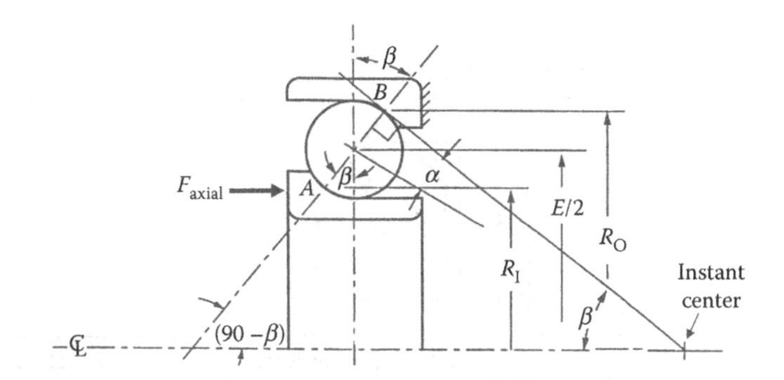

FIGURE 2.31
Thrust-load bearing kinematics.

Element speed ration:

$$\omega_r/\omega_I = 1/(2d)[1-(d/E)^2] \tag{2.10}$$

For a *thrust-load* bearing (Figure 2.31), similarly utilizing the pure rolling case, Figure 2.30 yields the following results for the thrust-load bearing.

$$V_A = R_I(\omega_C - \omega_I) = [E/2-(d/2)\cos\beta](\omega_I - \omega_C), \quad V_{A'} = (d/2)\omega_B$$
$$V_B = (d/2)\omega_B, \quad V_B' = R_O\omega_C = [(E/2)+(d/2)\cos\beta]\omega_C$$
$$\text{Setting } V_A = V_{A'} \text{ and } V_B = V_{B'} \text{ yields the following.}$$
$$\omega_C/\omega_I = 1/2[1-(d/E)\cos\beta] \tag{2.11}$$

Summarized in Table 2.1 is a compendium of various rolling element-bearing kinematic formulas (Jones 1946).

2.4 Computer-Aided Research and Development Analyses

A quite extensive list of commercially available computer codes for REB is given by Zaretsky and Poplawski (2009) and listed here in Table 2.2.

There is no single information source that summarizes the capability of all the REB software listed in Table 2.2. Following are the capability summaries for available REB computer codes extracted from readily available information sources.

The A. B. (Bert) Jones computer code was the first widely recognized general purpose rolling element-bearing code, initially developed in the early years of the modern digital computer (Jones 1960). Jones' computer code provided a completely general solution, where the elastic compliances of a system of any number of ball and radial roller bearings under any system of loads could be accurately simulated. Elastic deformation of the shaft and supporting structure are considered as well as centrifugal and gyroscopic loading of the rolling elements under high-speed operation. The simulation solution provides the loading and attitude of each rolling element in each bearing of the system as well as the displacement of the inner raceway with respect to the outer raceway. For ball bearings,

TABLE 2.1

Rolling Element-Bearing Kinematic Formulas

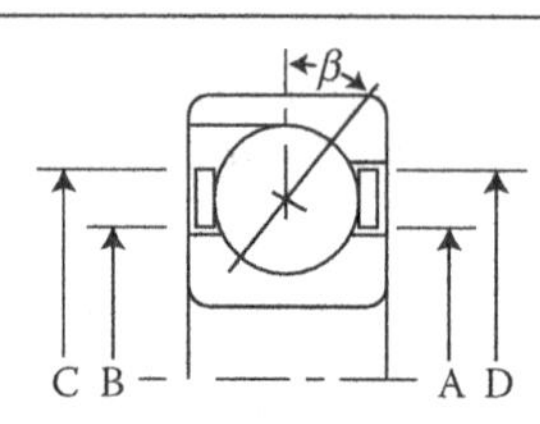

Quantity Required	Rotating Inner Race	Rotating Outer Race
Speed ratio, ρ Cage	$\rho_i = \frac{N_E}{N_i} = \frac{1}{2}\left[1 - \frac{d}{E}\mathrm{Cos}\,\beta\right]$	$\rho_o = \frac{N_E}{N_o} = \frac{1}{2}\left[1 + \frac{d}{E}\mathrm{Cos}\,\beta\right]$
Translation velocity of pitch circle, FPM	$V_E = \frac{\pi E N_i}{24}\left[1 - \frac{d}{E}\mathrm{Cos}\,\beta\right]$	$V_E = \frac{\pi E N_o}{24}\left[1 + \frac{d}{E}\mathrm{Cos}\,\beta\right]$
Angular velocity of ball about its center, RPM element	$N_B = \frac{E N_i}{2d}\left[1 - \left(\frac{d}{E}\right)^2 \mathrm{Cos}^2\,\beta\right]$	$N_B = \frac{E N_o}{2d}\left[1 - \left(\frac{d}{E}\right)^2 \mathrm{Cos}^2\,\beta\right]$
Velocity of rub of ball in pocket, FPM	$V_B = \frac{\pi E N_i}{24}\left[1 - \left(\frac{d}{E}\right)^2 \mathrm{Cos}^2\beta\right]$	$V_B = \frac{\pi E N_o}{24}\left[1 - \left(\frac{d}{E}\right)^2 \mathrm{Cos}^2\beta\right]$
Velocity of rub of C on D, FPM	$V_{CD} = \frac{\pi C N_i}{24}\left[1 - \frac{d}{E}\mathrm{Cos}\,\beta\right]$	$V_{CD} = \frac{\pi N_o}{12}\left[D - \frac{C}{2}\left(1 + \frac{d}{E}\mathrm{Cos}\,\beta\right)\right]$
Velocity of rub of B on A, FPM	$V_{BA} = \frac{\pi N_i}{12}\left[A - \frac{B}{2}\left(1 - \frac{d}{E}\mathrm{Cos}\,\beta\right)\right]$	$V_{BA} = \frac{\pi B N_o}{24}\left[1 + \frac{d}{E}\mathrm{Cos}\,\beta\right]$

Note: The formulas which apply to the various velocities are tabulated above.
N_E = rpm of pitch circle, N_i = rpm of rotating inner race, E = pitch diameter, N_o = rpm of rotating outer race, β = contact angle, d = ball diameter.

TABLE 2.2

Commercially Available RCB Analysis Computer Codes

	Source
I. Single-Bearing Analysis Codes	
1. CYBEAN	NASA Cosmic
2. TRBO	Wright-Paterson Development Center
3. CYFLEX	Naval Air PL
4. ROSDREL	Avco, Inc.
5. BASDREL	Avco, Inc.
6. BATHERM	Avco, Inc.
II. Up to 6 Bearings	
1. A. B. Jones	ABJ, Inc.
2. SHABERTH	NASA Cosmic
3. SHABERTH	SKF PC
4. COBRA	Poplawski Assoc.
5. GENROL	Franklin Institute Research Center. FIRC
III. Transient Dynamic Models	
1. ADORE	P. K. Gupta, Inc.
2. DREB	PKG, Inc.
3. RAPIDREB	PKG, Inc.

the precise positions of loading paths for each raceway are solved. Life estimates are thus more accurately made since the fatigue effects can be evaluated over known paths in the raceways. The solution is accomplished computationally through multiple nested interaction loops employing primarily the Newton–Raphson method. That is because the simultaneous multiplicity of Hertzian contact footprint loads are statically *indeterminate* and *nonlinear* (see *Force vs. Deflection* sketch in Figure 2.6). Subsequent REB general analysis computer codes such as those listed in Table 2.2 were all inspired by Bert Jones' original approach. Early in his career, during his professionally rewarding four years employment (1967–1971) as a research engineer at the Franklin Institute, the author worked on the initial development of one of those codes (Table 2.2, GENROL, organization then named the Franklin Institute Research Laboratories, FIRL), and worked as well on specialized computer code offshoots of GENROL.

CYBEAN (CYlindrical BEaring ANalysis) considers a flexible, variable geometry outer ring, EHL films, roller centrifugal and quasi-dynamic loads, roller tilt and skewing, mounting fits, cage and flange interactions. Kleckner et al. (1980) give the analytic foundation and software architecture for the computerized mathematical simulation of high-speed cylindrical rolling element-bearing behavior. This simulation includes both steady state and time-transient simulation of thermal interactions internal to and coupled with the surroundings of the bearing. They also provide a sample problem illustrating program usage.

BASDREL is a ball-bearing dynamics simulation computer code developed to rigorously model all the significant kinematic, structural, and dynamic effects. This computer code has the capability to analyze bearings of any material combination for the races, ball, and ball cage. The computer code also models the stresses and deflections of the loaded elements due to (1) preload, (2) external axial, radial, and moment loads, and (3) centrifugal and gyroscopic ball loads. BASDREL utilizes a six-degrees-of-freedom model of ball cage motion, formulated to analyze generalized ball and cage dynamics. BASDREL can also handle multiple simultaneous quasi-static ball-to-race load-deflection equations using a modified Newton–Raphson method. The Lawrence Livermore ODE numerical integration package (LSoDA) is employed for integration of the dynamic equations of motion. Analyzing ball and cage motions in the time domain, it predicts wear life, fatigue life, lubricant-film effects, ball-to-cage forces, torque noise, and many other bearing parameters.

SHABERTH was developed by NASA to simulate steady-state and transient performance of a shaft-bearing system including ball, cylindrical, and tapered roller bearings. The user's manual (Hadden et al. 1981) for this computer code is in the public domain, and is capable of simulating the thermomechanical performance of a load support system consisting of a flexible shaft supported by up to five rolling element bearings. Any combination of ball, cylindrical, and tapered roller bearings can be used to support the shaft. In calculating lubricant-film thickness and traction forces, the user can select from among many included model options. The formulation of the cage pocket-rolling element interaction model is formulated to optimize numerical convergence characteristics.

REBANS was developed by Aerojet-General Corporation under NASA sponsorship. It is a system of computer codes, offering improved capabilities for the analysis of high-speed rolling element bearings. REBANS is public domain, documented by Greenhill and Merchant (1994). The capabilities of the software packages in this system are unique to the analysis of rolling element bearings in that the effects of bearing ring and support structure flexibility are included directly in the calculations. A finite element representation, prepared using Version 4.4 of ANSYS, is used for modeling of bearing raceways and any support structure. The analysis system can determine the response of the bearing and surrounding structure to quasi-static loading conditions.

ADORE is the proprietary software of P. K. Gupta, Inc. The user's manual for it (Gupta 2014) is available on the Internet. In the opinion of rolling element-bearing research and development specialists, ADORE is perhaps the most advanced REB software, especially in simulating real-time dynamic motions for high-speed operation, with options to output video animations of such REB-motion simulations.

Computer code development work related to the ADORE, DREB, and RAPIDREB dynamic motion computer code simulations are compared against experimental data (Gupta et al. 1985). The general motion of the cage is predicted by the computer models in an angular-contact ball bearing operating up to a DN of 2 million. *Both the computer predictions and experimental data exhibited a critical shaft speed at which the cage mass center begins to whirl.* The predicted and measured whirl velocities and orbit shapes are in good agreement. Impressively, the computer models for the axial and radial velocities of the cage mass center also agree with test data within the tolerance band of the expected experimental error. Due to experimental difficulties, the cage angular velocity could not be reliably measured at high speeds. At low speeds, however, there is a fair agreement between the experimental data and the DREB/RAPIDREB computer model predictions.

2.5 Lubrication

For most heavy industrial and commercial applications, grease is the predominant choice for REB lubrication. Some of these applications are grease-packed and configured to be lubricated for the life of the bearing. Alternatively, for many applications there is provision in the design to facilitate relubrication at appropriate intervals. Relubrication facilitates keeping debris and water buildup from otherwise shortening bearing life. Higher speed applications employ oil bath, circulating oil, or oil mist lubrication which additionally cool the bearings. Figure 2.32 illustrates the required lubrication points for a REB.

Off-the-shelf greases with appropriate thickeners typically can function up to 350°F (176°C). It is recommended that for heavily loaded low-speed bearings an extreme-pressure (EP) additive should be in the grease. There is an extensive engineering technology underpinning

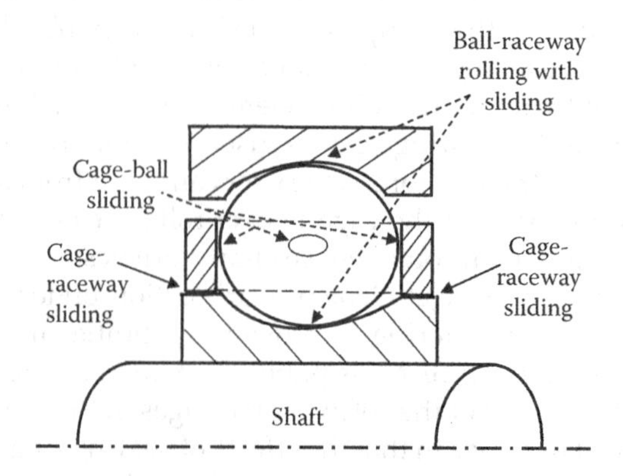

FIGURE 2.32
Required lubrication points in a rolling contact bearing.

for the influence and importance of lubrication parameters on successful REB operation, load capacity, and life, extensive enough that it should be a book its own. The remainder of this section is a condensation of that technology underpinning, with Zaretsky (1999) the primary information sources. Quoting Zaretsky (1999), a rolling bearing "lubricant has four major functions: (1) to provide a separating film between rolling and sliding surfaces in contact, thus preventing wear, (2) to act as a coolant for maintaining proper bearing temperature, (3) to prevent the bearing from becoming overtaken by a buildup of dirt, wear particles and other contaminants, and (4) to prevent corrosion of bearing surfaces."

Unlike the relatively thick hydrodynamic lubricating films, in rolling contact bearings (see Figure 1.25 and accompanying text) the very thin lubricant films are at much higher pressures, resulting in a quite large momentary oil viscosity increase and elastic contact surface deformations comparable to the film thickness. The Stribeck diagram as illustrated in Figure 1.1a is supplemented there with a region designated as EHL (elastohydrodynamic lubrication a la Zaretsky 1999) in keeping with Jones (1982), although the author feels that is a *conceptual misfit* since the abscissa of the Stribeck diagram is not a primary EHL parameter like it is for a sliding bearing. However, Jones (1982) supplements this with a useful correlation of the different lubrication regimes and severity of wear rate as illustrated in Figure 2.33. In the load zone OA (hydrodynamic and EHL), since there is no direct touching contact between the loaded-interacting surfaces, there is virtually no significant wear during normal operation, except perhaps accumulated from numerous starts and stops. For rolling bearings, this precludes element fatigue which can occur without touching contact between the mating surfaces.

The *mixed lubrication* zone AX is where surface contact touching begins to occur, with wear rates relatively low because of some partially separating lubricant film. In the boundary lubrication zone XY, wear rate becomes more significant since the only factor fighting wear is the long chemical molecules that are formed and rapidly wiped away by the sliding contact (see Figure 1.3 and associated text). The YZ zone in Figure 2.33 is obviously where bearings will not function successfully.

Zaretsky (1999) contains extensive tables of fluid and grease lubricants, listed by (1) MIL specification or SAE, ASTM, and AGMA numbers, (2) recommended temperature ranges,

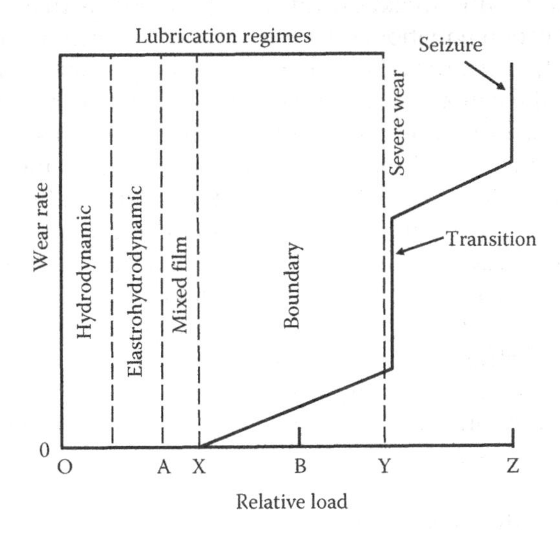

FIGURE 2.33
Wear rate-associated lubrication regime as a function of relative load.

and (3) composition. Rolling-bearing grease applications are advantaged because of the feasibility of simplified housings and seals. Whether in liquid or grease form, mineral oil is the most frequently employed lubricant, typically containing at least (1) an anti-wear extreme-pressure (EP) additive, (2) an anti-foam additive, and (3) an oxidation inhibitor. For specific applications, most notably the internal combustion engine, there are several other additives now utilized. Early in the lubrication field, viscosity was virtually the only perceived important lubricant parameter. In modern times, lubrication specialists jokingly suggest that the main function of modern lubricants is to carry around all the several additives. Of course, viscosity is always important as well as its longevity in use. Today oils *wear out*, not so much from viscosity degradation but usually first by breakdown of the additives that are so critical to modern engineered applications like internal combustion engines.

Synthetic lubricants, which have overcome some of the harmful effects of mineral oil oxidation, have become a primary choice for automobile new engine designs. Zaretsky (1999) recommends that synthetic lubricants should not be chosen over less expensive mineral oils if operating conditions do not require the advantages of a synthetic lubricant. Zaretsky further recommends that it is usually easier to incorporate synthetic lubricants in a new design than to convert an existing machine to their use. He emphasizes that when lubricant selection is an option, the first priority should be to determine anticipated bearing operating temperature, and then critically scrutinize the effects of temperature on the lubricant rheological properties. The second order of priority is the lubricant viscosity properties as a function of temperature and degradation in use.

The original identification of EHL oil films between rolling contact surfaces as a unique lubrication regime is generally attributed to Grubin (1949) but not widely appreciated until circa 1960. EHL is truly one of the major post WW II discoveries in the field of tribology, governing the reliability of highly stressed contacts between nonconforming surfaces like rolling contacts in bearings and rolling-with-sliding in gear tooth contacts. The introduction to EHL is given in Section 1.7 because of its incorporation of the Reynolds lubrication equation, with Figure 1.25 graphically illustrating EHL in the most import example, the ball bearing.

Calculations of EHL-oil-film thickness are based on models that couple the following: (1) the Reynolds lubrication equation, (2) the enormous momentary increase in viscosity (see Equation 1.17) only while the oil is momentarily trapped within the very thin rolling contact EHL film, (3) the contact-local elastic deformations at the contact footprint, and (4) the combined surface roughness of the two mating surfaces. Because of modeling the simultaneous confluence of these physically complex phenomena, computed results for EHL-film thickness do not readily accommodate normalization by one or two dimensionless numbers. That is, there is no simply definable dimensionless *"EHL number"* to add to that valuable family of other dimensionless numbers such as the Biot, Brinkman, Eckert, Fourier, Grashof, Knudsen, Lewis, Mach, Nusselt, Peclet (Reynolds x Schmidt), Prahl, Prandtl, Reynolds, Schmidt, Stanton numbers, plus a few others. Therefore, the published formulas for EHL-film thickness are based on elaborate curve fitting of multiple specific computed cases. The resulting formulas may understandably strike one as an unusual conglomerated multitude of random coefficients and exponents. Zaretsky (1999) conveniently summarizes such formulas of the main recognized contributors to EHL technology, as follows.

Grubin (1949), central film, line contact:

$$H = 1.95 G^{0.73} U^{0.73} W^{-0.091} \tag{2.12}$$

Dowson and Higginson (1966) minimum film, line contact:

$$H = 1.6G^{0.6}U^{0.7}W^{-0.13} \tag{2.13}$$

Archard-Cowking (1965–1966), central film, point contact:

$$H = 2.04\left(1+\frac{2R_x}{3R_y}\right)^{0.71} G^{0.74}U^{0.74}W^{-0.074} \tag{2.14}$$

Hamrock and Dowson (1977), minimum film, point contact:

$$H = 3.63\left(1-e^{-0.68k}\right)G^{0.49}U^{0.7}68W^{-0.073} \tag{2.15}$$

where
H = dimensionless-film thickness, h_c/R_X
h_c = film thickness, m (in.)
R_x = equivalent radius in rolling direction from $\frac{1}{R_x}=\frac{1}{R_{x1}}+\frac{1}{R_{x2}}$, m (in.), 1~body-1, 2~body-2
k = ellipticity ratio a/b (see Figure 2.25)
G = materials parameter, $\alpha E'$ with α pressure-viscosity exponent (Equation 1.18)
E' = reduced modulus of elasticity, from $1/E' = 1/2(((1-\upsilon_1^2)/E_1)+((1-\upsilon_2^2)/E_2))$
U = speed parameter, $\mu\eta_0 / E'R_x$, with
μ = coefficient of friction or traction
η_0 = absolute viscosity at temperature, N-s/m² (lbf-s/in.²)
W = load parameter, $F/E'R_x^2$, F = normal load, N (lbf)

Fortunately, Zaretsky (1999) provides Equation 2.15, a first order approximation of the various prior listed EHL-film thickness equations that should be valid for a maximum Hertzian stress between 1.4 and 2.4 GPa (200 and 350 ksi).

$$h_c = k_H(OD-ID)^{0.32}[N(OD+ID)]^{0.68}Z_0^{0.68}\bar{G} \tag{2.16}$$

where for OD and ID and h_c in inches, $k_H = 3.8\times10^{-11}$; for OD and ID in millimeters and h_c in inches, $k_H = 1.49\times10^{-12}$, N is speed in rpm. This last *odd-ball* unit mixture reflects that at one time all U.S. rolling element-bearing companies, except one, used millimeters for catalog listed dimensions. All use millimeters now.

Enhancement of EHL-film thickness with grease was explored by Dalmaz and Nantuo (1987). Their test results showed EHL-film thickness greater for grease than that calculated from Equation 2.15 using the viscosity properties of the base oil only, with the effective viscosity of the grease approximately 30% higher than the base oil alone. But Aihara and Dawson (1978) indicated that under long-term operation, separation of the base oil from the grease can starve the EHL film thus reducing it by 50%–70%, which Zaretsky (1999) states is almost always the case.

Additives in modern lubricating oils are a major contributor to the success of modern machinery. A prominent example is the modern automotive internal combustion engine. The recommended time between automotive engine oil changes is dictated by the chemical effective life of the additives such as additives for *anti-oxidation, anti-wear, extra pressure, cold start,* and *anti-foaming.* Specifically for rolling element bearings under heavy loads, high speed, and high temperature, anti-wear and extreme-pressure oil additives are extremely important. Not only are the contents of specific oil additive packages typically the producers' proprietary information, but even more so are the process steps for producing the additive packages price-wise competitive in large quantities. Correspondingly, additive packages and production processes are most often not patented since that comes with a time limit after which it all becomes public knowledge. Usually, lubricant additive producers opt for keeping everything forever secret, to provide the best long-term protection of their intellectual property. That makes a lot of common sense to the author since a patent only gives one a legal basis to pursue one of the costliest types of litigation, that is, patent infringement. Zaretsky (1999) reports on extensive research results for additive package testing by tribology researchers focusing on REB life.

2.6 Materials Selection and Processing

The evolution of modern rolling element-bearing steels follows with the history of the production of steels from iron, starting with the mid-nineteenth century *Bessemer* process for blowing air through molten iron to yield a relatively high grade steel. Shortly thereafter, *open-hearth melting* yielded an even higher quality steel and made steel increasingly accessible to industry. The evolution to modern steel production technology and production processes now provides a considerable variety of steel strength, deformation, and energy absorbing property options that are achieved through tight control of quench rate, carbon content, and several other alloy components. Pertinent to the evolution of REB steels, Harris and Kotzalas (2007) provide a detailed description of the evolution of steel production processes methods, including (1) melting methods, (2) basic electric furnace processes, (3) several vacuum degassing methods, and the SKF M-R process.

Exhaustive research testing (e.g., NASA, Parker and Zaretsky, 1972) showed that standard mechanical material strength properties obtained from static and dynamic tension/compression tests and rotating-beam fatigue tests did not correlate well enough with rolling element-bearing survival results. They showed that adequate assessment of steel/processing parameters for rolling element bear durability requires testing with actual bearings. Zaretsky (1999) and Zaretsky and Poplawski (2009) provide an overwhelming wealth of cutting edge test data far more than what bearing non-specialists could absorb or directly utilize. The intent of this section is to distill out from the wealth of such focused multiparameter testing the "kernels" potentially useful to the new generation of bearing engineers.

AISI 52100 bearing steel, first specified around 1920, is today still the most used bearing steel (Zaretsky 1999). Alloying components of AISI 52100 in weight by percent are as follows (C 0.89, Cr 1.60, Mn 0.20, Si 0.19, P 0.019). The rise and advancement of the modern gas turbine aircraft jet engine is what provided the primary driving force for a quantum leap beyond AISI 52100 in rolling element-bearing steels, as well as new non-ferrous bearing materials like ceramics. Zaretsky (1999) lists 26 through-hardened and carburizing-grade bearing steels and their compositions. Until the 1980s, AISI M-50's proven durability at temperatures up to 600°F (316°C) provided little motivation for further bearing material development. But gas turbine engine energy-efficiency performance-improvement advancements pushed operating turbine temperatures higher and higher. That led to needed bearing steel toughness improvements, achieved by chemistry modification beyond AISI M-50 and processing with heat treatment utilizing *carburizing* instead of through-hardening.

For optimum durability, *case hardening* (e.g., by *carburizing*) prior to final machining of rolling elements and raceways is a specific topic worthy of attention by anyone involved with high-tech machines like gas turbine aircraft engines. Harris and Kotzalas (2007) provide a comprehensive treatment of the methods for case hardening of REBs. Basically, *case hardening* produces a more durable REB than *through-hardening*. Case hardening allows the outer surface of the rolling contact surfaces to be hardened to only the depth needed to provide maximum durability against subsurface initiated fatigue and surface wear while retaining a softer interior core that does not have the unfavorable damage by through fracture failure that is a bearing failure mechanism more likely with through-hardened-bearing components. This can be easily visualized as illustrated in Figure 2.34, which the author devised for his senior undergraduate machine design class to explain this topic.

Figure 2.34 illustrates how sub-surfaced initiated fatigue at the maximum alternating shear stress depth leads to surface flaking, "spalling," on bearing rolling contact surfaces, and "pitting" on gear teeth contact surfaces. Consistent with basic material-fatigue knowledge, it is the formation of a micro-crack along the plane of maximum alternating shear stress that is the first step in a multistep crack-propagation process that culminates in a material fatigue failure. More is addressed on this topic in Section 2.7 (Failure Modes, Life, and Reliability). A concluding thought here is on the unhardened core of the rolling element, which is softer, preventing through fractures. The softer nature of the core means the case-hardened surface layer will possibly flex a bit more than with through-hardened components. Since the depth of the case-hardened surface layers need only be thin to

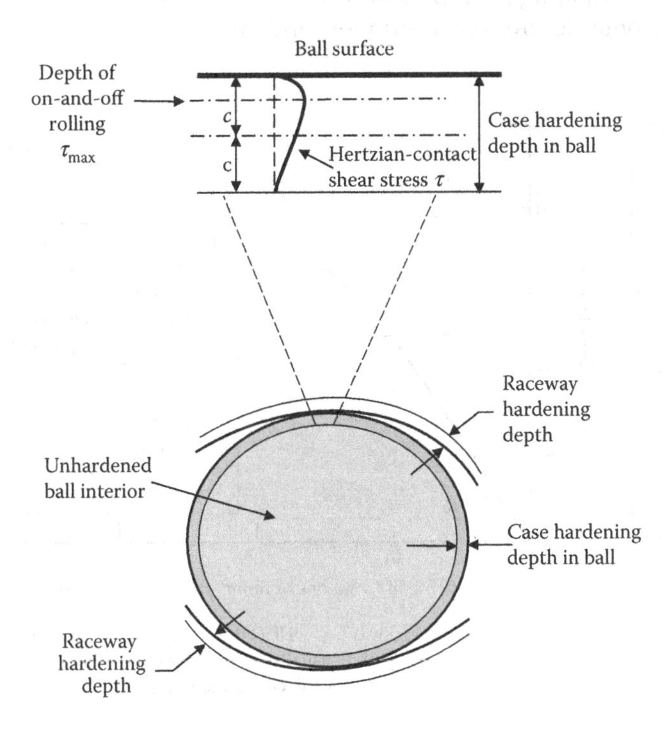

FIGURE 2.34
Case-hardened-bearing components encapsulate the alternating stress.

combat subsurface initiated fatigue at the rolling τ_{max}, their flexing a la beam bending stress is minimized by virtue of the hardened layers' thinness, that is, beam bending stress $\sigma = Mc/I$, where c in this case is nominally only one-half of the hardened-layer depth.

2.7 Failure Modes, Life, and Reliability

If everything in the life of a rolling element bearing were perfect, for example, flawless steel, ideal lubrication, zero contamination, totally non-corrosive environment, proper handling, no dimensional discrepancies, and so on, the bearing would ultimately be life limited due to *subsurface initiated fatigue*, leading to *spalling* as briefly introduced in the previous section. Correspondingly, REBs are *life rated* statistically by manufacturers based on extensive laboratory test data.

An L_{10} life rating of a particular bearing model means that 90% of the bearings of that model number will last that long before spalling occurs, as illustrated in Figure 2.35. Following that life rating information from bearing manufacturers' catalogs, *every undergraduate machine design textbook's* chapter on REB attributes bearing life rating to be primarily from fatigue. So, when the author teaches machine design, he prominently describes the information provided in Figure 2.36, which clearly shows that the life of REBs is almost always dictated by some other failure cause that happens sooner than spalling from fatigue. As Figure 2.37 illustrates, fatigue spalling can initiate not only below the surface but can also initiate on the surface. Zaretsky (1999) provides considerable depth on the many factors influencing bearing fatigue life, including (1) steel inclusions and carbides, (2) bearing overload, (3) crack propagation, (4) geometric stress concentration, and (5) microspalling.

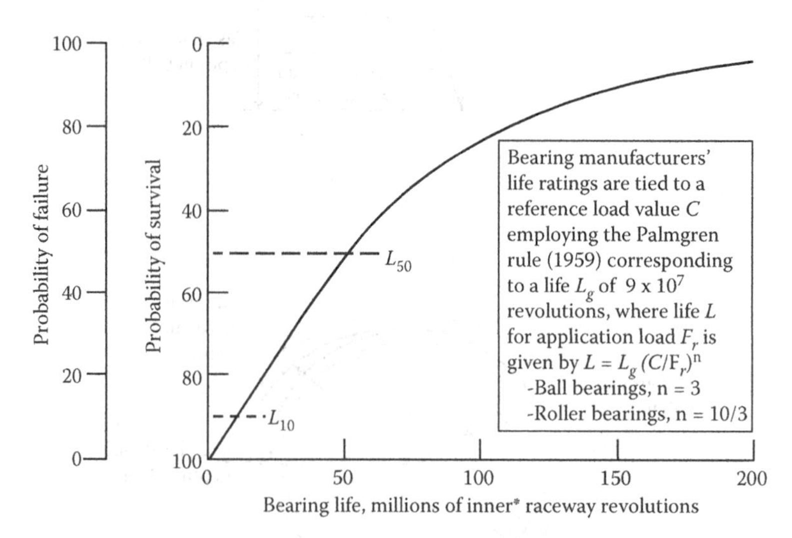

FIGURE 2.35
Statistically documented rolling bearing spalling fatigue-life ratings.

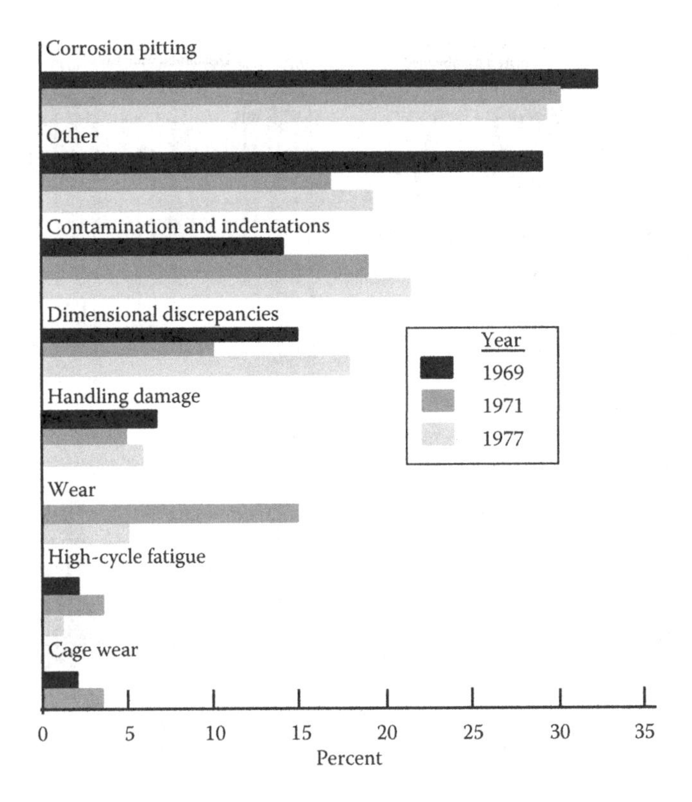

FIGURE 2.36
Rolling bearing life limiting causes. (From Zaretsky, E. V. and Poplawski, J. V., *Rolling Element Bearing Technology*, Handout for 3-day short course, Case Western Reserve University, Cleveland, Ohio, p. 973, June 23–25, 2009.)

A REB failure means it ceases to function normally (e.g., becomes rough running and/or noisy), which fortunately usually happens before the bearing stops working. As compared to hydrodynamic-film bearings, REBs are more likely to give forewarning that they are in trouble before they cease to function. A monitored forewarning used in modern aircraft gas turbine engines is by magnetic sensor detection of any bearing ferrous debris in the bearing

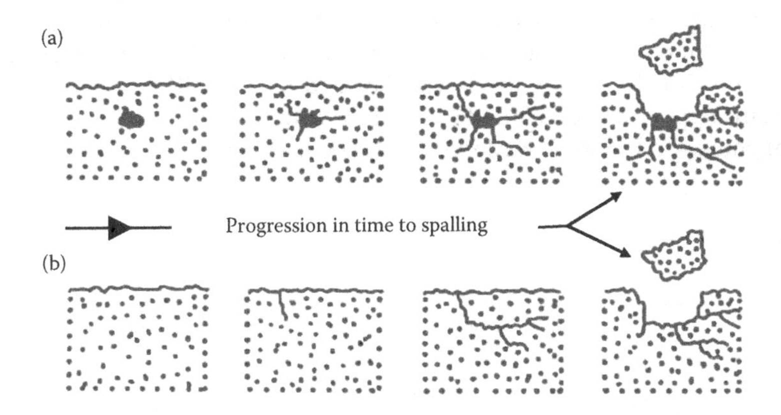

FIGURE 2.37
Bearing fatigue leading to spalling (a) subsurface, (b) surface.

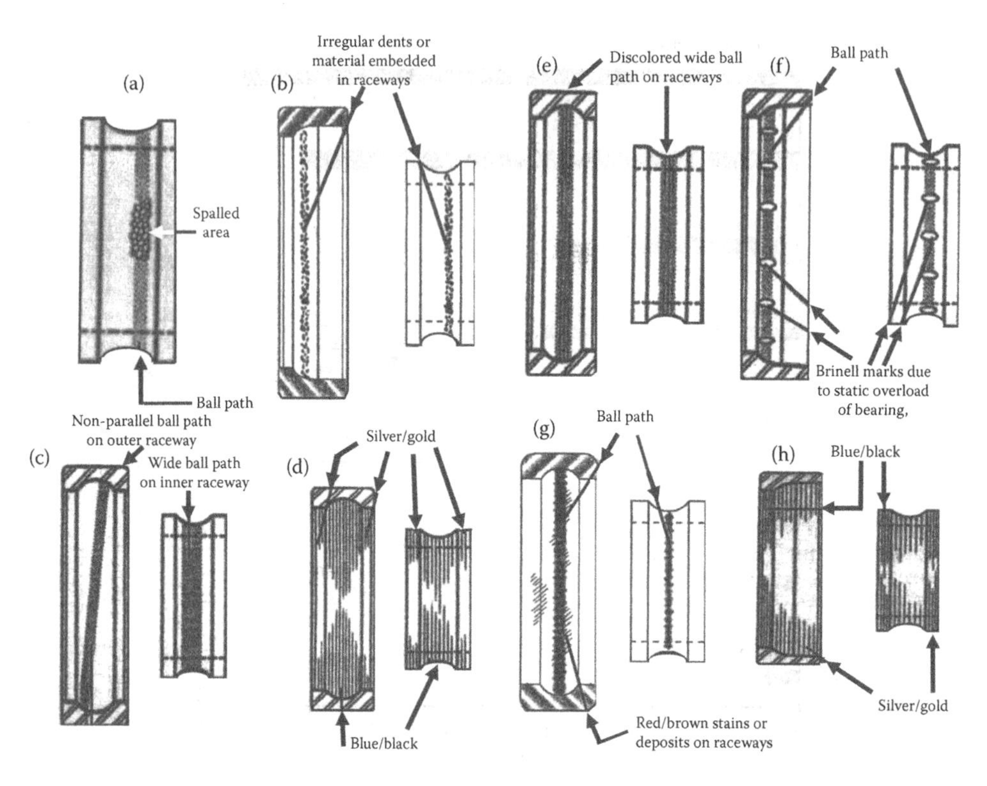

FIGURE 2.38
Sketches of various bearing damage causes; (a) spalling, (b) contamination, (c) misalignment, (d) lubrication failure, (e) tight fit, (f) brinelling, (g) corrosion, (h) overheating.

oil circulation system. Clearly, REBs "won out" over hydrodynamic-oil-film bearings when gas turbine jet engines were initially developed because a total interruption of lubrication, while perhaps degrading a REB and shortening its life, will not cause the bearing(s) to immediately cease working, as would be the result for an operating hydrodynamic-film bearing subjected to a lubricant supply total interruption.

The category of *Other* in Figure 2.36 includes the common failure modes: (1) misalignment, (2) true and false brinelling, (3) excessive axial thrust, (4) lubrication deficiency, (5) heat and thermal distress, (6) roller edge stresses, (7) cage fracture, (8) element or raceway fracture, (9) skidding, and (10) electric-arc discharge. For end users of REBs it is clearly useful to deduce from the appearance of a failed bearing the probable root cause(s) of the failure. To assist in that endeavor, Figure 2.38 illustrates the physical appearances for several distinct failure causes.

References

Adams, M. L., *Rotating Machinery Research and Development Test Rigs*, CRC Press/Taylor & Francis Boca Raton, FL, p. 217, 2017.

Aihara, S. and Dawson D., A Study of Film Thickness in Grease Lubricated Elastohydrodynamic Contacts, *Proceedings: Elastohydrodynamic Lubrication Institution of Mechanical Engineers*, pp. 104–115, 1978.

Archard, J. F. and Cowking, E. W., Elastohydrodynamic Lubrication at Point Contacts, *Proceedings: Elastohydrodynamic Lubrication*, Institution of Mechanical Engineers, London, 180(B3), pp. 1965–1966, 1965–66.

Conrad, R., British patent no. 12,206, 1903; U.S. patent no. 822,723, 1906.

Greenhill, L. M. and Merchant, D. H., *Modeling of Rolling Element Bearing Mechanics.* Computer Program (REBANS) User's Manual, NASA-CR-196557 Final Report, 1994.

Dalmaz, G. and Nantuo, N., An Evaluation of Grease Behavior in Rolling Bearing Contacts, *Lubrication Engineering*, Vol. 43 (12), pp. 905–915, 1987.

Dowson, D. and Higginson, G. R., *Elastohydrodynamic Lubrication: The Fundamentals of Roller and Gear Lubrication*, Pergamon Press, New York, 1966.

Grubin, A. N., Fundamentals of the Hydrodynamic Theory of Lubrication of Heavily Loaded Cylindrical Surfaces, In: Ketova Kh. F. (editor) *Investigation of the Contact Machine Components*, 1949. Translation of Russian Book No. 30, Central Scientific Institute for Technology and Mechanical Engineering, Moscow, Chapter-2, (Available from Department of Scientific and Industrial Research, London, Great Britain, Trans. CTS-235 and Special Libraries Association, Trans. R-3554).

Gupta, P. K., *Advanced Dynamics of Rolling Elements, ADORE User's Manual*, 2014.

Gupta, P. K., Dill, J. F. and Bandow, H. E., Dynamics of Rolling Element Bearings; Experimental Validation of the DREB and RAPIDREB Computer Programs, *ASME Journal of Tribology*, Vol. 107(1), pp. 132–137, 1985.

Hadden, G. B., Kleckner, R. J., Ragen, M. A. and Sheynin, L., Research report: User's manual for computer program AT81y003 SHABERTH. *Steady state and transient thermal analysis of a shaft bearing system including ball, cylindrical and tapered roller bearings, NASA*, 1981.

Hamrock, B. J. and Dawson, D., Isothermal Elastohydrodynamic Lubrication of Point Contacts, Part-III—Fully Flooded Results, *ASME Journal of Lubrication Technology*, Vol. 99(2) pp. 264–276, 1977.

Harris, T. A. and Kotzalas, M. N., *Rolling Bearing Analysis*, 5th ed., CRC Press/Taylor & Francis, Boca Raton, FL, 2007.

Hertz, H., *On the Contact of Rigid Elastic Solids*, Miscellaneous papers, MacMikkiiN, London, pp. 163–183, 1896.

Jones, A. B., *Analysis of Stresses and Deflections*, New Departure Engineering Data, Bristol, CT, pp. 161–170, 1946.

Jones, A. B., A General Theory for Elastically Constrained Ball and Radial Roller Bearings Under Arbitrary Load and Speed Conditions, *ASME, Journal of Basic Engineering*, Vol. 82(2), pp. 309–320, 1960.

Jones, W. R., *Boundary Lubrication Revisited, NASA TM-82858*, 1982.

Kleckner, R. J., Pirvics, J. and Castelli, V., High Speed Cylindrical Rolling Element Bearing Analysis "CYBEAN"—Analytic Formulation, *ASME Journal of Lubrication Technology*, Vol. 102(3), pp. 380, 1980.

Lundberg, G. and Palmgren, A., *Dynamic Capacity of Rolling Bearings*, Acta Polytech. Mech. Eng. Ser. 1, R.S.A.E.E., No.3, 7, 1947.

Palmgren, A., *Ball and Roller Bearing Engineering*, SKF Industries, Inc., Philadelphia, 1959.

Parker, R. J. and Zaretsky, E., Rolling-Element Fatigue Lives of Through-Hardened Bearing Materials, *ASME Journal of Tribology*, Vol. 94 (2), pp. 165–173, 1972.

Zaretsky, E. V. (editor), *Life Factors for Rolling Bearings*, Society of Tribologists and Lubrication Engineers, STLE Publication SP-34, 2nd ed., p. 314, 1999.

Zaretsky, E. V. and Poplawski, J. V., *Rolling Element Bearing Technology, Handout for 3-day short course*, Case Western Reserve University, Cleveland, Ohio, p. 973, 2009.

3

Magnetic Bearings

The generic configuration of an *actively controlled magnetic* bearing is shown in Figure 3.1, which schematically illustrates the essential components. The main feature of active magnetic bearings which has attracted the attention of some practical minded rotating machinery designers is that they are *oil-free bearings*. This means, for example, that with large pipe line compressor rotors supported on oil-free bearings, the elimination of oil precludes the eventual coating of pipe line interior surfaces with always-present machinery lost oil that otherwise must be periodically cleaned out of the interior pipe line, at considerable service and downtime costs. Not surprisingly to the author, this feature was not even in the minds of magnetic-bearing conceivers, who for the most part were academicians with a particular focus on control theory. They conceived the modern active magnetic bearing as an electromechanical actuator device that utilizes rotor position feedback to a controller in order for the magnetic bearing to provide electromagnetic non-contacting rotor levitation with attributes naturally occurring in conventional bearings, that is, static-load capacity along with stiffness and damping. Magnetic-bearing technologists focused their attention on the fact that the rotor dynamic properties of magnetic bearings are freely prescribed by the control law designed into the feedback control system, and thus can also be programmed to adjust in real time to best suite a machine's current operating needs, such as *extra damping* to (1) attenuate unbalance excited rotor-bearing resonant modes (i.e., at critical speeds) or (2) active tuning around critical speeds, as well as (3) to suppress dynamic instability self-excited rotor vibration. That was *Mana from heaven* for the rotor-vibration-technology clan.

3.1 Unique Operating Features of Active Magnetic Bearings

Magnetic-bearing systems can routinely be configured with impressive versatility not achievable with conventional bearings. In addition to providing real-time controllable-load support, stiffness, and damping, they can simultaneously provide feed-forward-based dynamic-bearing forces to partially negate rotor vibrations from other anticipated inherent sources. They can also employ notch filtering strategies to isolate the machine's stator from specific rotor vibration frequency forces, primarily from the ever-present once-per-revolution synchronous residual rotor mass unbalance. In this last feature, notch filtering out the unbalanced forces from the rotor–stator interaction bearing forces would appear quite desirable but then the rotor will tend to spin about its principal polar inertia axis through its mass center, and thus its surfaces will accordingly have runout, meaning that rotor–stator rubs and/or impacts at small rotor–stator radial clearances have an increased likelihood of occurring, such as at radial-dynamic seals.

A natural extension of magnetic-bearing systems is their integration with machinery condition monitoring strategies. Not only do active magnetic-bearing systems possess the displacement sensors inherent in modern rotating machinery conditioning monitoring

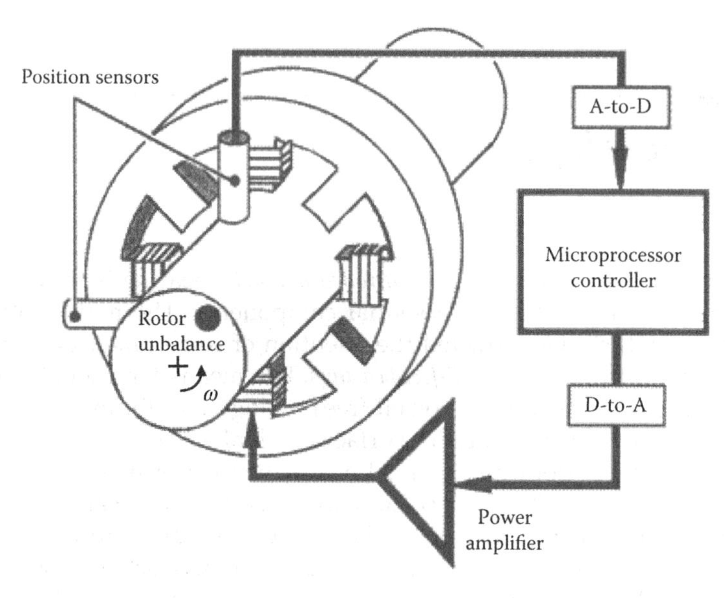

FIGURE 3.1
Active magnetic-bearing schematic.

systems, but they automatically provide the capability of *real-time monitoring of bearing forces*, a long wished-for feature of rotating machinery problem diagnosticians. As alluded to earlier, magnetic bearings being real-time-controlled force actuators can also be programmed to impose static and dynamic-bearing-load signals that can be "intelligently" composed to alleviate (at least partially and temporarily) a wide array of machine operating difficulties such as excessive vibrations and rotor-stator rubbing initiated by transient thermal distortions of the stator or other components. Clearly, the concept of a so-called *smart machine* for rotating machinery is not difficult to conceptualize when active magnetic bearings are employed for rotor support. A review of substantive magnetic-bearing publications is given by Allaire and Trumper (1998). Schweitzer (1998) focuses on *smart rotating machinery.*

3.2 Shortcomings of Magnetic Bearings

Magnetic-bearing systems are relatively quite expensive, encompassing a system with position sensors, A-to-D and D-to-A multichannel signal converters, multichannel power amplifiers, and a microprocessor, each with its own reliability factors. So this lack of basic simplicity with such a multicomponent electromechanical system surely translates into concerns about reliability and thus the need for component redundancy (e.g., sensors).

The most obvious manifestation of the reliability/redundancy factor is that magnetic bearings in actual applications require a backup set of *catcher bearings* (typically ball bearings) onto which the rotor drops when the magnetic-bearing operation is interrupted, such as by a power interruption or failure of a primary non-redundant component, or the magnetic bearing is overloaded. The dynamical behavior of the rotor when the catcher bearings take over was initially not properly evaluated by magnetic-bearing technologists. But in rigorous application testing, it was found that severe large amplitude nonlinear rotor vibration can occur when the rotor falls through the catcher-bearing clearance gap and hits the catcher bearings (see the Chapter 11 failure case study). Depending on the catcher-bearing radial clearance, damage to close-running clearance components like radial seals is a strong possibility.

Fluid-film bearings and rolling contact bearings both possess considerable reserve capacities for non-sustained momentary overloads, for example, impact loads. Since these conventional bearings completely permeate the modern industrial world, their high capacities for momentary overloads are essentially taken for granted since they *do that job and keep on running.* On the other hand, magnetic bearings electrically *saturate* when loads are pushed to their limits, and thus provide little capacity for large-load increases that momentarily exceed the bearing's design-load capacity by substantial amounts. This is a serious limitation for many applications. For static-load and lower-frequency stiffness and damping properties, magnetic-bearing force capacity is limited by the *saturation flux density* of the magnetic iron, as illustrated in Figure 3.2a. A further limitation is set by the maximum rate at which the control system can change the current in the windings. The magnets have an inherently high inductance which resists a change in current, thus the maximum *slew rate* depends on the voltage available from the power amplifier. In practical terms, the required slew rate is a function of the frequency and amplitude of rotor vibration experienced at the bearing. Figure 3.2b illustrates the combined effects of magnetic saturation and slew-rate limitation on magnetic-bearing-load limits.

Conventional bearings are not normally feedback-controlled devices, that is, they achieve their load capacity and other natural characteristics through mechanical design features grounded in fundamental mechanics principles as illuminated in Chapters 1 and 2. Conversely, the basic operation of active magnetic bearings relies on feedback of rotor position signals to adjust instantaneous bearing forces. As a result, a generic shortcoming of active magnetic bearings stems from this fundamental reliance on feedback control. It is referred to with the control theory terms *spillover* and *collocation error.* Feedback control design is traditionally viewed as a compromise between *response and stability.* Whenever a feedback loop is closed, there is the potential for instability, as is well known. Specifically for active magnetic bearings, collocation error arises from the sensors not being placed exactly where the bearing force signals are applied to the rotor, and this can produce rotor dynamical instabilities (spillover) that would not otherwise occur. Surely no rotating

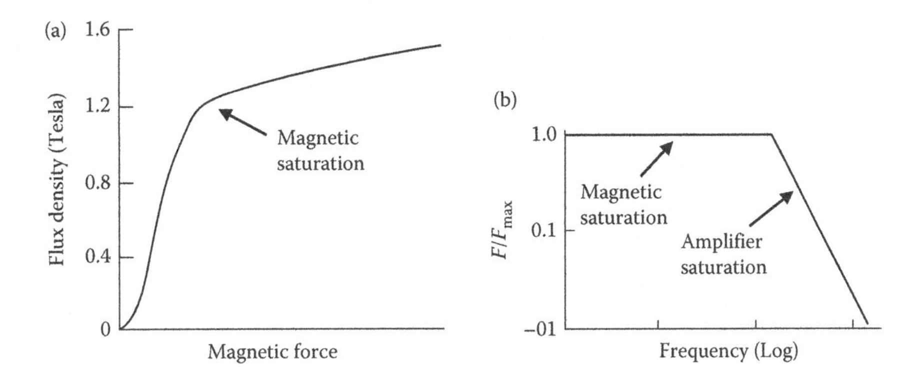

FIGURE 3.2
Magnetic and amplifier saturation effects on magnetic-bearing-load limits. (a) flux density characteristic, (b) frequency effect. (From Fleming, D. P., Magnetic Bearings-State of the Art, *NASA Technical Memorandum*, p. 104465, July 1991.)

machinery vibration specialist will be enthused about this, since other traditionally recognized and unavoidable rotor dynamical instability mechanisms are always lurking, especially in turbomachinery.

The magnetic-bearing technologists' answer to this fundamental shortcoming was to have programmed into the control law a very accurate dynamics model of the rotor system. However, machines constantly change their dynamic characteristics in response to operating point changes and as a result of normal seasoning and wear over time. This approach would necessitate continuous automatic real-time recalibration of the dynamics model residing in the programmed control law. Under some well-defined operating modes, rotor dynamical systems can be quite nonlinear, and then having an accurate model in the control law for the actual rotor system becomes a formidable challenge. It prompts one to sarcastically ask *"how can all those non-smart 'stupid' oil-film bearings and ball bearings do all the things they do?"* The answer simply is, *"their non-academician designers are smart."*

References

Allaire, P. and Trumper, D. L. (editors), Review of publications on magnetic bearings, *Proc. 6th International Symposium on Magnetic Bearings*, Boston, August 1998.

Fleming, D. P., Magnetic Bearings-State of the Art, *NASA Technical Memorandum*, p. 104465, July 1991.

Schweitzer, G., Magnetic Bearings as a Component of Smart Rotating Machinery, *Proceedings. IFToMM 5th International Conference on Rotor Dynamics*, Darmstadt, pp. 3–15, September 1998.

4

Synovial-Fluid Joints—The Body's Bearings

The human body contains several joints as illustrated in Figure 4.1. These joints can be categorized according to their kinematic function: (1) hinge (fingers and toes), (2) ball and socket (hip and shoulder), (3) pivot (neck), and (4) gliding (wrist). The *freely moving joints* are lubricated by *synovial fluid* to yield low-friction relative-sliding motion, similar to machines and mechanisms that have parts connected to each other that are in relative motion while transmitting loads. That is, *synovial joints are bearings*, the topic of this book. *Synovial-fluid joint* also (1) absorbs shock loading (squeeze-film lubrication) as the synovial fluid becomes more viscous under applied pressure; (2) provides nutrient and waste transportation; (3) serves as a molecular sieve as the synovial-fluid pressure acts as a barrier against migrating cells into or out of the joint space, and (4) contains lubricin known as PRG4 which acts to provide boundary lubrication (see Figure 1.1a). Synovial fluid (means "like egg white") is similar to egg white and is non-Newtonian.

4.1 Early Work on Synovial Joint Tribology

Synovial joints have a fluid-filled cavity between the articulating bones, as illustrated in Figure 4.2, which enables articulating connected bones to move freely with respect to each other. The bone ends are covered with articular cartilage within the synovial joint. The synovial joint is sealed from the surrounding tissues by the synovial membranes and the cavity is filled with synovial fluid. This creation generates very good lubrication between the connected bone ends that move relative to each other, producing very low frictional resistance and preventing bone-to-bone contact wear. Scientific engineering study of human joint lubrication is of course hampered by the absence of a true test apparatus other than a living human body, thus precluding the totality of rigorous experimental freedoms afforded other machinery research test rigs such as those covered by Adams (2017). Lubrication and bearing technology significantly advanced in the second half of the twentieth century with some of the well-recognized bearing and lubrication researchers having human joint lubrication within their total scope of lubrication research investigations. Today their work would be claimed by that relatively new field named *biomedical engineering*. Reaching a clear scientific understanding of synovial joint lubrication had some *false starts*, clearly reflecting the absence of *living specimen test rigs* with which to rigorously test, sharpen, and validate theoretical mathematical models (Dawson 1981; Medley et al. 1984).

From force measurements, Swanson (1979) estimated *joint friction coefficient* values to be in the range of 0.01–0.001, indicating the presence of some type of fluid-film lubrication, but did not identify the specific type of lubricant film (see Figure 1.1a). The explanations of Higginson (1971) and Medley et al. (1984) of how such films can be generated and

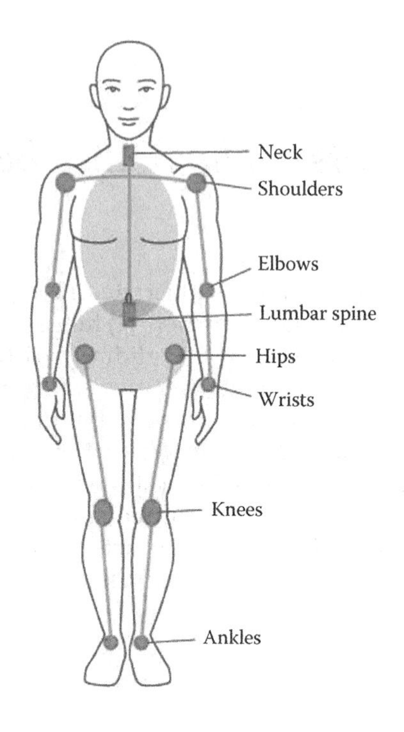

FIGURE 4.1
Human joints.

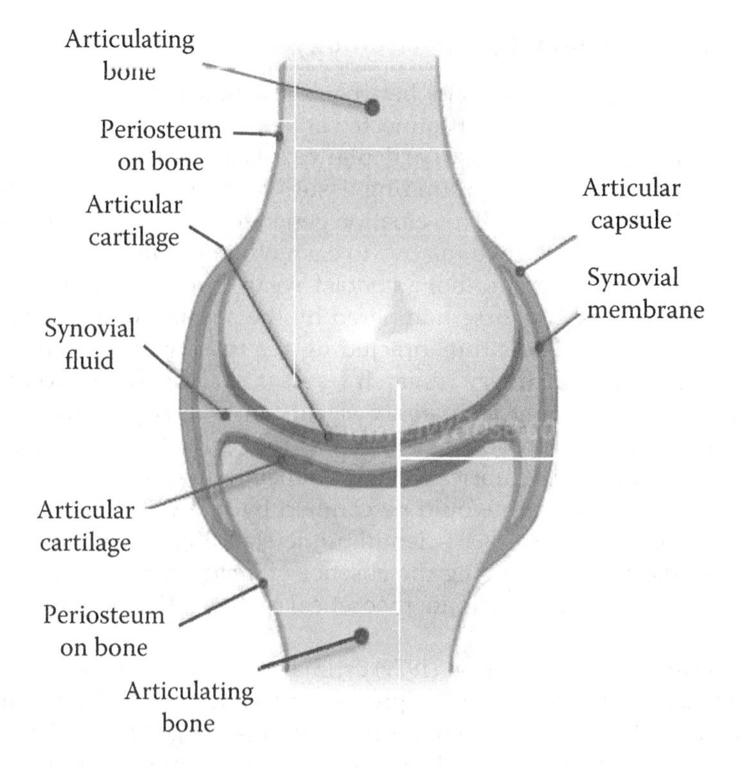

FIGURE 4.2
Structure of synovial joints.

maintained were not considered to be convincing by Dawson and Jin (1986), who provided their mathematical model based on a microvariation of the then already well-developed elastohydrodynamic lubrication (EHL) analysis approach covered in Chapters 1 and 2. They suggest that micro-EHL analysis is the approach most likely to correctly characterize joint lubrication, but without relying on the rolling-element-bearing and gear-set guidelines proven necessary for the generation and maintenance of an EHL film. That is, the need in hard-surface bearings and gears for the combined surface roughness to be substantially smaller than the predicted EHL-film thickness. Because based on the work of Gardner et al. (1981), synovial joints EHL-film thickness was estimated to be in the range of 0.1–1 μm, whereas healthy cartilage appears to have an unloaded combined surface roughness in the range of 2–5 μm. Dawson and Jin (1986) appear to have successfully overcome this inconsistency by coupling local micro-EHL associated with surface asperities. Their theoretical results predicted that locally generated EHL pressures have a remarkable ability to "smooth" initially rough cartilage surfaces as they pass through the loaded region. Their resulting analysis predicts film thicknesses several times greater than the locally reduced effective roughness of the opposing cartilage surfaces. Their investigation on synovial joint lubrication thus offered a convincing explanation of *nature's remarkable bearings*.

4.2 More Recent Developments on Synovial Joint Tribolgy

Jin (2009) provides a comprehensive review of synovial joint lubrication, citing 28 published works through to 2008. Eight of these works of most engineering interest include the following: a review of *human joint tribology* (Unsworth 1991), *influences on friction coefficient* (Forster and Fisher 1996), *boundary friction* (Ateshian 1997 and Hills 2000), *role of cartilage* (Kumer 2001 and Graindorge 2006), *role of lubricin* (Jay 2007), and *load variation on friction* (Katta 2007).

Tandon et al. (1994) present an approach to synovial joint lubrication analysis that uses a Bingham fluid model as the lubricant between the approaching porous cartilaginous surfaces. They postulate that the strength of the core formed, due to thickly concentrated hyaluronic acid molecules, increases as the surfaces come closer. They explain that this is due to the withdrawal of the base fluid through a boosted lubrication mechanism, leading to the formation of lubricating gel. They further clarify this by explaining that the gel ultimately acts as a boundary lubricant which very briefly prevents cartilage-to-cartilage contacts during gait-cycle impact loading.

There is in any non-bone biological tissue a degree of local deflection compliance under local load. Similarly there are such machinery bearings as *compliant–surface bearings* (e.g., Figure 1.24). However, the chemistry of the human joint lubricant is considerably different than that of machinery lubricants. And the tissue in human joints is not at all similar to conventional machinery bearing materials, like metal, plastic, rubber, and others. These differences warrant taking a more biological approach to understanding how synovial joints work. To that end, Bera (2013) approached the characterization of synovial joint functioning from a more biological point of view, albeit while still retaining conventional tribology concepts as encapsulated in Figure 1.1a.

Bera points out that during rolling motion of a wheel on a rough surface, the friction force generated at the wheel-road interface produces rotation of the wheel with a motive drive torque applied to the wheel that yields the vehicle motion. But at the very slippery synovial joint cartilage contact interface, such conventional rolling motion does not occur. Instead, cross-connected ligaments and muscles produce rolling motion (articulation) consistent with the classical kinematic four-bar linkage, which happens prominently in the knee joint. Bera contends that the highly slippery lubrication is possible only by boundary lubrication from the synovial fluid. Of course, it is hard to imagine that Bera's use of the label *boundary lubrication* means the same thing as what occurs in conventional boundary lubricating machinery oils, as briefly described in Section 1.1 with Figure 1.3.

Bera explains that synovial joints are so slippery and almost frictionless because of the adsorption of *albumin over gamma globulin* after formation of globulin coating on the hydrophilic cartilage surface, producing extremely low friction, enhanced by the effect of phospholipids. Of course, this terminology challenges engineers, including the author, but perhaps not the biomedical engineer. And thus neither *abrasive* nor *adhesive* sliding cartilage surface wear occurs. This explanation of synovial joint lubrication is clearly a bit outside a mechanical engineer's realm to say the least, but nevertheless interesting. And it is indicative of the author's contention in the prior section here that fully *understanding synovial joint lubrication is hampered by the lack of a test rig* (living human) on which to perform rigorous experimental research.

References

Adams, M. L., *Rotating Machinery Research and Development Test Rigs*, CRC Press/Taylor and Francis, Boca Raton, FL, p. 217, 2017.

Ateshian, G. A., *A Theoretical Formulation for Boundary Friction in Articular Cartilage, J. Biomech. Eng.*, Vol. 119(1), pp. 81–86, 1997.

Bera, B., Mechanism of Boundary Lubrication and Wear of Frictionless Synovial Joint, *International Journal of Engineering Research and Applications (IJERA)*, Vol. 3(4), pp. 1593–1597, 2013.

Dawson, D., *Lubrication of joints: A—Natural Joints, in Introduction to the Biomechanics of Joints and Joint Replacements*, Mechanical Engineering Publications, London, Ch. 13, pp. 120–133, 1981.

Dawson, D. and Jin, Z., Micro-Elastohydrodynamic Lubrication of Synovial joints, *Engineering in Medicine*, Vol. 15(2), pp. 63–65, 1986.

Forster, H. and Fisher, J., The Influence of Loading Time and Lubricant on the Friction of Articular Cartilage, *Proceedings, Inst. Mech. Eng. [H].*, Vol. 210, pp. 109–119, 1996.

Gardner, D. L., O'Connor, P., and Oates, K., Low Temperature Scanning Electron Microscopy of Dog and Guinea-Pig Hyaline Articular Cartilage, *J. Anat.*, Vol. 132, 267–282, 1981.

Graindorge, S., The Role of the Surface Amorphous Layer of Articular Cartilage in Joint Lubrication, *Proceedings, Inst. Mech. Eng. [H]*, Vol. 220(5), pp. 597–607, 2006.

Higginson, G. R., Elastohydrodynamic Lubrication in Human Joints, *Proc. Instn Mech. Engrs*, Vol. 191, pp. 217–223, 1971.

Hills, B. A., Boundary Lubrication in Vivo, *Proceedings, Inst Mech Eng [H].*, Vol. 214(1), pp. 83–94, 2000.

Jay, G. D., The Role of Lubricin in the Mechanical Behaviour of Synovial Fluid, *Proc., Natl. Acad. Sci. USA*, Vol. 104(15), pp. 6194–6199, 2007.

Jin, Z., Lubrication of Synovial Joints, In: Luo J., Meng Y., Shao T., and Zhao Q. (editors) *Advanced Tribology*. Springer, Berlin, Heidelberg, pp. 871–872, 2009.

Katta, J., Effect of Load Variation on the Friction Properties of Articular Cartilage, *Proceedings, Inst. Mech. Eng.[L]*, Vol. 221(3), pp. 175–181, 2007.

Kumar, P., Role of Uppermost Superficial Surface layer of Articular Cartilage in the Lubrication Mechanism of Joints, *J. Anat.*, Vol. 199(3), pp. 241–250, 2001.

Medley, J. B., Dowson, D., and Wright, V., Transient Elastohydrodynamic Lubrication Models for the Human Ankle Joint, *Engng. Med.*, Vol. 13, pp. 137–151, 1984.
Swanson, S. A. V., *Lubrication, in Adult Articular Cartilage*. Pitman Medical, Bath, UK, pp. 415–460, 1979.
Tandon, P. N., Bong, N. H., and Kushwawa, K., A New Model for Synovial Joint Lubrication, *International Journal of Bio-Medical Computing*, Vol. 35(2), pp. 125–140, Elsevier, 1994.
Unsworth, A., Tribology of Human and Artificial Joints, Proceedings, *Inst. Mech. Eng. [H]*, Vol. 205(3), pp. 163–172, 1991.

5

Bearing Monitoring and Diagnostics

The development of modern sensors and real-time monitoring/diagnostics (MD) of operating machinery co-evolved significantly in the second half of the twentieth century. Modern machinery components in (1) *power plants*, (2) *automobiles*, (3) *aircraft*, (4) *marine vessels*, (5) *railway*, (6) *manufacturing, mining, and construction equipment*, (7) *HVAC systems*, (8) *home appliances*, (9) *medical service devices*, and (10) *communication systems* are but prominent application examples of today's wide presence of real-time MD. Monitoring of bearing-operating parameters specifically has become essential for real-time assessment of bearing health, wear-and-tear, remaining life, and alerts of impending bearing failures. At the same time, operating-parameter monitoring at the bearings has also become an invaluable provider of symptomatic information directly relatable to several specific machinery health and potential impending failure mechanisms. Two prominent examples of this include: (1) rotor vibration spectra give symptomatic indicators for a wide variety of identifiable health factors in operating machinery, and (2) predictive maintenance (Adams 2010, 2017a).

5.1 Primer and Examples for Monitoring and Diagnostics (MaD)

The field of modern condition monitoring is now over 60 years into its development and thus is truly a mature technical subject. However, it continues to evolve and advance in response to continuous demand for further reductions in machinery downtime and maintenance costs.

Vibration is the most regularly monitored condition parameter in modern rotating machinery, and it is usually measured at or near the bearings. In addition, bearing temperatures and acoustic emissions as well as machine-internal pressures/fluctuations are also commonly monitored for the real-time diagnostic information therein contained. The power plant boiler feed water multistage centrifugal pump illustrated in Figure 5.1 is a prominent example of critically needed MaD real-time signals that are now typically monitored on such high-energy-density machines that must operate 24 hours a day, for example, in order for its electric power generating steam turbine generator unit or chemical process system to operate without interruption. The largest capacity pumps of this type that are in service require 80,000 hp to drive (with an auxiliary steam turbine) (Adams 2017a). The now widely appreciated need for all these time-base signals to be continuously monitored in this application evolved through the many years of pioneering power plant pump troubleshooting by the late Dr. Elemer Makay, as widely reported, for example, Makay (1978, 1980), Makay and Szamody (1980), and Makay and Barrett (1984).

Figure 5.2 shows the typical fundamental use of vibration monitoring in rotating machinery, to provide warning of gradually approached and/or suddenly encountered excessively *high-vibration levels* that could potentially damage the machinery. Trending a machine's vibration levels over an extended period of time can provide early warning of

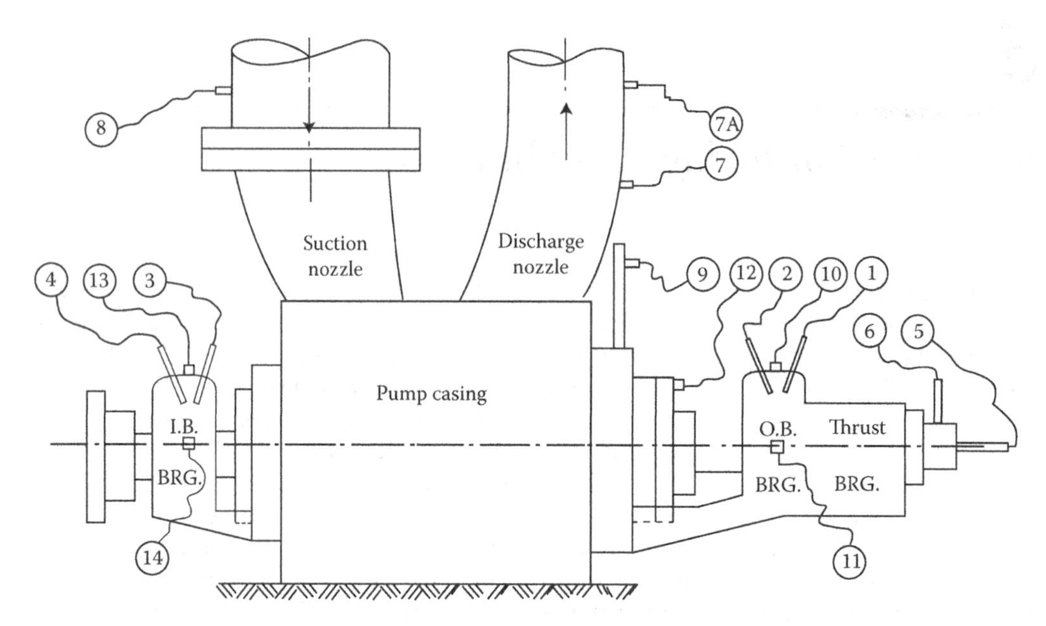

FIGURE 5.1
Time-based condition monitoring measurements of a multistage high-power centrifugal pump. Channels 1 to 6 are shaft-relative-to-bearing vibration displacement by eddy-current-inductance non-contacting proximity probes; channels 7 to 9 are pressure transducers; channels 10 to 14 are accelerometers.

impending problems other than just excessive vibration, providing machinery operators with valuable diagnostic information for critical decision making in order to schedule a timely shutdown of a problem machine for corrective action like rebalancing the rotor.

MaD has become the essential component in the development and utilization of *predictive maintenance*. As illustrated in Figure 5.3 for one version of predictive maintenance, each machine of a given group is provided with specific maintenance actions based on the machine's monitored condition instead of a fixed-time maintenance cycle. In principle this makes a lot of sense, but as most practitioners know "the devil is in the details." This effort has been driven by both industrial and government organizations efforts to drastically reduce maintenance costs, primarily by making large reductions in maintenance and technical support staff. This prevailing "bean counter" mentality has created new

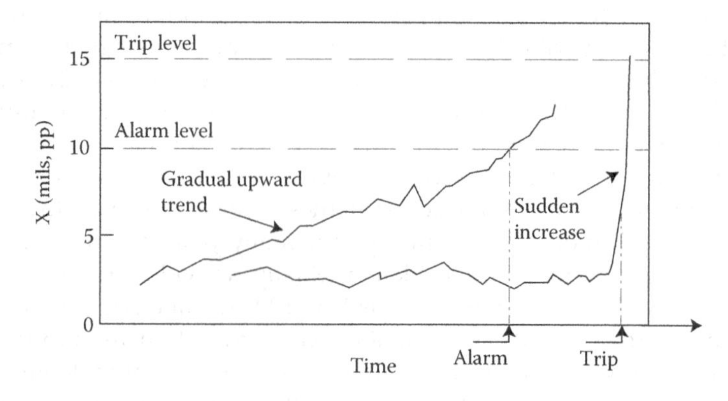

FIGURE 5.2
Tracking a representative vibration-peak amplitude over time.

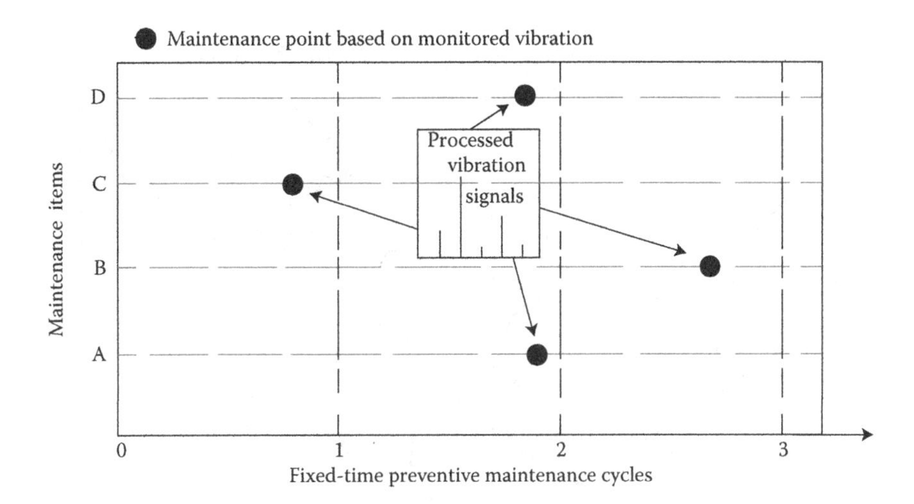

FIGURE 5.3
Predictive maintenance contrasted to *preventive* maintenance.

business opportunities for suppliers of machinery condition monitoring products and for new approaches to glean increased diagnostic information from already continuously monitored machine parameters.

5.1.1 Signal Processing and Utilization

The development of the *fast Fourier transformation* (FFT) in the mid-1960s is one of the most significant factors in the evolution of modern MaD. Fourier *transformation of time-base signals from the time domain into the frequency domain* has long been a somewhat confusing process to machinery maintenance personnel because of its imbedding within integral calculus, as follows.

$$X(\omega) = \int_{-\infty}^{\infty} x(t)e^{-i\omega t}dt \tag{5.1}$$

However, the author has been told by several persons that their first view of the illustration in Figure 5.4 (Adams 2010) was the first time they really understood what an FFT does. As clearly observable, viewing a time-varying signal such as vibration in the frequency domain is far more delineating of its contents than viewing it in the time domain. Not shown in Figure 5.4 is the phase angle for each harmonic. But the peak amplitude for each of the harmonics as illustrated are the FFT outputs that are directly relatable to specific characteristic mechanisms in the monitored machine.

The rotor-vibration example of such an output for a rotating machine illustrated in Figure 5.5 is a fairly common case. (1) It shows a synchronous once-per-revolution (1N) harmonic component that typically dominates the vibration signal, being indicative of an ever-present residual rotor mass unbalance. (2) The sub-synchronous harmonics typically show the presence of fluid-rotor interaction such as from (*i*) bearings, (*ii*) pump hydraulic rotor forces, and (*iii*) present but tolerable dynamic instability phenomena. (3) The progressively diminishing higher harmonics (2N, 3N, ...) that are always present because real dynamic systems are never perfectly linear, so the 1N rotor unbalance spills a little

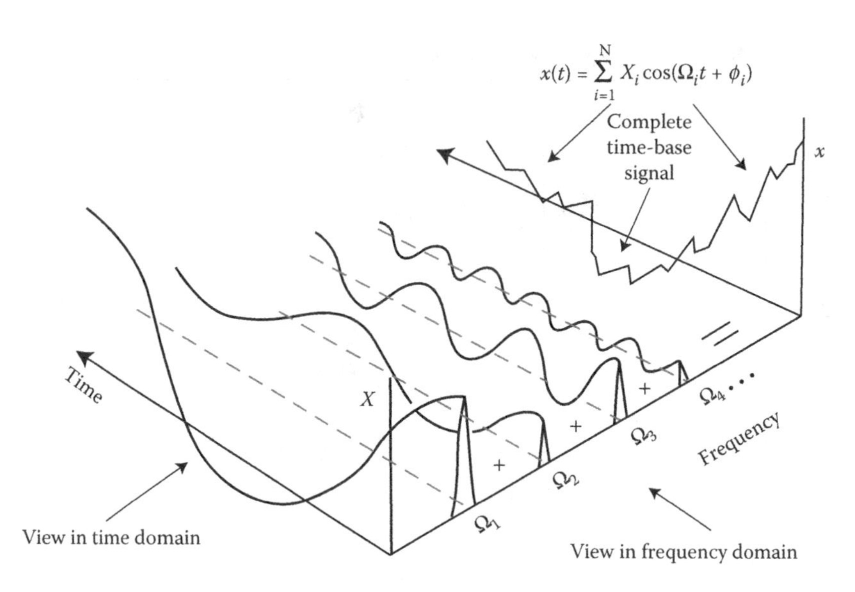

FIGURE 5.4
Graphical depiction of a time-base signal's frequency spectrum.

energy into its higher frequency integer harmonics. In emergency situations such as where a large blade of a turbine detaches producing a very large rotor unbalance, the resulting large rotor vibration becomes quite nonlinear, yielding vibration spectrum then exhibiting large higher harmonics of 1N as well as integer fractional harmonics (N/2, N/3, ...). See the last paragraph of Section 5.2.

Wavelet analysis has emerged over the last 35 years as essentially an advanced brother of FFT. The collection of theory and computational methods now known by the label *wavelets* is a mature topic in some cutting-edge applications. Some specific applications include: (1) *computer vision* systems that process variations in light intensity at several resolution levels, similar to how animal and human vision is now postulated to function; (2) digital

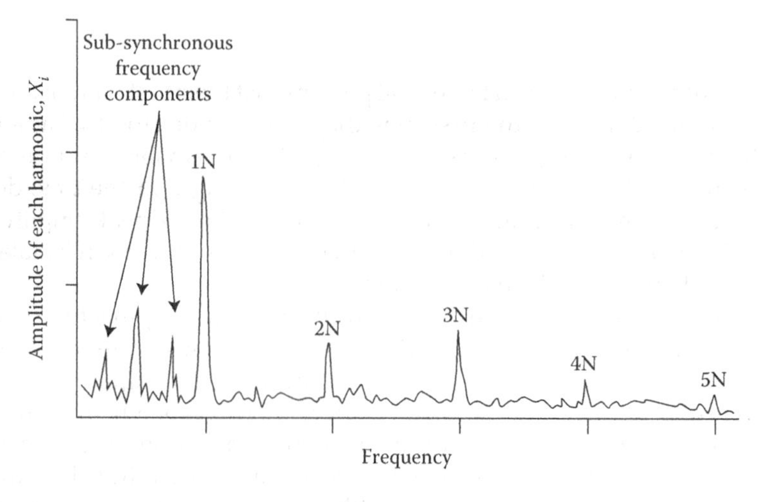

FIGURE 5.5
Example spectrum of a rotating machinery vibration signal.

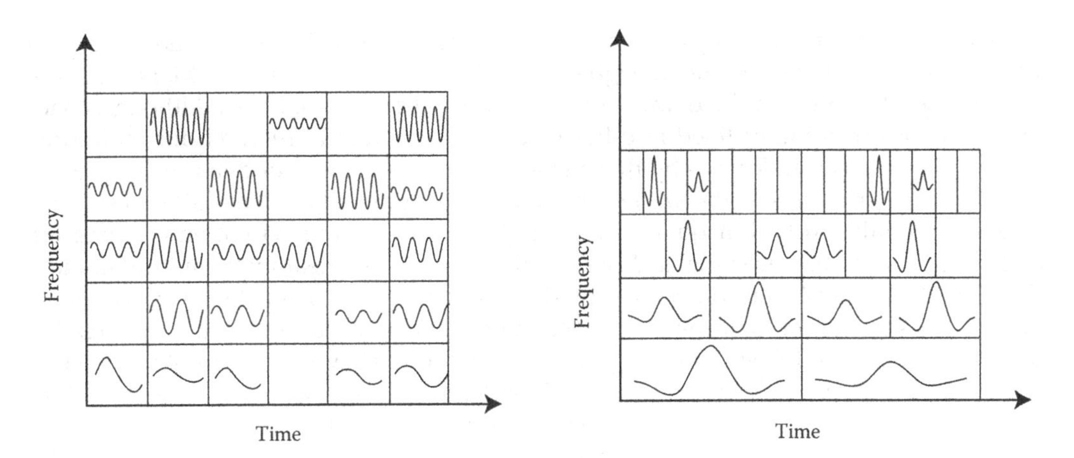

FIGURE 5.6
Time-frequency plane, comparing (a) FT and (b) WT.

data compression of *human fingerprint* images; (3) *de-noising* contaminated time-base signals; (4) detecting *self-similar behavior patterns* in time-base signals over a wide range of time scales; (5) *sound synthesis*; and (6) *photo image* enhancements.

Wavelet transform (WT) is a powerful extension to the Fourier transform (FT). Computationally fast algorithms are readily available for WTs, similar in details to FFT algorithms. These two transforms are also similar in their mathematical details. With FTs, just as the various harmonic frequency components are difficult to see at a glance of the time signal $x(t)$ (Figure 5.4) the time-base information contained in $X(\omega)$ (Equation 5.1) is difficult to see because it is hidden in the phase angle $\varphi(\omega)$ of $X(\omega)$. Adams (2010) details the mathematical foundations of wavelet transformations. The practical benefit of wavelet transforms over Fourier transforms is succinctly illustrated in Figure 5.6, visually presenting the ability of WTs to simultaneously acquire both *time and frequency localization*. The well-known intermittent nature (i.e., non-stationary) of significant vibration signal content, often at high frequencies, is symptomatic of a number of incipient machine failure phenomena. To capture such intermittent high-frequency components means that with FFT as the detection tool, a very small time window must be employed, and that low-pass filters out the lower part of the spectrum. Similarly, with a relatively long FFT time window, intermittent high-frequency components will be inherently time-average filtered out. Figure 5.6 shows that a real-time WT essentially functions like an FFT that would simultaneously be able to employ multiple-duration time windows. Another feature that distinguishes WTs from FTs is as follows. FT employs only a family of sinusoidal functions to decompose a time-base signal into its harmonic components. In contrast, while WTs can also employ a family of sinusoidal functions/waves, it is free to use wave families of other preferred shapes, into which to decompose the time-base signal. The wave function is referred to as the *mother wave*. For particular applications, *wavelet specialists* have devised particular mother waves usually named after the originator, for example, the Daubechies (1993) mother wave, found to yield more quickly convergent series than sinusoidal functions.

5.1.2 Advanced Diagnostics Using Chaos Theory and Probability Statistics

Adams and Abu-Mahfouz (1994) introduced to rotating machinery vibration specialists the application of *chaos theory* as a tool for extracting diagnostic information from time-base signals that is not extractable by other means thus far divulged. The use of chaos theory for the analysis of machinery vibration signals has not yet found its way into MaD products, being still pretty much in the domain of a few researchers in academia. Although chaos analysis tools are being utilized in other fields such as medical research for monitored heart signals. As is well known by those who have studied chaos of dynamical systems, the *necessary ingredient is nonlinearity* of the dynamical system. Of course, all real systems have some nonlinearity, but quite often it is tame enough that vibration engineering analyses in general achieve reliable design analyses based on linear mathematical models. However, Adams and Abu-Mahfouz provide examples where chaos theory-based signal analysis tools can detect some important rotating machinery conditions that are not readily detectable by any of the standard signal analysis tools. They expose an abundance of interesting possibilities for machinery condition feature detection using signal mappings that are regularly employed by chaos specialists. Just a few of these are presented here, and in Adams (2010), for the two highly nonlinear dynamic systems shown in Figure 5.7.

Figure 5.8 presents simulation results of the rotor unbalance excited rub-impact model illustrated in Figure 5.7a, showing a confluence of rotor orbit trajectories and their signal chaos-content mappings using some typical chaos signal processing tools. The central portion of Figure 5.8 is a *bifurcation diagram*, plotting the orbit's x-position coordinate (with a dot) once for each shaft revolution as a reference mark (keyphaser) on the rotor passes the same fixed position point on the stator. If the rotor orbit is strictly rotational-speed synchronous, then the same dot appears once per rev, repeatedly. If a half-synchronous (N/2) subharmonic component is superimposed, then only the same two dots repeatedly appear. Similarly, the *Poincaré maps* in Figure 5.8 contain a dot deposited for the rotor orbit's (x, y) position at each shaft revolution as the reference keyphaser mark fixed on the rotor passes the same fixed stator point. The term quasiperiodic is used by chaos specialists and others to label non-periodic signals that are comprised of incommensurate (non-integer-related) periodic signals.

Spanning the range of non-dimesional unbalance shown in Figure 5.8, the orbital motion goes from quasiperiodic to period 5 (5 shaft revolutions to complete one cycle), bifurcates into

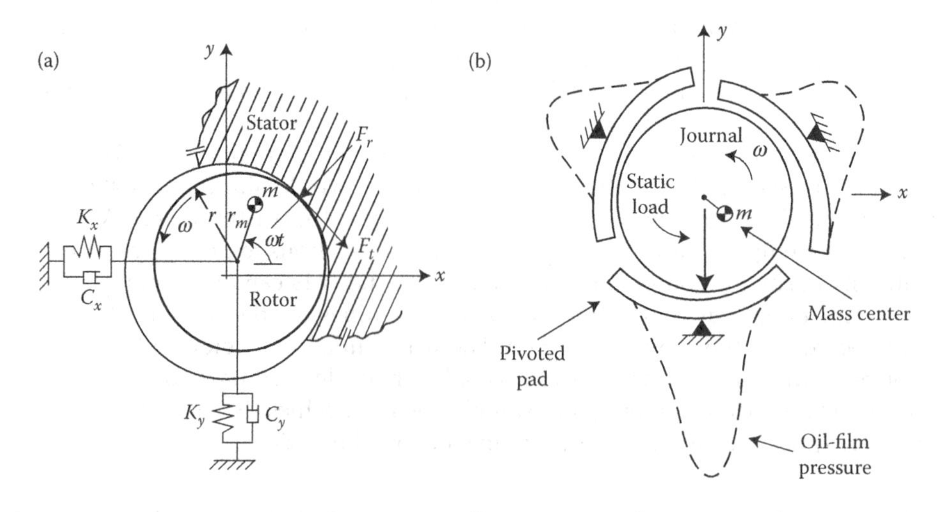

FIGURE 5.7
Rotor dynamics models for chaos studies; (a) unbalance excited rub-impact and (b) 3-pad tilting-pad journal bearing. (From Adams, M. L., *Rotating Machinery Vibration–From Analysis to Troubleshooting*, 2nd ed., CRC Press/Taylor & Francis, Boca Raton, FL, 2010.)

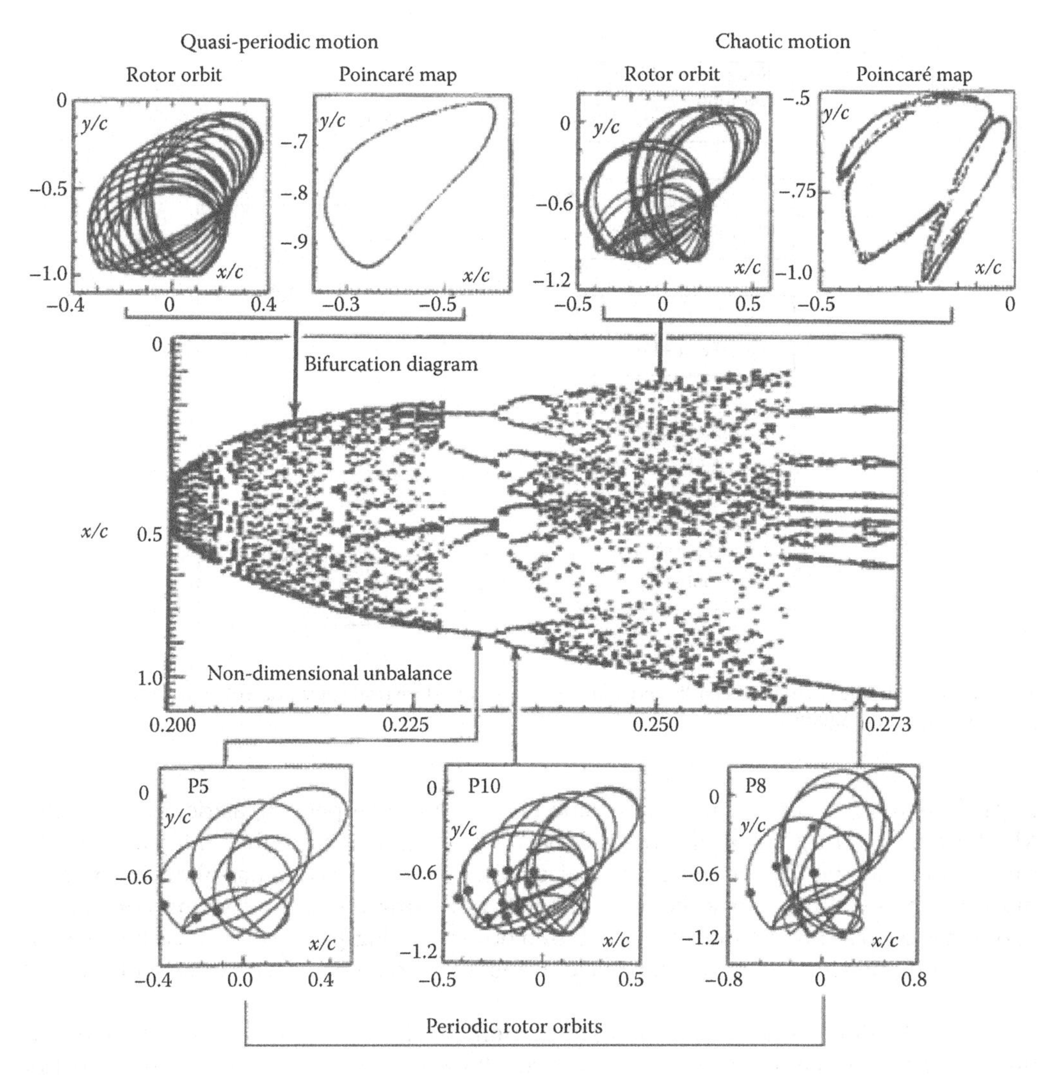

FIGURE 5.8
Rotor orbits and chaos-tool mappings for the unbalance-excited rub-impact simulation model in Figure 5.7a.

period 10 motion, becomes chaotic (non-periodic and deterministic but not random), finally emerging from the chaos zone as period 8 motion. To the author this is quite intriguing. On the periodic rotor orbits shown at the bottom of Figure 5.8, a fat dot is deposited at each keyphaser mark, thus period 5, period 10, and period 8 orbits have 5, 10, and 8 marks, respectively. These dots deposited on the periodic orbits are, by themselves, the *Poincaré* maps of their respective periodic orbits. Thus, for any *periodic motion,* the Poincaré map is a limited number of dots equal in number to the number of revolutions per period of motion. For a *quasiperiodic motion,* the dots on the Poincaré map, over time, fill in one or more closed loops, the number of loops equal to the integer number M of superimposed incommensurate periodic components minus one, that is, M-1. Thus, the quasiperiodic orbit shown in Figure 5.8 has two incommensurate periodic components, thus one loop.

For chaotic motion, the Poincaré map has a *fractal* nature to it and therefore has a fuzzy appearance, as displayed for the chaotic orbit in Figure 5.8. There are mathematical

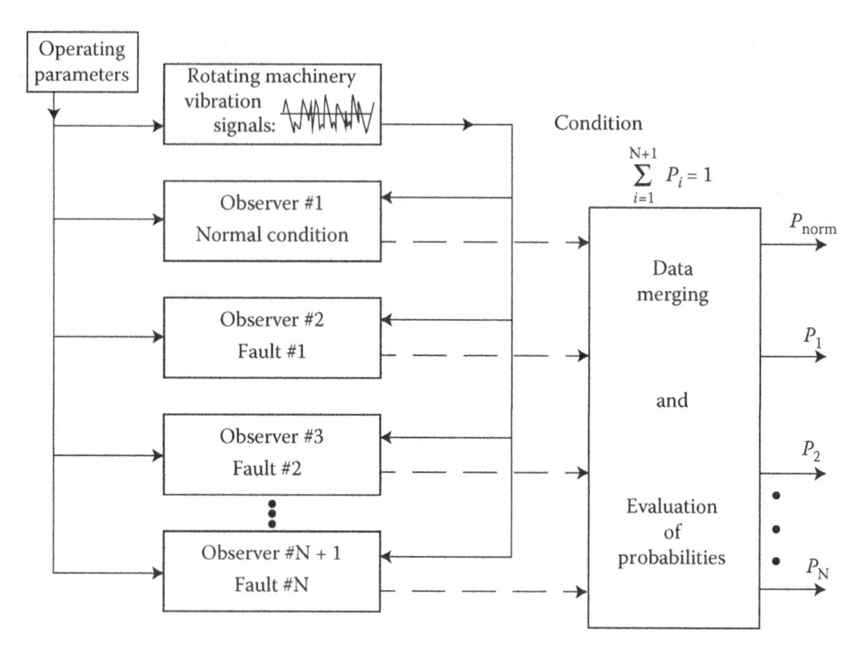

FIGURE 5.9
Real-time probabilities for defined faults and severity levels from statistical correlation of monitored and model-predicted vibration signals.

algorithms to compute a scalar dimension of such a fractal pattern, as detailed by Abu-Mahfouz (1993). In general, the fussier the Poincaré map, the higher the fractal scalar dimension and the higher the degree of chaos. Thus, special signal-filtering methods must be employed to remove the noise without removing the chaos content. For this, the author has used in his laboratory model-based observers (Figure 5.9) combined with signal-threshold de-noising in the signal wavelet transforms to reconstruct de-noised time-base signals.

Figure 5.10 shows some additional simulation results for the unbalance-excited rub-impact model shown in Figure 5.7a. These results were generated to study the detection of small losses in system damping capacity at off-resonance conditions. Since the well-known attenuation effect of damping on vibration amplitudes is primarily at or near a resonance, with off-resonance vibration amplitudes not significantly affected by reductions in damping. As the Figure 5.10 results clearly show, the fractal nature of the associated Poincaré maps for 8% and 11% critical damping cases provided a quite measurable parameter of a small deterioration in damping. However, the FFT signatures for the two compared cases show virtually no monitorable differences. It is relevant and insight enhancing to point out at this point that constructing such a Poincaré map requires the acquisition of many consecutive cycles of real-time data, in contrast to a windowed FFT. One can thus intuitively appreciate how chaos monitoring tools can be so amazing in deciphering time-base signals. That is, because the data utilized covers a relatively long time period, sort of like trending. Accordingly, comparing a monitoring tool like windowed FFT to a chaos tool is analogous to comparing a photograph to a video. It is not difficult to appreciate the practical implication of early detection of damping loss to rotor vibration. For example, in a large steam turbine generator set, through some progressive journal-bearing deterioration process, one can readily imagine a slowly progressing loss

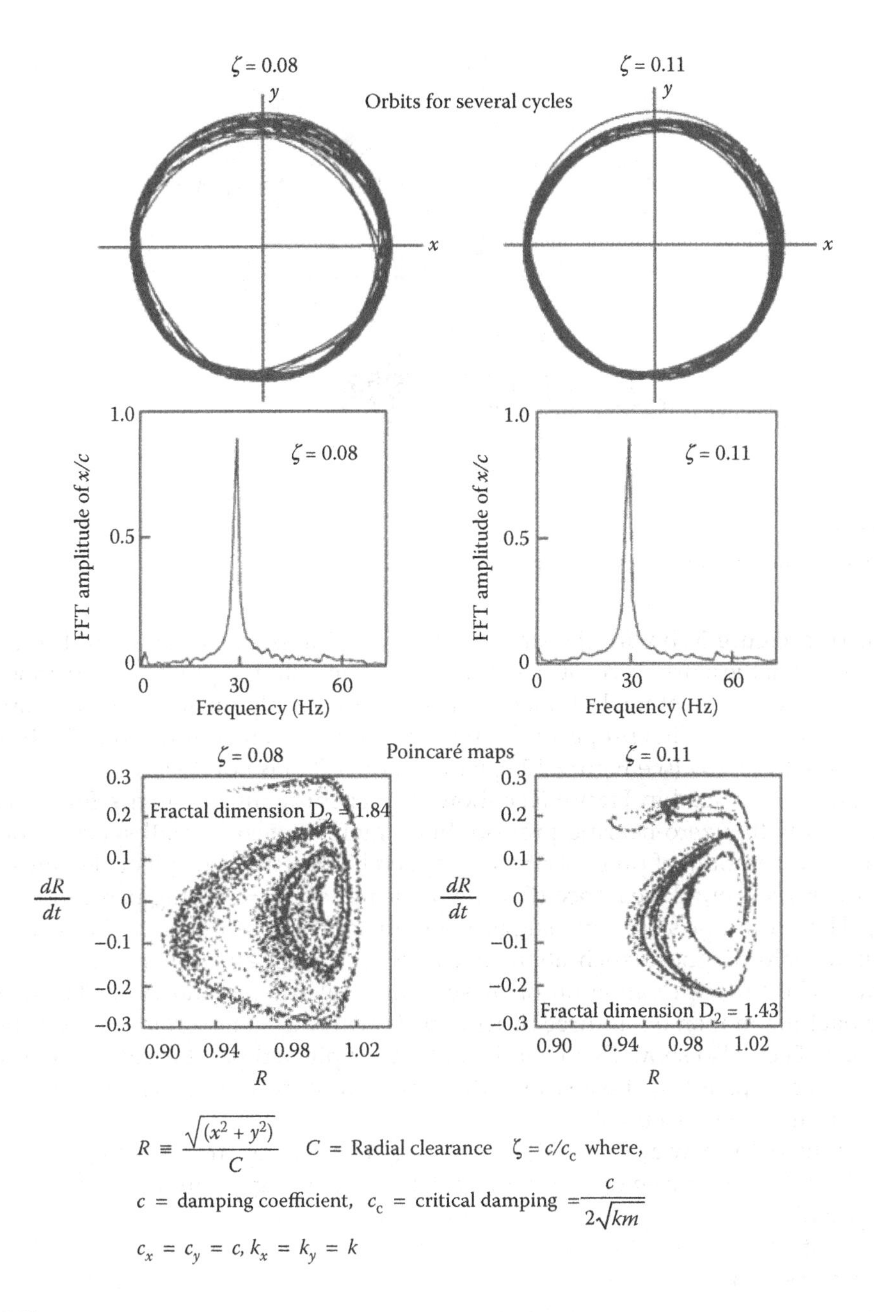

$$R \equiv \frac{\sqrt{(x^2 + y^2)}}{C} \quad C = \text{Radial clearance} \quad \zeta = c/c_c \text{ where,}$$

$$c = \text{damping coefficient}, \ c_c = \text{critical damping} = \frac{c}{2\sqrt{km}}$$

$$c_x = c_y = c, k_x = k_y = k$$

FIGURE 5.10
Poincaré mapping of chaotic response reveals small loss in damping at off-resonance condition; from simulation model illustrated in Figure 5.7a.

of bearing fluid-film damping that would not result in detectable increasing vibration levels at operating speed but could result in dangerously high-rotor-vibration levels on machine shutdown while coasting down through the machine's natural-frequency-mode(s) resonant critical speed(s).

Figure 5.11 shows one of several interesting results on chaotic motion with pivoted-pad journal bearings presented by Adams and Abu-Mahfouz (1994). For the 3-pad bearing

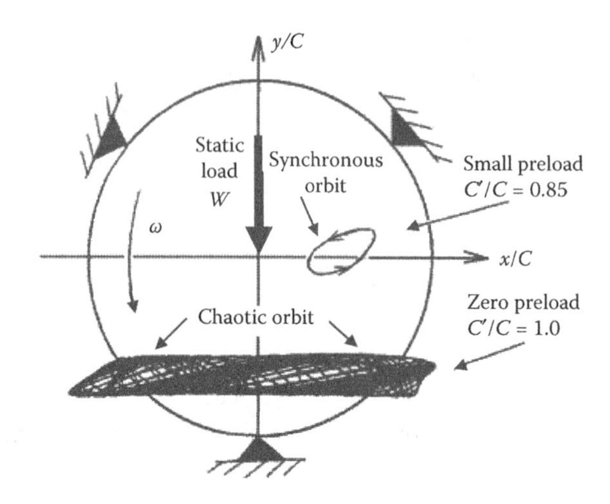

FIGURE 5.11
Chaotic rotor vibration originating in a tilting-pad journal bearing.

illustrated in Figure 5.7b with the static load directed into a pivot location (bottom), it is well known that the journal eccentricity will find a stable static equilibrium position on one side of the pivot or the other, but not exactly on the statically unstable pivot position (see Figures 7.23 and 7.29). This property can be mitigated or even eliminated if the bearing is assembled with *preload* (see Figure 7.25), which more often is not done.

The results illustrated in Figure 5.11 show the synchronous unbalance force causing a chaotic orbit with a zero-bearing preload, but a more expected small synchronous orbit when a modest amount of preload (15%) is applied by adjusting the pivot clearance to 85% of the bearing's ground clearance (C' = pivot clearance, C = bearing-pad radius-journal radius). These examples strongly suggest some useful applications of chaos-tool signal analysis to help diagnose such abnormal rotor vibration and other related operating problems. The fuller presentation of these results by Adams and Abu-Mahfouz show chaotic pitching motion of all three bearing pads with the chaotic rotor motion shown in Figure 5.11. They also show that chaotic motion for pivoted-pad bearings is possible for other numbers of pads (e.g., 4 pads) and other operating conditions where the static-bearing load is not directed into a pivot.

Application of the new evolving MD technology is driven by the persisting environment to greatly *reduce maintenance personnel* and the shrinking population of *true experts*. Development of new MD systems was a topic of extensive ongoing research in the author's laboratory at Case Western Reserve University (CWRU) throughout the 1990s. The CWRU team developed model-based MD software, incorporating an array of machine-specific vibration simulation models, specific to an extensive array of operating modes as well as fault types and fault severity levels. As illustrated in Figure 5.9, each model (called an "observer") is run in real-time and its simulated vibration signals are continuously combined with the machine's actual monitored vibration signals and correlated through a novel set of statistical algorithms and model-based filters, as summarized by Loparo and Adams (1998). Probabilities are generated for each fault-type and severity level potentially in progress. The vibration models in the observers also remove signal noise which does not statistically correlate with the models. In contrast to conventional signal noise filtering techniques, such model-based statistical-correlation filtering retains components such as signal chaos content for on-line or off-line analysis.

One of the many interesting findings by the CWRU team is that the various fault and fault-level-specific observer vibration models do not have to be as "nearly perfect" as one might suspect if thinking in the time-signal domain and/or frequency domain framework. Because the sum of the probabilities is constrained to = 1. So a model (observer) only has to be representative enough of its respective operating mode to "win the probability race" among all the observers when its fault (or fault-combination)-type and severity level are in fact the dominant condition. Compared to the rule-based approach inherent in the so-called expert systems, this physical model-based statistical approach is fundamentally much more open to correct and early diagnosis, especially of infrequently encountered failure and maintenance-related phenomena and especially of conditions not readily covered within a rule-based "expert system." An additional benefit of a model-based diagnostics approach is the ability to combine measured vibration signals with vibration simulation model outputs to make real-time determinations of rotor-vibration signals at locations where no sensors can be installed, that is, by *virtual sensors* (Section 5.1.5).

5.1.3 Machinery Vibration Sensors and Safe Vibration-Level Guidelines

As any mechanical engineering undergraduate knows, the three elementary dynamics/kinematics parameters are *displacement, velocity,* and *acceleration,* directly relatable to each other through elementary calculus. And these are the *three vibration parameters* that are regularly measured and monitored on plant operating machinery. The *sensors* for these three vibration signals are schematically illustrated in Figure 5.12. For accelerometers, the accurate useable frequency range must be considerably below the accelerometer's own one-degree-of-freedom (DOF) natural frequency. For a velocity sensor, the useable frequency range must be sufficiently above the sensor's own one-DOF natural frequency. Because velocity sensors are intrinsically fragile, they are primarily for laboratory use. So, an industrial velocity sensor is in fact an accelerometer with a built-in signal integrator. For non-contacting position sensing, the inductance proximity probe must be selected for the shaft material. Because of residual magnetism in a shaft, there will be some indicated runout that is not mechanical. So in extra high-accuracy applications, the rotating target is chrome plated to filter out the electrical run out (Horattas et al. 1997).

Adams (2010) gives detailed technical information for selecting sensor specifications as well as the industry experience-based vibration-severity guideline shown in Figure 5.13. Although both parts of this guideline contain the same guideline information, one clearly sees the appeal of using the *velocity severity levels* since a particular velocity peak value has the same severity interpretation over the entire frequency range of concern for most machinery. Table 5.1 provides *vibration-displacement severity criteria* based on the ratio of journal vibration displacement-to-bearing-clearance ratio, as measured by proximity probes (Eshleman 1999; Adams 2010).

5.1.4 Monitoring and Diagnostics in High-Energy Centrifugal Pumps

In addition to vibration monitoring, Figures 5.1, 5.14, and 5.15 (Adams 2017a) show additional important parameters to diagnostically real-time monitor in real-time: (1) pressure pulsations and (2) balance drum leak-off flow in high-energy-density pumps. These parameters are good relative measures of *pump hydraulic discomfort* as operation necessarily intrudes into off-design low flow operation. In addition, these parameters also provide good relative measures of the associated life-shorting of pump internals. Figure 5.16 is a diagnostic guideline for high-energy centrifugal pumps that succinctly combines vibration

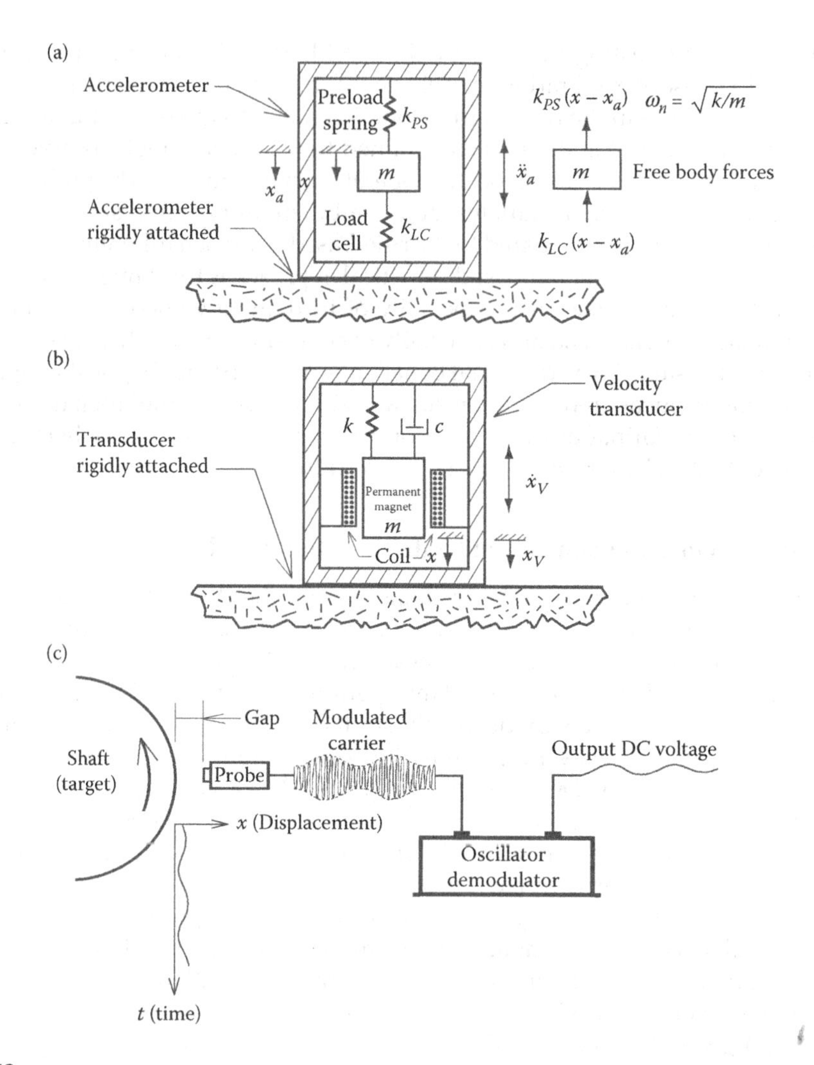

FIGURE 5.12
Vibration-measurement sensors; (a) accelerometer, (b) velocity transducer, and (c) eddy-current non-contacting probe position sensing.

spectral characteristics and root causes as functions of the percent of design best-efficiency operating flow. Table 5.2 summarizes recommended centrifugal pump parameters to monitor and record (Florjancic 2008; Adams 2017a). Depending on the specific configuration and application of a pump, combinations of monitored parameters can further provide early warning of emerging operating problems.

5.1.5 Model-Based Condition Monitoring through Virtual Sensors

Figure 5.17 is a photograph of the Case Western Reserve University (CWRU) multistage centrifugal pump test loop recently developed for research in *model-based conditioning monitoring* for power plant pumps (Adams 2017a). Its initial development was in response to pump-condition monitoring in general, but specifically in response to plant services where the pumps are submerged and thus impractical for periodic condition inspections, for example, river pumps for nuclear power plants. Figure 5.18 shows the submerged

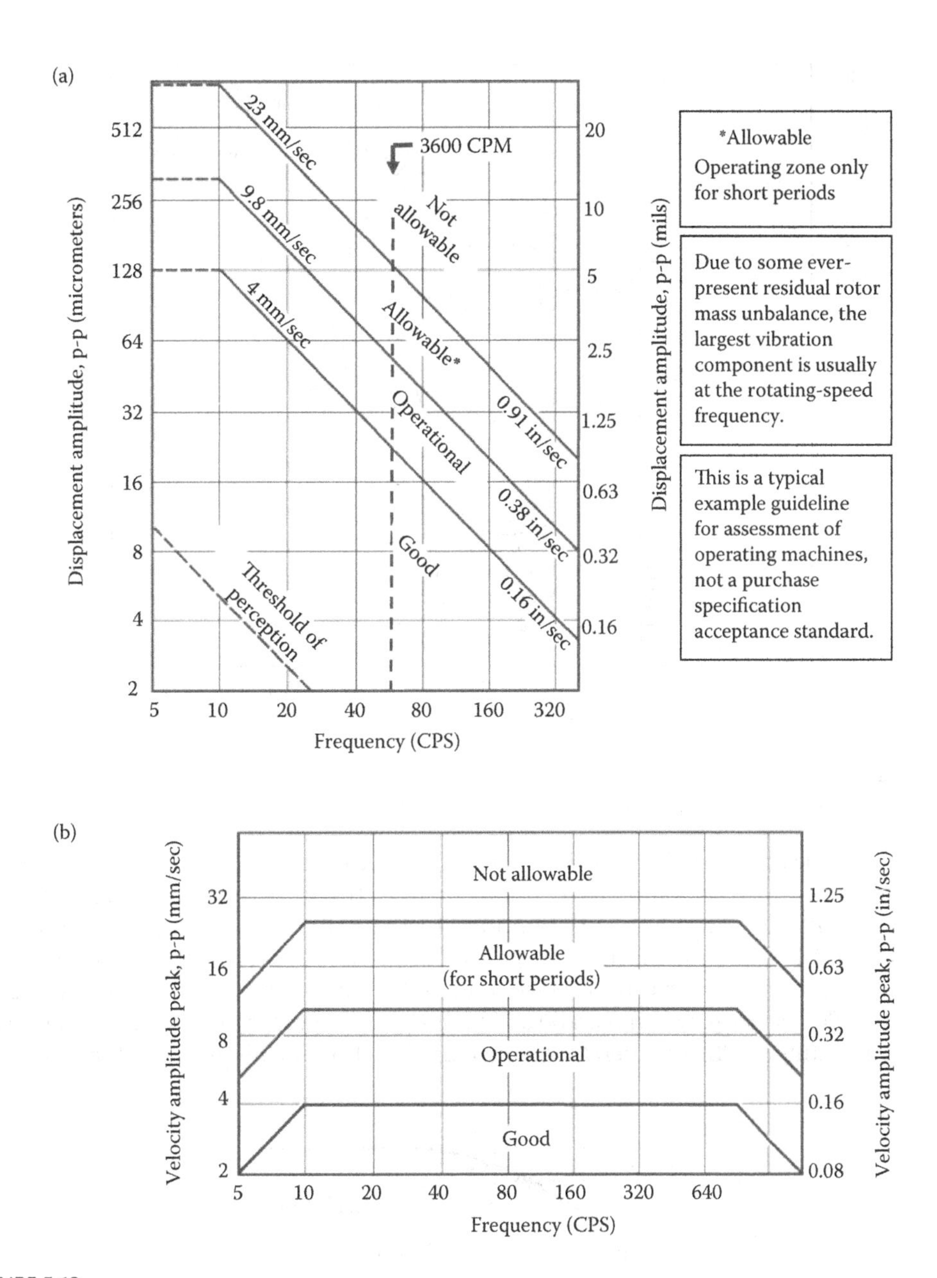

FIGURE 5.13
Bearing-cap vibration acceleration and velocity amplitude guideline; (a) acceleration p-p, (b) velocity s-p, (a) and (b) contain the same information.

TABLE 5.1
Guideline: Journal-Vibration Displacement Relative to Journal Bearing

Speed	Normal	Surveillance	Plan Shutdown	Immediate Shutdown
3600 rpm	$R/C < 0.3$	$0.3 < R/C < 0.5$	$0.5 < R/C < 0.7$	$R/C > 0.7$
10,000 rpm	$R/C < 0.2$	$0.2 < R/C < 0.4$	$0.4 < R/C < 0.6$	$R/C > 0.6$

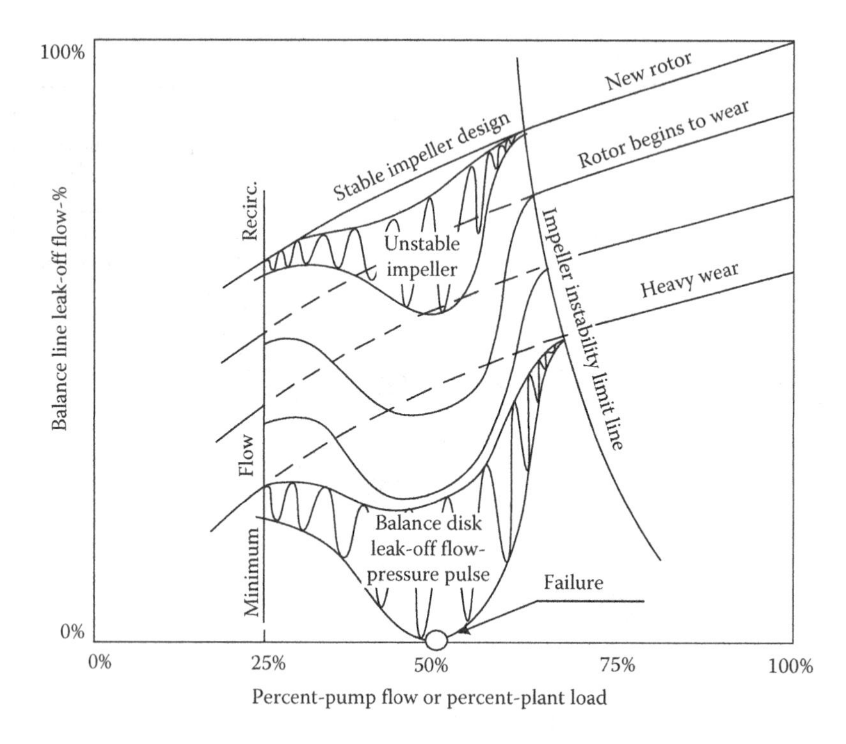

FIGURE 5.14
Pressure pulsations warn of pump discomfort/damage at low flows.

accelerometer locations on the CWRU test pump inner casing surface. Figure 5.19 shows sensor locations on the test pump as it is submerged in the transparent test-loop outer can. A fairly new "player" in pump condition monitoring is the Robertson efficiency probes (Robertson and Baird 2015). A matched pair of these probes accurately measures pump efficiency in real-time using ultra-precise temperature difference measurements (Figure 5.20). Pump *efficiency degradations* can be a valuable complementary parameter in detecting pump deteriorating health and thereby giving fault detection warnings to *avoid forced outages*.

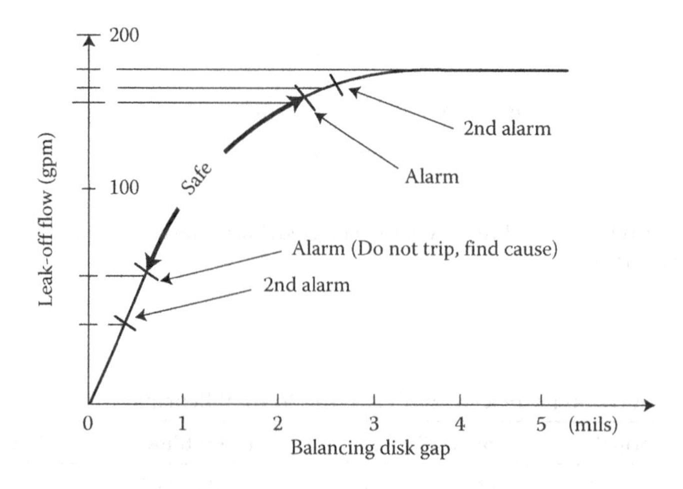

FIGURE 5.15
Balancing disk leak-off flow as an indicator of pump health.

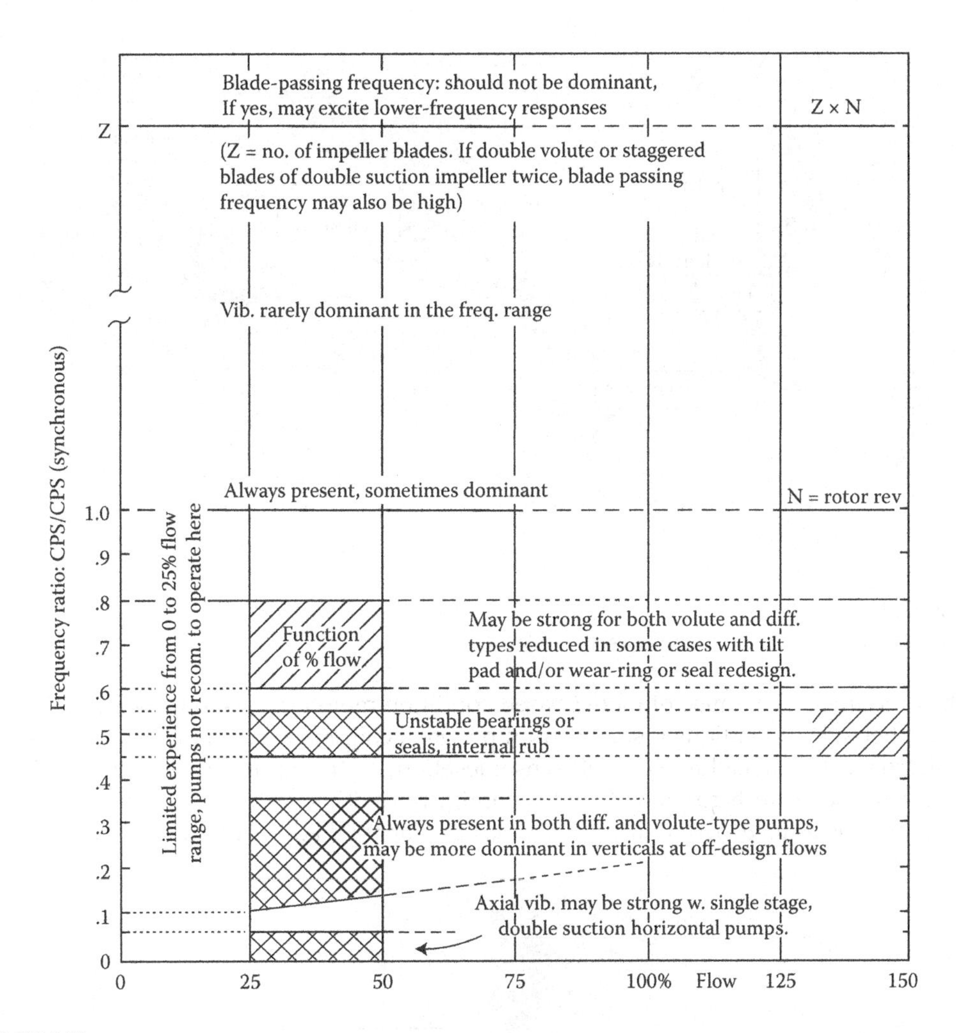

FIGURE 5.16
Sources and description of pump-vibration forces versus flow range.

TABLE 5.2
Recommended Monitoring for High-Speed and Multistage Pumps

Recommended for Recording	An Indication of
Pump flow, pump speed	Internal wear
Shaft vibration (amplitude vs. frequency)	Internal wear
Balance drum water flow	Drum piston clearance wear[a]
Thrust-bearing temperature	Change in axial thrust
Suction pressure and temperature	Cavitation[a]
Discharge pressure	Internal wear
Leak-off flow	Overheating of pump
Radial-bearing temperature	Overload/wear[a]
Seal-drain temperature	Breakdown of seal
Barrel temperature (top and bottom)	Casing distortion, insulation

[a] Recommended for small and simple pumps.

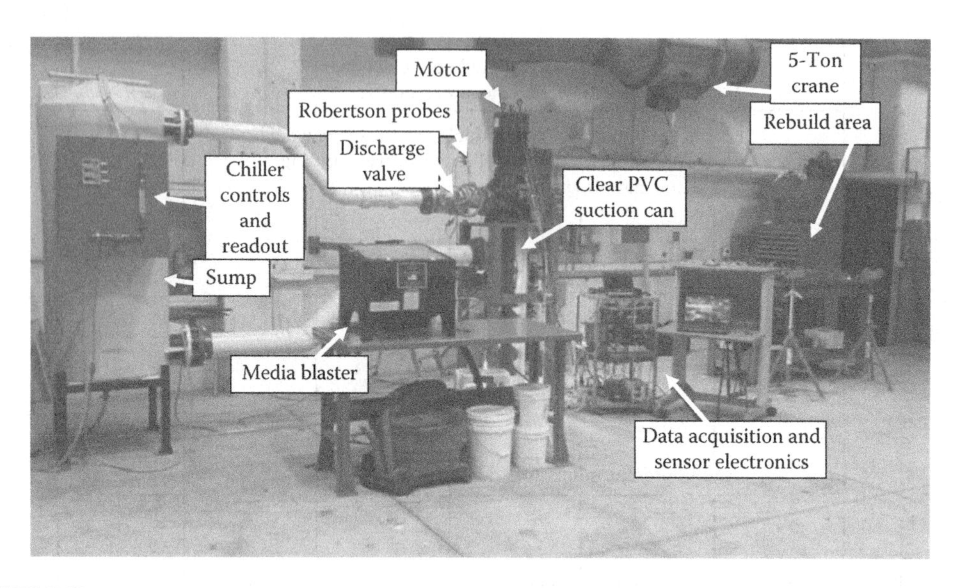

FIGURE 5.17
CWRU multistage centrifugal pump research test loop.

The fundamental premise of model-based condition monitoring is to reconstruct in real-time the behavior conditions inside the pump from a computer model driven by measured signals from external readily accessible sensor locations on the pump-driver unit. The CWRU test facility shown in Figures 5.17–5.20 is configured with sensors inside the unit, where sensors are not feasible to normally place in power plant operating pumps. And additionally,

FIGURE 5.18
Submerged accelerometer locations on CWRU 3-stage test pump.

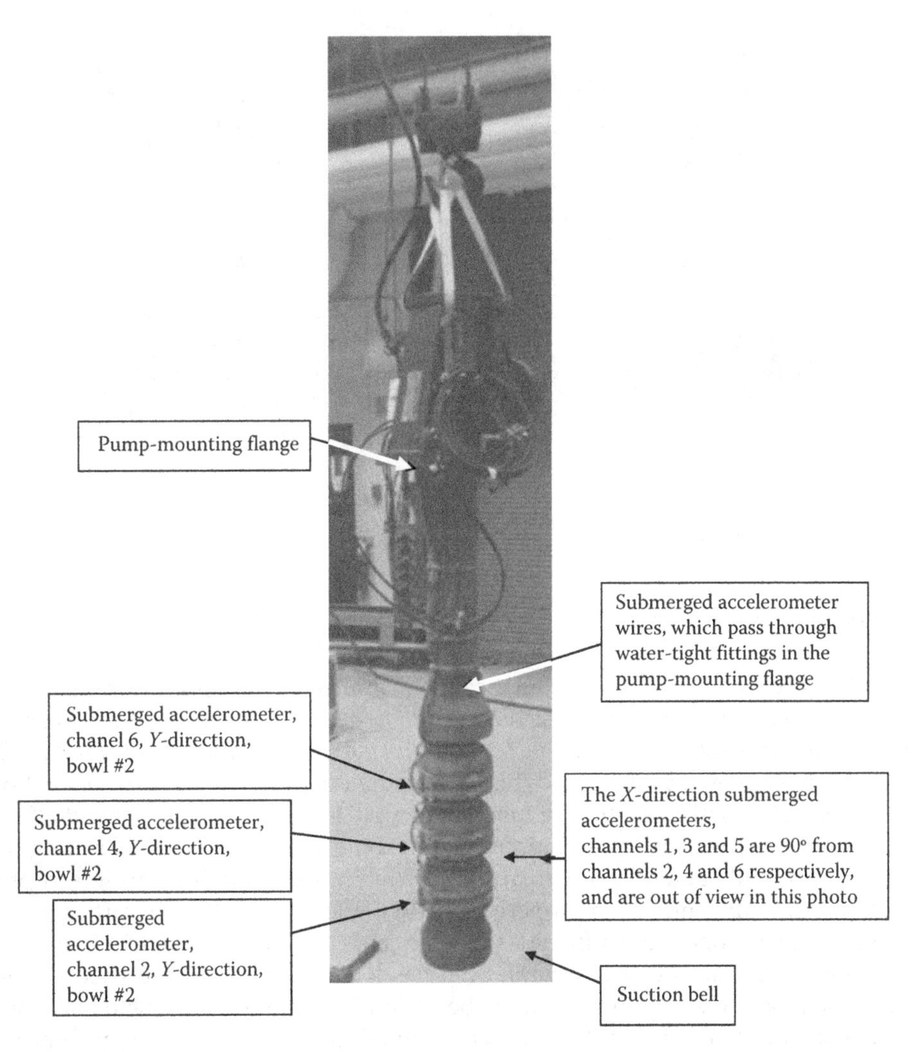

FIGURE 5.19
Sensor locations on installed CWRU 3-stage test pump.

sensors are located at readily accessible external locations. In this research, *competing online real-time computer models* are "tested" to determine their adequacy in replicating the internal measurements from only the signals of the external sensors. This ongoing research is a promising pursuit likely as well to ultimately benefit many other types of rotating machinery.

5.2 Hydrodynamic Fluid-Film Journal and Thrust-Bearing MaD

Under normal circumstances, running a hydrodynamic fluid-film-bearing (HDFFB) functions with no bearing-to-journal (or thrust collar) contact, except for boundary-lubricated starts and stops. Thus, HDFFBs are unlike rolling element bearings that are life-limited by subsurface initiated fatigue, but even more likely to first dysfunction from another cause (see

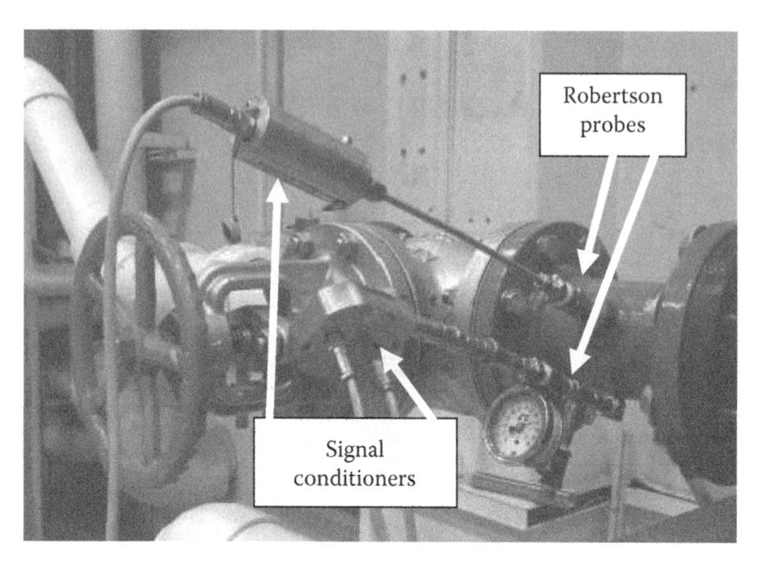

FIGURE 5.20
Robertson pump-efficiency probes.

Figure 2.37). Since HDFFBs are not automatically life limited by any specific mechanism, they have often been loosely characterized as having *infinite life*, which of course is an exaggeration to put it mildly. *The most sudden and complete way an HDFFB can fail in operation is if the flow of lubricant supply is interrupted.* A hydrodynamic bearing is essentially a *viscous pump*. So with an interruption of a continuous flow of lubricant supply to the bearing, it *pumps itself dry* in a very short period of time. That is why aircraft gas turbine jet engines use only rolling contact bearings, which while potentially incurring accelerated degradation by an interruption of lubricant supply, will continue to properly function at least until the aircraft safely lands. There are a number of operating difficulties that can cause degradation of HDFFBs to an extent necessitating *refurbishment* or *replacement*. So it has long been recognized that monitoring and assessment of operating HDFFBs must be included within the complete real-time MD assessment of critical machinery. A prominent example is large power plant machinery such as main steam turbine generator sets, major pumps, fans, and auxiliary machinery.

The most important parameter to continuously monitor in HDFFBs is the *operating temperature* in the lubricating film, which is substantially heated by the inherent viscous energy dissipation within the sheared oil film, as detailed in Chapter 7. Depending on the bearing size and unit load (see Figure 1.1a), minimum-film thickness is typically 0.5–5 thousands of an inch (0.013–0.13 mm). So, measurement of fluid-film temperature directly is not practical because that would put the temperature measurement thermocouple tips much too close to the high-speed rotating surface which would easily wipe out the thermocouples when contacting them during starts, stops, and significant vibration. Figure 5.21 illustrates a commonly used approach to get a thermocouple as close as possible to the oil film without being directly within the film, thus registering a temperature nearly the same as the oil film. This arrangement shows that it is desirable to have the thermocouple tip securely touching a quite thin section of the bearing white metal approximately 0.1 inch (2.5 mm) from the oil film. The diameter of the thermocouple recess hole is made as small as possible to prevent any significant radial deformation (dimple) of the bearing white-metal liner under the hydrodynamic-film pressure which can be well over 1000 psi (70 bar). Maximum HDFFB allowable operating temperature is typically set by the bearing liner material, conservatively

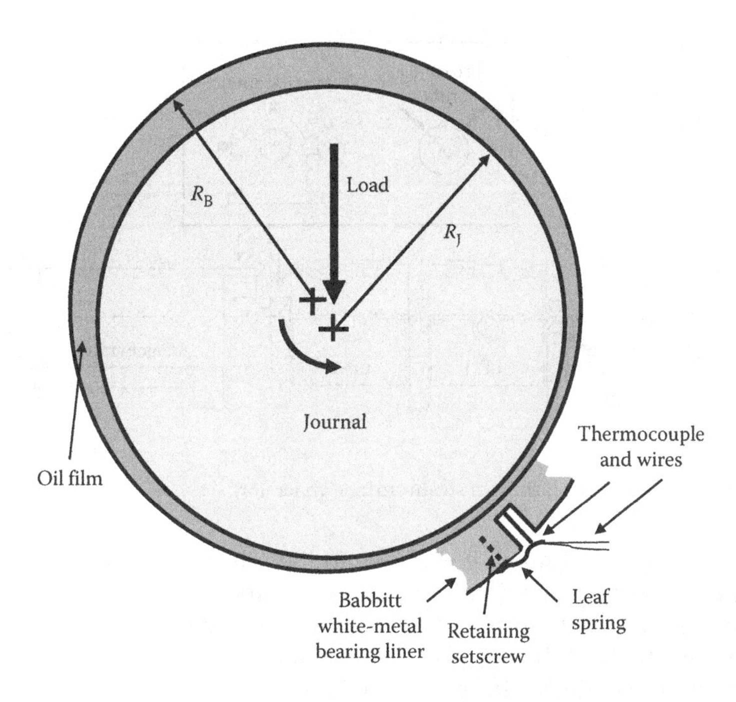

FIGURE 5.21
Journal bearing with leaf-spring-retained thermocouple. Oil-film and white-metal thicknesses exaggerated.

around 200°F (93°C) for bearing white metal. Overload of the bearing or lubricant supply partial starvation will cause a significant rise in bearing-film temperature, detectable as illustrated in Figure 5.21.

Inadequate lubricant filtration of course always increases the risk of progressive surface damage to the bearing liner and turning surfaces. An automotive engine has several radial HDFFBs; *main-crank-shaft, connecting-rod* and *wrist-pin bearings* (see Figure 7.1a). It should be well known by all automobile owners, although it is not, that when the motor oil is routinely changed within the designated mileage/time interval, changing the oil filter at the same time is even more important than changing the oil. The importance of proper oil filtration is equally important for industrial machinery. Debris content in the lubricant is a monitored parameter in critical applications, particularly in aircraft engines as covered in the next section.

Rotor vibration at HDFFBs is always present since there is always residual rotor unbalance as well as other possible sources of vibration. *Excessive vibration* for a given application is a potential source of progressive bearing deterioration as well as indicative of other operating problems occurring in a machine. Figure 5.22 illustrates the vibration monitoring at the radial HDFFBs of a multi-bearing rigidly coupled-rotor large 500 MW electric power steam turbine generator unit. Such units are typically equipped with two orthogonal x and y proximity probes (Figure 5.12c), bearing-mounted to continuously measuring shaft vibration orbit relative to each radial bearing. And accelerometers (Figure 5.12a) are mounted on each bearing housing, with signals twice-integrated in time to get real-time-bearing displacement signals that are combined with the proximity-probe displacement signals to obtain real-time total-rotor vibration displacement with respect to the ground. Figure 5.23 shows a typical installation of the sensors.

The typical angular orientation of a pair of proximity probes is at 45° and 135° because in most applications the major axis of the rotor vibration orbit is near 45°-to-horizontal due

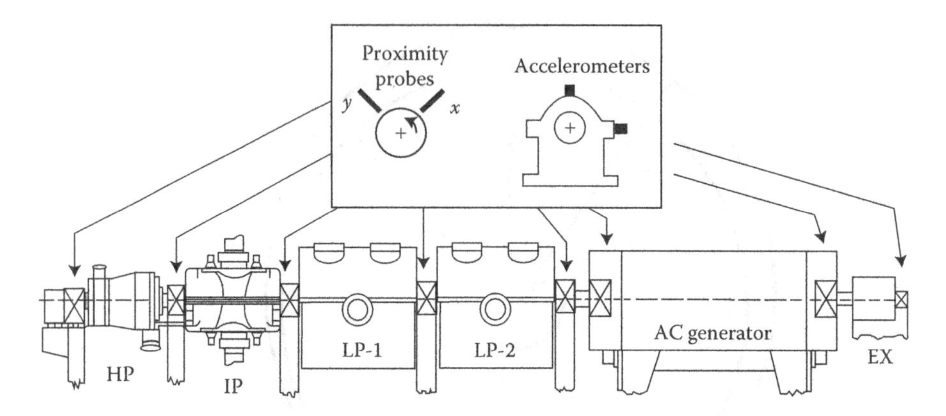

FIGURE 5.22
Vibration monitoring channels at bearings of steam-turbine generator.

to the journal-bearing oil film being stiffest into the minimum film thickness direction, that is, along the line-of-centers joining the rotor centerline and the bearing centerline. The 45° channel is therefore normally selected as the rotor-vibration channel used for rotor balancing in the field, having the highest signal-to-noise ratio if it closely aligns with the rotor-vibration orbit's ellipse major axis. Total rotor-vibration displacement level at the bearings is then obtained by vectorially adding the conditioned outputs of the integrated accelerometer transducer measurement and the proximity probe displacement measurement. An approach not easily employed in plants but commonly employed in laboratory rotor test rigs (Adams 2017b) is simply to mount the proximity probes to an essentially non-vibrating fixture.

In addition to the vibration diagnostic signal processing methods treated in Section 5.1, the specific simultaneous real-time monitoring of x–y rotor-radial vibration orbital trajectories at the bearings yields a real-time picture of the rotor-vibration orbit, which by itself contains informative diagnostic information. Figure 5.24 illustrates two typical rotor-vibration synchronous orbits. The dot on the orbits is placed where (when) the rotor keyphaser fixed mark on the rotor passes a fixed detection point as illustrated in Figure 5.25. Orbit plots augment multifrequency vibration FFT spectra. A simple example of this is illustrated in Figure 5.26, showing both the orbit and FFT of a periodic orbital vibration that contains a dominant N/2 harmonic. Figure 5.27 shows an important quite interesting

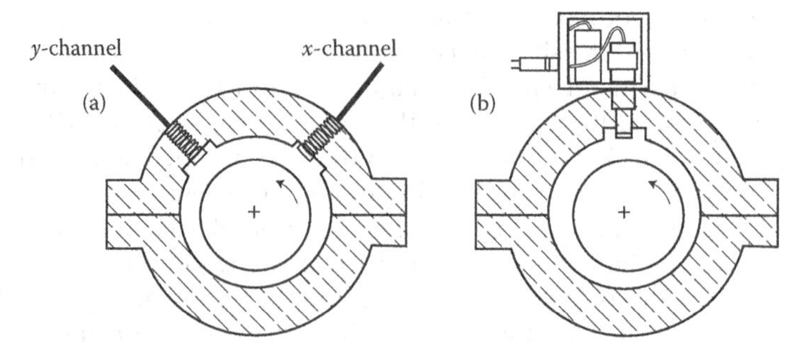

FIGURE 5.23
Bearing-mounted vibration monitoring sensors; (a) typical 2-proximity probes at 90° relative placement and (b) probe with accelerometer.

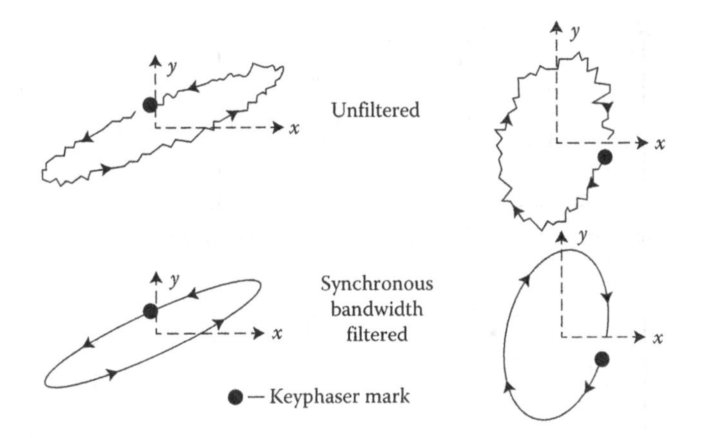

FIGURE 5.24
Examples of measured rotor synchronous orbits; (a) forward co-rotational whirl and (b) backward counter rotational whirl.

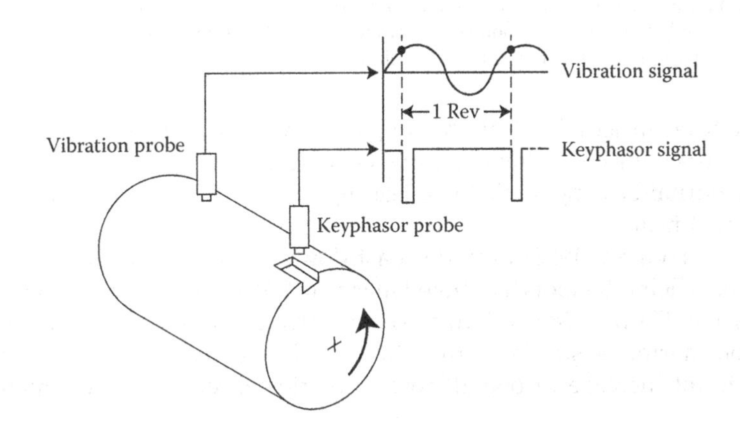

FIGURE 5.25
Monitored rotor vibration signals are referenced to a single keyphaser.

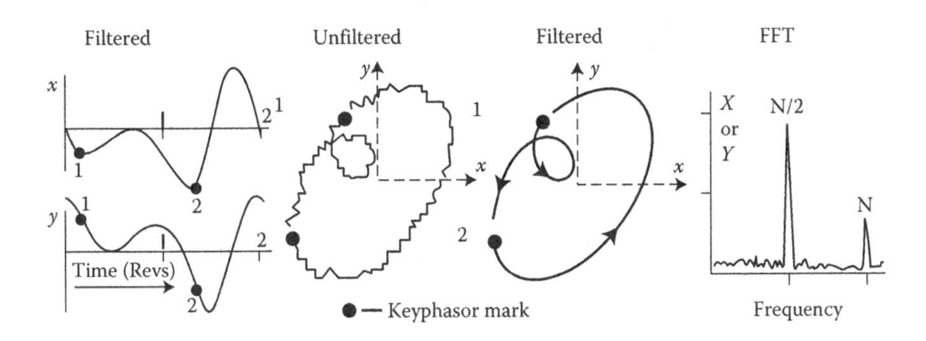

FIGURE 5.26
Measured orbit with synchronous and half-synchronous harmonics.

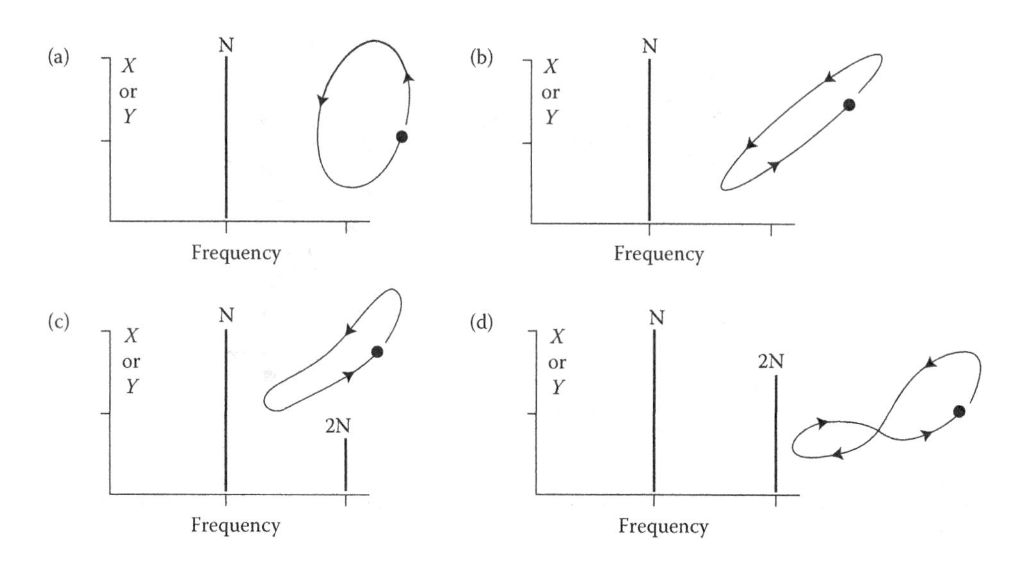

FIGURE 5.27
Journal orbit and FFT for increasing bearing radial load. (a) Nominal radial load; synchronous linear motion, (b) Moderate radial-load increase; synchronous linear motion, (c) Substantial radial load; nonlinear motion with some $2N$, and (d) Very high-radial load; nonlinear motion with high $2N$.

example taken from an actual troubleshooting case (Adams 2010) where after a cold start, a progressively worsening condition in a turbo centrifugal compressor caused a progressive increase on a journal-bearing static load, leading to the rotor orbital vibration becoming considerably nonlinear.

Bently and Muszynska (1996) provide coast down orbit and cascade plots for a quite similar machine. Their orbit plots illustrated in Figure 5.28 start at running speed (5413 rpm). The orbit shown in Figure 5.28a is characteristic with a 2N harmonic superimposed on the 1N synchronous harmonic similar to that shown in Figure 5.27d. As this machine coasted down a significant increase in overall rotor vibration levels was encountered between

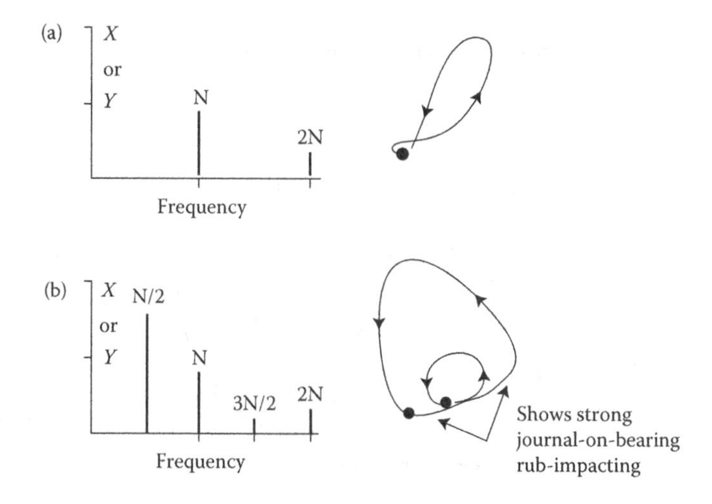

FIGURE 5.28
FFT and journal orbit changes; coast down with high-radial load.

4500 and 4100 rpm as shown, Figure 5.28b. The orbit and its FFT spectrum are a clear picture of what was occurring in the 4500–4100 speed range, as follows. At 4264 rpm, the spin speed traversed twice the 2132 cpm critical-speed resonance frequency at half the spin speed. Significant bearing dynamic nonlinearity from the high-radial static-bearing load or misalignment plus the inherent characteristics of the journal bearings to have lowered damping for subsynchronous frequencies combined to produce a dominant N/2 subharmonic vibration component in the speed neighborhood of 4264 rpm. As the orbit plot in Figure 5.28b clearly shows, the significant increase in the overall rotor-vibration level near this speed caused pronounced journal-on-bearing rub-impacting at the monitored bearing. Once the rub-impacting occurs, the dynamic nonlinearity increases even further, synergistically working to maximize the N/2 harmonic through what is tantamount to a so-called *nonlinear jump phenomenon*, similar to that analyzed by Adams and McCloskey (1984) and covered in Chapter 11. As detailed by Adams (2010), each rotor elliptical orbit harmonic component as shown in Figure 5.24 for the 1N component can be decomposed into a *circular forward whirl* and a *circular backward whirl* component as illustrated in Figure 5.29. The cascade plot from field measurements for the Figure 5.28 coast down case are shown in Figure 5.30, separating the forward-whirl circular orbital portion from the circular backward-whirl portion for each harmonic's elliptical orbit. The period of the vibration around 4264 rpm is two revolutions of the shaft because of the dynamic nonlinearity making possible the 1N unbalance synchronous excitation force substantially exciting the

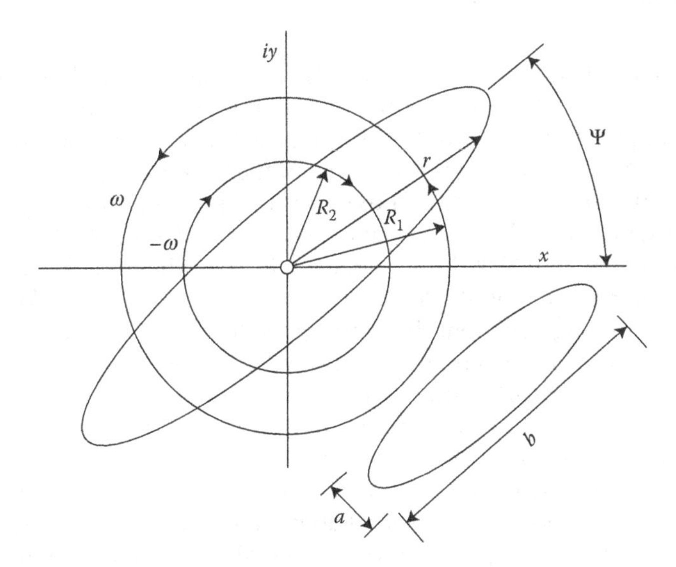

Elliptic orbits are from $x(t)$ and $y(t)$, each harmonic at same frequency
Where, $x = X\cos(\omega t + \phi_x)$ and $y = Y\cos(\omega t + \phi_y)$
Forward-whirl elliptical orbit when $0 < (\phi_x - \phi_y) < 180°$
Backward-whirl elliptical orbit when $-180° < (\phi_x - \phi_y) < 0$
Straight-line orbit when $(\phi_x - \phi_y) = 0$, a = 0
$a = |R_1 - R_2|$, $b = R_1 + R_2$, $\Psi = \dfrac{\beta_1 + \beta_2}{2}$,
Where β_1 and β_2 are the angles of R_1 and R_2 w.r.t. x-axis @ $t = 0$

FIGURE 5.29
Elliptical orbit: the sum of two counter-rotating circular orbits.

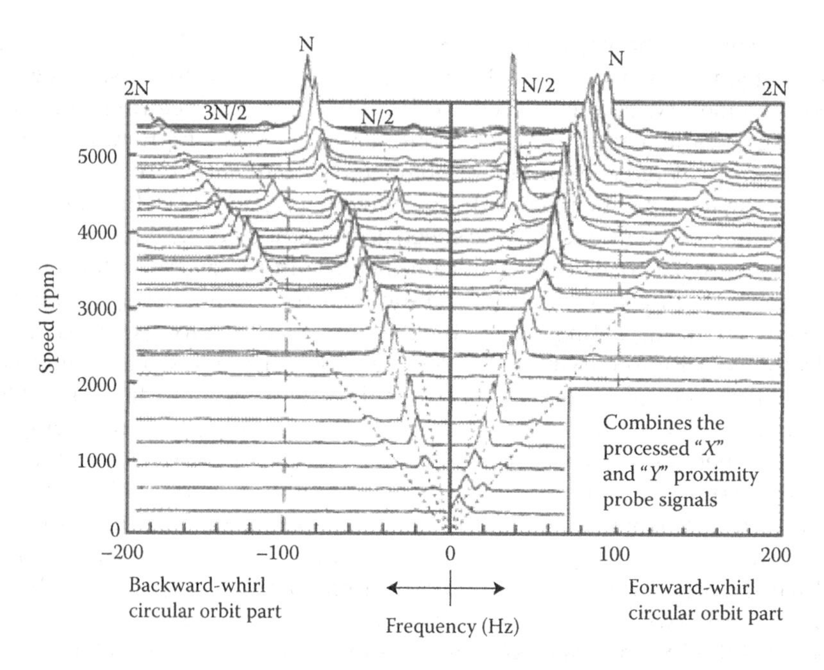

FIGURE 5.30
Cascade for coast-down of case in Figure 5.29.

N/2 subharmonic natural frequency mode. Thus, as Figure 5.30 shows at the coast down speed around 4264 rpm the period of the vibration is two revolutions because of the N/2 harmonic. So higher harmonics of the N/2 frequency are possible. And sure enough, a relatively small 3N/2 component is detectable in the backward circular orbit component.

The theoretical connection between dynamical system nonlinearity and the resulting potential for excitation of higher integer harmonics of 1N (2N 3N ...) and fractional-integer harmonics of 1N (N/2, N/3 ...) relates to a mathematically rigorous rule from elementary ordinary differential (ODE) equations for linear systems. Where a linear ODE has a sinusoidal forcing function, the *particular* solution can contain only the excitation frequency(s). For example, the one-degree-of-freedom equation $m\ddot{x} + c\dot{x} + kx = F\sin\omega t$ must have a steady-state (*particular*) solution for $x(t)$ which contains only the excitation frequency ω, because the equation is linear. That is, $\ddot{x}$, $\dot{x}$ and x are all to the first power and the coefficients m, c, and k are all constant as well.

To summarize, monitoring and diagnostics pertaining to fluid-film bearings are today quite essential components within the total domain of health monitoring for machinery, far more than just for the bearings' health per se, but also for the overall health of a running machine as detectable from vibration signal analysis.

5.3 Rolling Element-Bearing MaD

In contrast to monitoring vibration signals at fluid-film bearings, MaD of rolling element bearings (REB) is focused almost solely on assessing the condition of the bearing, but not additionally on monitoring the overall health of the machine. This can be fully appreciated

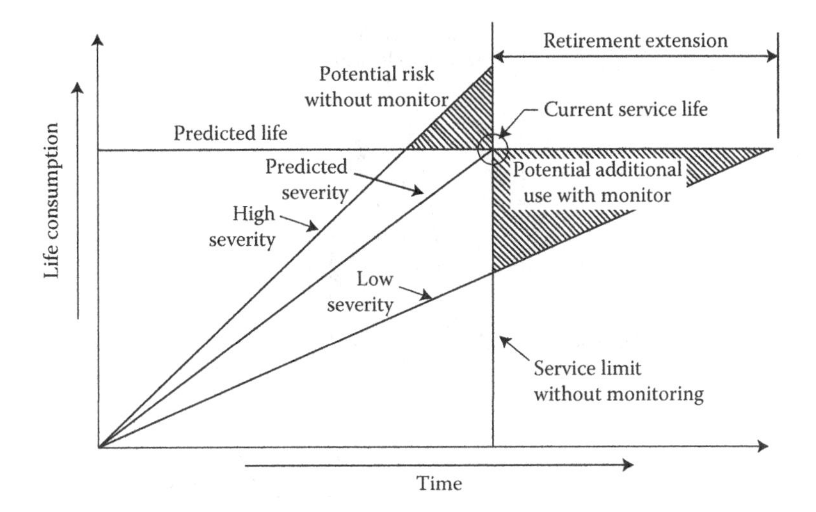

FIGURE 5.31
REB-life consumption in: actual service compared to design analysis.

by scrutinizing the information contained in Figures 2.37 and 2.38, which delineate the many different ways in which an REB reaches the end of its inevitable finite life. Specifically, REBs do not have a "thick-film" space between rolling and stationary components, thus there is not the same potential for vibration measurements at an REB that also provide the useful diagnostic information of overall machine health to the considerable extent provided from rotor-bearing vibration monitoring at fluid-film bearings.

However, overall machine vibration levels are surely influenced by REBs, primarily: (1) as part of a machine's structural stiffness makeup, (2) as a producer of vibration from the intrinsic way the load distribution between the rolling elements and raceways cyclically varies, and (3) as caused by geometrical imperfections from manufacturing, installation, wear, and damage in service. Good auto mechanics and mechanically inclined drivers know that when a wheel bearing becomes noisy, it is advisable to stop at the next service area hopefully before the noisy wheel bearing seizes. *Acoustic noise emission* and *vibration* are the main REB-time-based parameters monitored to detect REB-emerging faults and insipient failures. That practice has become a considerable cost saver over the last quarter century in manufacturing where a convenient time-chosen prefailure intervention on critical-path machinery can avoid otherwise untimely and costly production line stoppages. Additionally, Harris and Kotzalas (2007) review a then fairly new approach called condition-based maintenance (CBM) that goes a step farther. They show a development example for *component life consumption* of helicopter mechanical components that incorporates prediction of remaining REB life based on monitored/recorded *actual accumulated operating conditions*, not on design-analysis-assumed operating conditions, as exampled in Figure 5.31. Harris and Kotzalas also delineate *noise-sensitive* from *vibration-sensitive* machinery applications. Time-frequency analyses such as with fast Fourier transform (FFT) and wavelet transform (WT), Section 5.1.1, are the two major monitoring tools within the numerous identified signal processing tools.

The importance of REB-MaD has fostered continuous research aimed specifically at REB-MaD. As a result, the literature on MaD for REBs is overwhelmed with numerous signal-analysis methodologies in large measure by academics, buried in mathematically heavy analysis approaches. At the same time, laboratory running of REBs to their end-of-life for different loads, speeds, defects, design variations, and so on, has always been an enormously

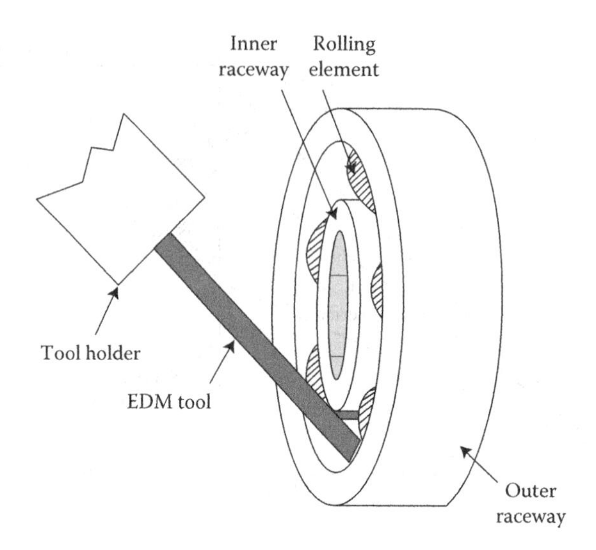

FIGURE 5.32
Schematic of EDM fault seeding process.

costly endeavor, to say the least. Given this considerable gap between (1) blue-sky math-intensive theories and (2) cost-prohibitive laboratory-results-driven approaches, there is a strong motivation for developing REB-MaD methods that combine the following components:

1. REB-fault dynamics modeling
2. Overall machinery dynamics modeling
3. Useful time-frequency analyses
4. Real-world machinery outside the confines of laboratory test rigs

That is where the most promising research deliverables are to be found. An early effort at this overall approach is reported by Adams (author's #2 son) and Loparo (2004). They conducted the testing part of their effort on a mass-produced 2-horsepower, 3-phase induction motor. Testing was conducted with various defect-seeded ball bearings of the same catalog number used in the production motor. Vibration of the motor was measured using magnetically attached accelerometers at the drive end of the motor housing at the 12-o'clock position directly above the drive-end bearing.

Since most REB-failure modes result in a spall on the bearing raceways, single spall-like faults were seeded onto the inner and outer raceways by electro-discharge machining (EDM) as illustrated in Figure 5.32. The shape of a spall is roughly triangular with one triangle point located at the initiation point of the initial surface crack resulting from the subsurface initiated fatigue crack, with the remainder of the spall out in front of this point in the direction of rolling.

A 29-degree-of-freedom nonlinear model was developed to simulate the monitored-vibration acceleration of the 2-horsepower motor as tested. This simulation model was formulated as a time-transient dynamically nonlinear system incorporating the nonlinear characteristics of both the seeded-raceway faults (spalls) and the inherently nonlinear ball-raceway Herztian contacts. The model included motor housing, bearing elements, and rotor, illustrated in Figure 5.33.

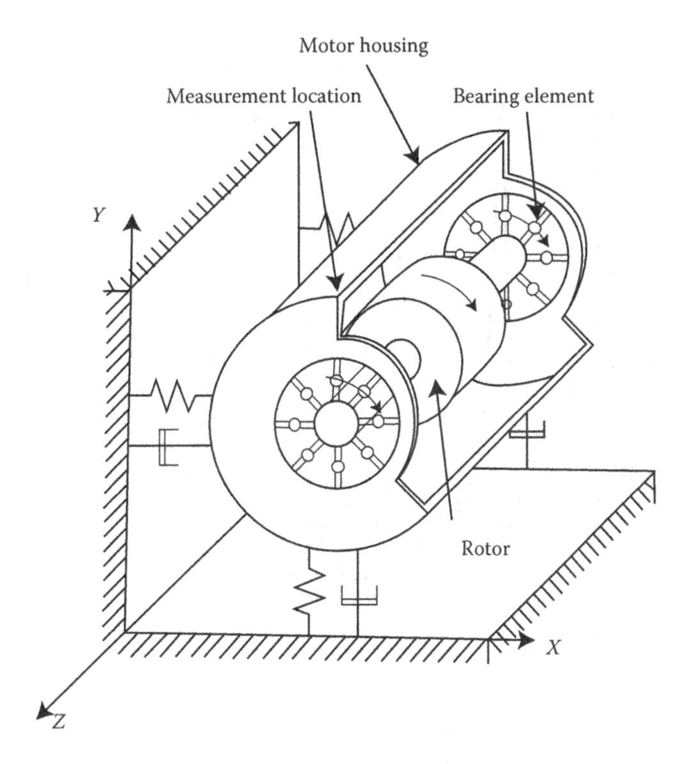

FIGURE 5.33
Vibration model of 2-horsepower motor as tested with REBs.

Adams and Loparo then used the well-known signal processing technique known as *envelope analysis*. A sample of their results comparing measured and model-based simulation results are shown in Figure 5.34. As is well known, envelope analysis is a 3-step data reduction process as follows: (1) the time-base signal is band-width filtered around a system resonance frequency, followed by (2) half-wave rectification, followed by (3) low-pass filtering and then FFT into the frequency domain, sometimes referred to as the *high-frequency resonance technique* (HFRT), graphically illustrated in Figure 5.35. Each time a defect strikes its mating element, a pulse of short duration is generated that excites the resonances periodically at the characteristic frequency related to the defect location. The resonances are thus amplitude-modulated at the characteristic defect frequency. Demodulating one of these resonances yields a time-based signal directly relatable to the bearing condition. The signal is bandpass-filtered around one of the resonant frequencies, thus eliminating most of the unwanted vibration signals from other sources. This bandpass-filtered signal is then demodulated by an envelope detector in which the signal is rectified and smoothed by low-pass filtering to eliminate the carrier or bandpass-filtered resonant frequency. The spectrum of the envelope signal in the low-frequency range is then obtained to get the characteristic defect frequency spectrum of the bearing, as exampled in Figure 5.35.

The Figure 5.34 FFT results of the enveloped signal yield the *acceleration impulse train*. The close agreement between their measured and model-based results provides strong promise for reliable low-cost model-based assessment of REB faults from a near-bearing real-time vibration signal for industrial operating machinery. Specifically in this example, a vibration model that couples RCB *fault dynamics* to the overall *machine dynamics model* was shown to

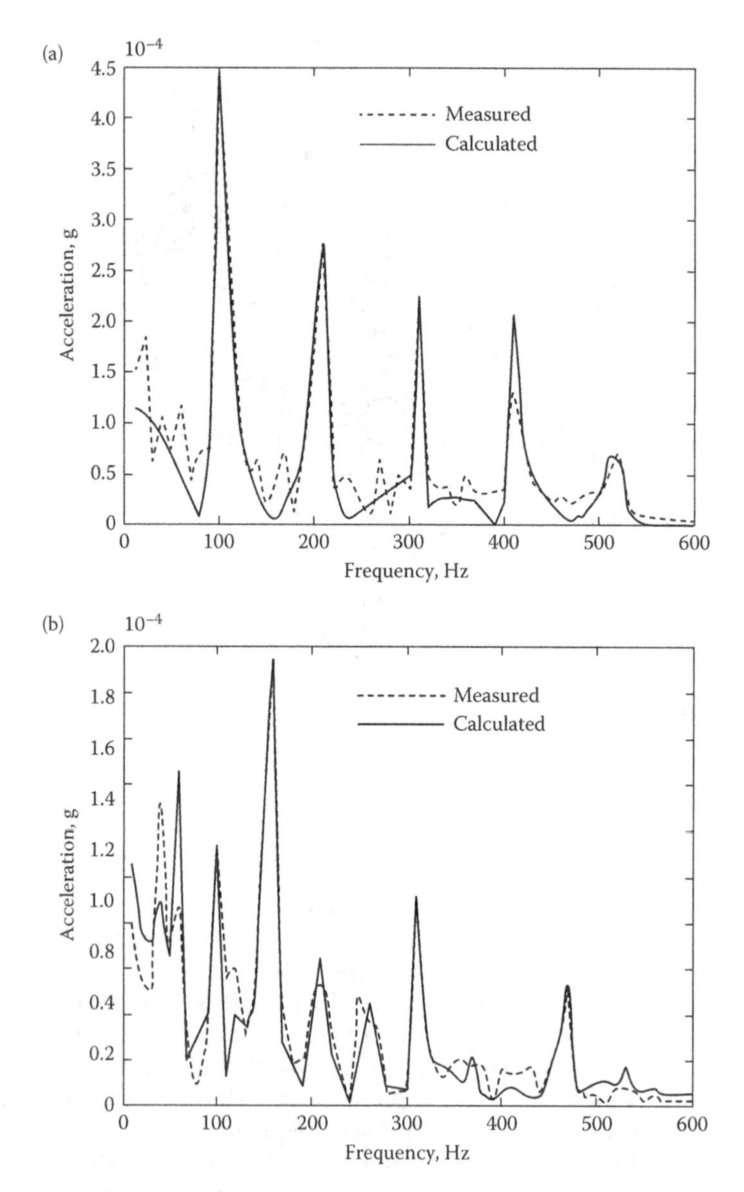

FIGURE 5.34
Experimental and calculated results; (a) outer-raceway defect and (b) inner-raceway defect.

yield a reliable low-cost MD approach for REB-spall detection and assessment. Since many of the frequencies resulting from distributed defects coincide with those due to localized defects, the difficulty in discriminating distributed defects from localized defects from frequency information alone is also a good selling feature for the model-based approach of Adams and Loparo (2004).

Tandona and Choudhuryb (1999) provide a thorough review of REB-MaD technology as summarized here. The WT method has been used to extract very weak signals for which FFT becomes ineffective as described in Section 5.1.1 and Figure 5.6. WT provides a variable-resolution time-frequency distribution from which periodic structural ringing

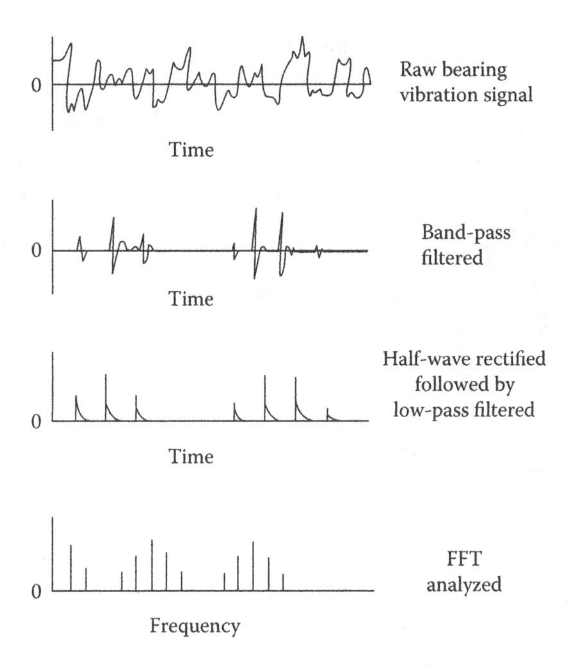

FIGURE 5.35
Extraction process of demodulated spectra by HFRT.

due to repetitive force impulses, generated on the passing of each rolling element over the defect, are detected. With both *time and frequency localization*, WT is superior to windowed FFT, thus potentially able to detect a defect even at the pre-spalling stage.

Accelerometers (Figure 5.12a) are by far the most frequently used vibration sensors for REB-MaD. Additionally, eddy-current non-contacting displacement proximity probes (Figure 5.12c) are also widely employed to capture the radial displacement of the outer raceway directly as the rolling elements pass under it. The signal-to-noise ratio can be improved by reduced or eliminated extraneous vibrations of the housing structure. However, the installation of these probes for REB-MaD is difficult, requiring not only drilling and tapping of the bearing housing but also fine adjustment of the gap between the probe tip and the outer race (Adams 2010). And the gap adjustment can drift due to vibration, thermal expansion, and dirt. Advantageously, since the outer raceway does not rotate, there is no *electrical runout* corrupting the displacement signal as it does in the typical application where the probe targets a rotating journal. Similarly, *fiber-optic sensors* have likewise been used to measure the radial displacement of the outer raceway directly as the rolling elements pass under it. Bearing cage-to-shaft speed ratio has also been found to be an informative parameter for REB-health monitoring (see Equation 2.7).

Automated interpretation in fault diagnosis schemes based on pattern recognition employing short-time-signal processing techniques has been in use for about 30 years. More recently, the establishment of identifiable patterns for specific fault types and severity has dominated the published research. Artificial neural networks have emerged as a popular tool for signal processing and pattern classification tasks, and are suitable for REB-condition-monitoring programs. An artificial neural network has the ability to "learn" how to solve a problem, instead of having to be preprogrammed with a precise algorithm.

In addition to spalls, distributed REB defects include *surface irregularities* like *roughness*, *waviness*, and *off-size rolling elements*, for which the vibration responses have been studied

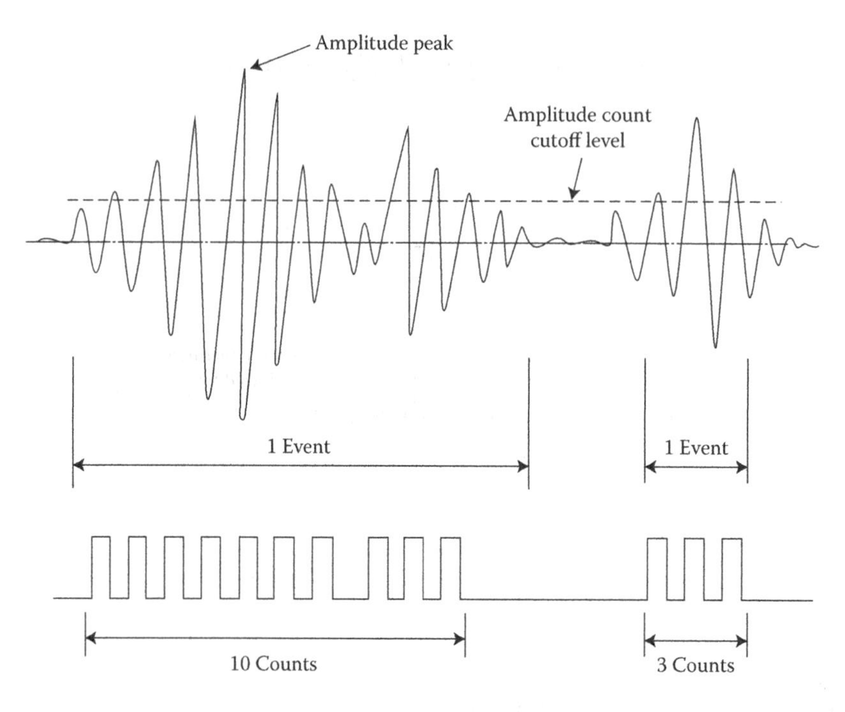

FIGURE 5.36
Time-based AE burst signal. (From Tandona, N. and Choudhuryb, A., A Review of Vibration and Acoustic Measurement Methods for the Detection of Defects in Rolling Element Bearings, *Tribology International*, Vol. 32(8), pp. 469–480, August 1999.)

primarily in the frequency domain. Surface roughness or similar features have been found to be significant only when the asperities break through the lubricant film and contact the opposing surface (see Figure 1.25). However, longer-wavelength features (waviness) or varying rolling element diameter have a more detectable effect on the measurable-vibration levels. The frequency range within which waviness produces a significant-vibration level has been found to be below 60 times the rotational speed.

Transient elastic wave generation due to a rapid release of strain energy caused by a structural alteration in a solid material under mechanical or thermal stresses is the cause of significant *acoustic emissions* (AE). Among the primary sources of AE are the generation and propagation of cracks from cyclic plastic deformation. A piezoelectric transducer with its preamplifier and signal conditioner is the primary sensor used for AE time-based signals. Because of the very high stiffness of piezoelectric crystals, they have a very high natural frequency. Ringdown counts, events, and peak amplitude of the signal are the most frequently measured AE parameters. Figure 5.36 demonstrates these for an example AE signal. Ringdown counts are the number of times the amplitude exceeds a preset voltage threshold level in a specified time period, giving a simple integer AE signal measure. AE monitoring can detect growth of subsurface cracks before they emerge at the REB surface, a distinct advantage over vibration monitoring that can detect cracks only after they grow into surface defects. Also, the energy radiated by various neighboring components up to 50 kHz, that easily mask vibration energy released from a defective REB, does not alter AE signal content in the very high-frequency range. Acoustic-noise sound pressure (ANSP) measurement has also been found to be a useful diagnostic parameter for REB-MaD in laboratory testing.

References

Abu-Mahfouz, I., *Routes to Chaos In Rotor Dynamics*, Ph.D. Thesis, Case Western Reserve University, 1993.

Adams, M. L., *Rotating Machinery Vibration–From Analysis to Troubleshooting*, 2nd ed., CRC Press/Taylor & Francis, Boca Raton, FL, 2010.

Adams, M. L., *Power Plant Centrifugal Pumps–Problem Analysis and Troubleshooting*, CRC Press/Taylor & Francis, Boca Raton, FL, 2017a.

Adams, M. L., *Rotating Machinery Research and Development Test Rigs*, CRC Press/Taylor & Francis, Boca Raton, FL, 2017b.

Adams, M. L. and Abu-Mahfouz, I., Exploratory Research on Chaos Concepts as Diagnostic Tools for Assessing Rotating Machinery Vibration Signatures, *Proceedings, IFTOMM 4th International Conference on Rotor Dynamics*, Chicago, September 1994.

Adams, M. L. and Loparo, K. A., Analysis of Rolling Element Bearing Faults in Rotating Machinery: Experiments, Modelling, Fault Detection, and Diagnosis, *Proceedings of IMechE 8th International Conference on Vibration in Rotating Machinery*, Swansea, Wales, September 2004.

Adams, M. L. and McCloskey, T. H., Large Unbalance Vibration in Steam Turbine-Generator Sets, *Proceedings, I. Mech. E. 3rd International Conference on Vibrations in Rotating Machinery*, pp. 491–497, York, England, September 1984.

Bently, D. E. and Muszynska, A., Vibration monitoring and analysis for rotating machinery, *Noise and Vibration 1995 Conference*, Pretoria, South Africa, Keynote Address Paper, pp. 1–24, November 7–9, 1996.

Daubechies, I., *Different Perspectives on Wavelets*, American Mathematical Society, Short Course 1993.

Eshleman, R. L., *Machinery Vibration Analysis II, Short Course*, 1999.

Florjancic, D., *Trouble-Shooting Handbook for Centrifugal Pumps*, Turbo Institute, Ljubljana, Slovenia, p. 414, 2008.

Harris, T. A. and Kotzalas, M. N., *Rolling Bearing Analysis*, 5th ed., CRC Press/Taylor & Francis, Boca Raton, FL, 2007.

Horattas G. A., Adams, M. L., and Dimofte, F., Mechanical and Electrical Run- Out Removal on a Precision Rotor-Vibration Research Spindle, *ASME Journal of Acoustics and Vibration*, Vol. 119(2), 1997, pp. 216–220.

Loparo, K. A. and Adams, M. L., *Development of Machinery Monitoring and Diagnostics Methods, Proceedings, 52nd Meeting of the Society for Machinery Failure Prevention*, Virginia Beach, April 1998.

Makay, E., *Survey of Feed Pump Outages*, EPRI Research Project Rp-641, Final Report FP-754, 1978 (edited by M. L. Adams).

Makay, E., How Close are Your Feed Pumps to Instability-Caused Disaster, *Power Magazine*, pp. 69–71, 1980.

Makay, E. and Barrett, J., Changes in Hydraulic Component Geometries Greatly Increased Power Plant Availability and Reduced Maintenance Costs: Case Histories, *Proceedings of Texas A & M International Pump Symposium*, 1984.

Makay, E. and Szamody, O., *Recommended Design Guidelines for Feedwater Pumps in Large Generating Units*, EPRI Report CS-1512, September 1980.

Roberson, M. and Baird, A., (www.robertson.technology) *Thermodynamic Pump Performance Monitoring in Power Stations*, I Mech E seminar, Fluid Machinery in the Power Industry, Bristol, UK, June 10, 2015.

Tandona, N. and Choudhuryb, A., A Review of Vibration and Acoustic Measurement Methods for the Detection of Defects in Rolling Element Bearings, *Tribology International*, Vol. 32(8), pp. 469–480, August 1999.

Section II

Bearing Design and Application

6

Bearings without Thick-Fluid-Film Lubrication

There are numerous applications where bearing *loads* and/or relative *speeds* between the load-transmitting bearing parts are sufficiently small that neither full-film lubrication nor rolling elements are necessary. Such applications permeate both industrial and consumer products. The group of bearings that function without a full film separating the bearing load-transmitting parts in relative sliding motion include those generally identified by the following designations: (1) dry, (2) mixed film, and (3) thin film. Table 6.1 (Raimondi et al. 1968) classifies all bearing types based on the means of transmitting load between the bearing parts in relative motion.

6.1 Dry and Solid Film

In unlubricated surface-to-surface contacts between bearing parts, the load transmitted between the parts is carried by a sufficient number asperity-to-asperity tip contacts with asperities of the softer material flattened by yielding, giving the following expression relating transmitted load to the flattened asperities.

$$W = S_y \sum_{n=1}^{N} A_n \tag{6.1}$$

where

W = transmitted load

S_y = yield strength of softer material

A_n = flattened area of nth asperity

N = number of flattened asperities

The total of all the flattened asperity-to-asperity contact areas, which varies roughly proportional to the bearing load W, is miniscule compared to the apparent geometric contact area between the two load-transmitting bearing parts. The associated sliding force needed to slide one bearing part relative to the other is well known to entail what is called *cold welding* between contacting asperities and that sliding force is roughly proportional to the total of all flattened asperity-to-asperity contact areas. Thus, the standard assumption that the *Coulomb sliding friction coefficient* is a constant is usually a good approximation. As explained in Section 1.1 and illustrated in Figure 1.2, *adhesive wear* debris is theorized as the result of cold welded asperity tips staying attached to the other contacting asperity, then becoming a wear debris particle. As illustrated in Figure 6.1, skidding friction and adhesive wear is greatly reduced by interposing a solid lubricant like (1) carbon graphite, molybdenum disulfide, or PTFE plastic, or by a (2) sulfur-rich or phosphorous-rich layer

TABLE 6.1
Classification of Bearings Based on Means of Transmitting Load

Bearing Type	Means of Carrying Load between Moving Parts
Full film (liquid and gas)	Film pressure in comparatively thick film separating the parts in relative sliding motion
Thin film	Fluid pressure in comparatively thin film or resistive force in solid-film separating parts
Mixed film	Fluid pressure in comparatively thin film or resistive force in solid-film separating parts, in addition to solid-to-solid contact between parts
Dry	Solid-to-solid contact between parts
Filmless (flexural and torsional)	Flexure or torsion of an elastic member between parts
Rolling	Resistive force of rolling elements between parts

formed by adsorption from lubricating oil. Even without interposing such ingredients, natural oxides when automatically formed on the surfaces provides nature's way of reducing sliding friction.

Unlike fluid-film bearings, dry sliding bearings are not amenable to fundamental first-principle characterization, that is, RLE (Equation 1.10) for fluid-film bearings, because the governing phenomena entail the physical-chemical interactions between the sliding parts and any naturally present or interposed agents (Figure 6.1). Therefore, reliable design information/guidelines for dry friction sliding bearings rest entirely on empirical data for specific bearing materials and interposed agents. Two fundamental criteria govern the bearing design: (1) allowable *operating temperature* of the rubbing bearing surface material and (2) allowable bearing surface *wear*. Both of these criteria are governed by the product of *bearing load* and *sliding velocity*. In consequence, the bearing design parameter that emerged many decades ago is the bearing material PV factor, where P is the unit load (units of pressure) on the bearing, $P = W/A$, where $W =$ bearing load, $A =$ load-carrying apparent bearing area, and $V =$ sliding velocity. From extensive laboratory test data, commercially available bearing materials are tabulated with their PV rating value. A PV rating is also employed in sizing mixed-film and thin-film bearings.

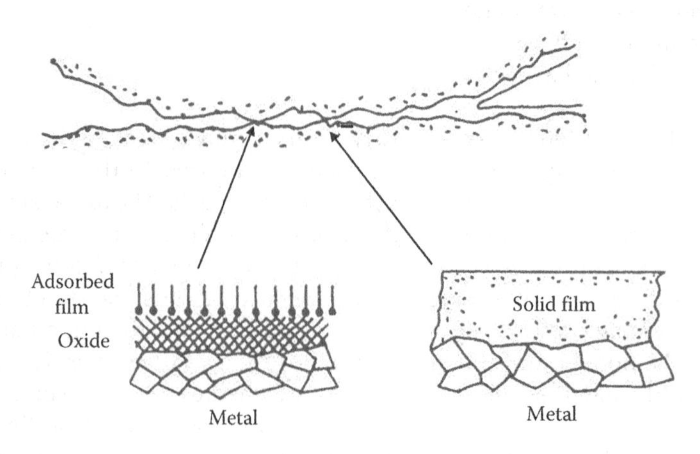

FIGURE 6.1
Asperity contact between rubbing surfaces.

6.1.1 PV Determined by Allowable Operating Temperature

PV as related to allowable temperature is governed by the rate of mechanical energy dissipated into heat from the sliding friction between the mating sliding surfaces and thus is dictated by the following elementary equation.

$$\text{Power} = \text{Force} \times \text{Velocity} \tag{6.2}$$

The conversion from mechanical energy can be expressed as follows (Raimondi et al. 1968).

$$H_d = \frac{FV}{J} \tag{6.3}$$

where
- H_d = heat dissipated
- $F = fW$
- f = coefficient of friction
- W = bearing load
- V = sliding velocity
- J = mechanical heat equivalent

Assuming that conduction is the governing-type of heat transfer yields the following:

$$H_d = kA(t - t_0) = k\Delta t \tag{6.4}$$

where
- k = coefficient of heat conduction
- A = apparent contact area
- Δt = temperature rise

Combining Equations 6.3 and 6.4 to equate heat generated to heat dissipated yields the predicted temperature rise as follows.

$$\Delta t = \frac{fWV}{JkA} = \frac{f}{kJ}PV \tag{6.5}$$

For a specified allowable temperature rise (i.e., Δt = constant) and an assumed value for sliding friction coefficient, the following relationship for the *PV* factor is obtained.

$$PV = \frac{Jk\Delta t}{f} = \text{constant (based on allowable temperature rise)} \tag{6.6}$$

6.1.2 PV Determined by Allowable Wear

Based on the adhesive wear mechanism covered in Section 1.1 and illustrated in Figure 1.2, the volume of material worn from sliding surfaces in contact is expressed as follows.

$$V_L = k_w A_0 L = \frac{k_w WL}{H} = Ah \tag{6.7}$$

Therefore,

$$\frac{V_L}{AT} = \frac{Ah}{AT} = \frac{k_w WL}{HAT} = \frac{k_w PV}{H} \tag{6.8}$$

where

$P = W/A$ bearing load/apparent load-carrying area
V_L = volume of material removed
k_w = adhesive wear coefficient
A_0 = real asperity-tip contact area = W/H
H = material hardness
L = sliding distance traveled
A = apparent area
T = time
h = average thickness of material layer removed

For h/T the allowable wear rate as well as k_w and H assumed constant yields the following.

$$PV = \frac{hH}{Tk_w} = \text{constant (based on allowable wear)} \tag{6.9}$$

Raimondi et al. (1968) report that the PV value computed for *allowable wear* is typically a smaller value than that computed for *allowable temperature rise*. However, wear rate may be significantly influenced by temperature rise and vice versa, that is, wear rate and temperature rise significantly coupled. So, employing one or the other of these two limiting PV criteria could in some atypical cases be a flawed design criterion in some atypical applications. In such a case, wear test data taken over a sufficient range of elevated temperatures would be needed to achieve a reliable design criteria that couples wear rate and temperature rise.

6.1.3 PV Values of Various Materials

The bearing material manufacturers have a long list of their PV factors for units P (psi) and V (ft/min). It is not easy to rigorously compare the various manufacturers' listed PV values since their respective test conditions are not detailed. For example, the test temperature range used and if temperature has a significant influence on the published PV value may not be listed by the material manufacturer. It is surely reasonable to suspect that temperature is a significant influence since the PV value is usually based on allowable temperature rise, not on allowable wear rate. A comparison of advertised PV ratings of a wide range of bearing materials is plotted in Figure 6.2 (Raimondi et al. 1968) based on the temperature rise criteria. Manufacturers also often place limits on *maximum pressure* and *maximum velocity* independent of the PV rating. Of course, as with some uncertainty in any material's published strengths (yield, ultimate, endurance limit), there should be an appropriate factor-of-safety > 1 imposed on a bearing design.

6.1.4 PV Applied to Journal Bearings

Applying the PV approach to journal bearings can be facilitated by converting the formulation into journal-bearing nomenclature, employing journal rotational speed N and nominal journal diameter D, as follows.

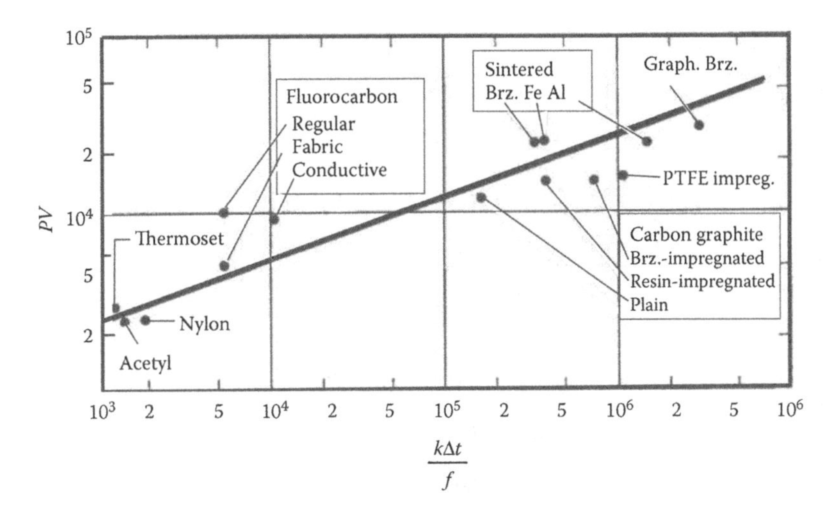

FIGURE 6.2
PV versus $k\Delta t/f$ for various commonly used bearing materials.

$$\Delta t = \frac{f}{kJ} PV = \frac{\pi D f}{Jk} PN \tag{6.10}$$

This clearly shows that the temperature rise in a journal bearing of specified diameter is proportional to f and the product PN. Again, a suitable factor of safety should be employed, especially since the manufacturer's material test conditions may be significantly different from those of the application.

6.1.5 Solid Films

Application for lubrication at temperature extremes, both high and low, has fostered the use of solid lubricants such as carbon graphite, molybdenum disulfide, and PTFE plastic particles. Such extreme temperature requirements while existing in a few industrial applications are critical in outer-space applications, especially the near absolute zero outer-space temperatures. Solid lubricants can either be in powdered form or mixed as additives with organic oil, synthetic oil, grease solvents, or water. Long-chain lubricants such as metal soaps are melted or destroyed above 400°F (204°C), necessitating solid lubrication at above this temperature to have adequately higher melting points. For the very low outer-space temperatures, lubricants are needed that do not freeze into solid chunks.

Carbon graphite is especially affective at both very high and very low operating temperatures. In gas turbine aircraft engines, the rolling element-bearing temperatures can reach well over 1000°F (538°C). Molybdenum disulfide competes with carbon graphite, effective at temperature up to 800°F (427°C) (Larson and Larson 1968). Solid lubricants derive their lubricity from the very low strength of their crystalline boundaries, as illustrated in Figure 1.4.

6.2 Filmless, Mixed Film, and Thin Film

The transition of a bearing's classification can occur during a normal operating cycle. As a common example, consider an oil-lubricated journal bearing designed to run

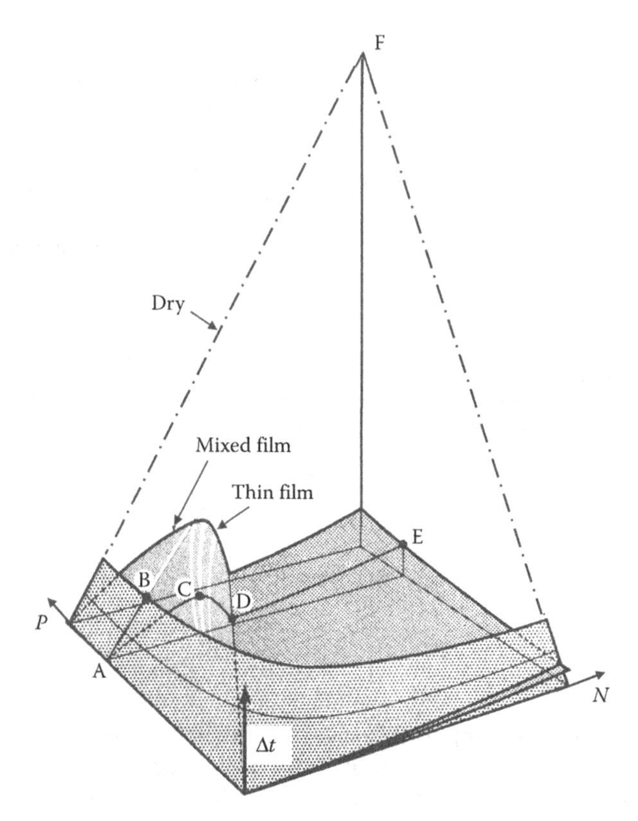

FIGURE 6.3
Transition under load during startup; dry-to mixed film to full film.

hydrodynamically with a relatively thick oil film. At zero rpm dead start under full bearing load, the bearing operates momentarily as filmless or boundary-lubricated bearing, as illustrated for startup in Figure 1.1(b). Figure 6.3 illustrates the *speed-contact pressure-temperature* locus as the bearing; (1) starts under full load at zero rpm, point A, (2) progresses to mixed-film lubrication, point B, (3) then to thin-film lubrication, point C, (4) then to the hydrodynamic regime point D, and (5) on up to operating speed, point E. If there were no lubricant present thus forcing the bearing to operate dry, the operating point F would result in a very high operating temperature rise. On coast down from operating speed, the trajectory of lubrication classification is typically milder near the stopping point since a fluid film is already established and thus tends to "hang in there" as the rotor comes to a stop at zero rpm.

On some critical applications, the hydrodynamic bearing has within its load-bearing surface a *small hydrostatic bearing recess* (see Figure 1.18) into which oil is pumped to "float" the rotor on a fluid film at startup. One prominent example of this is illustrated in Figure 1.22, showing a primary reactor coolant pump for a nuclear power plant pressurized-water reactor (PWR). In these machines, the double-acting tilting-pad thrust bearing (see Figure 1.12) at the top of the motor carries all the axial loads including the full weight of the rotor. Another prominent example is large steam turbine generator sets in electric power plants. The journal-bearing designs frequently have *small hydrostatic lift pads* in the bottom of the journal-bearing cylindrical-load-bearing surface, either as original

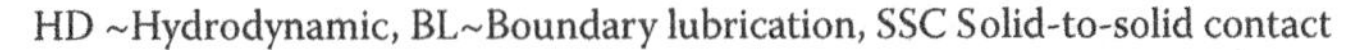

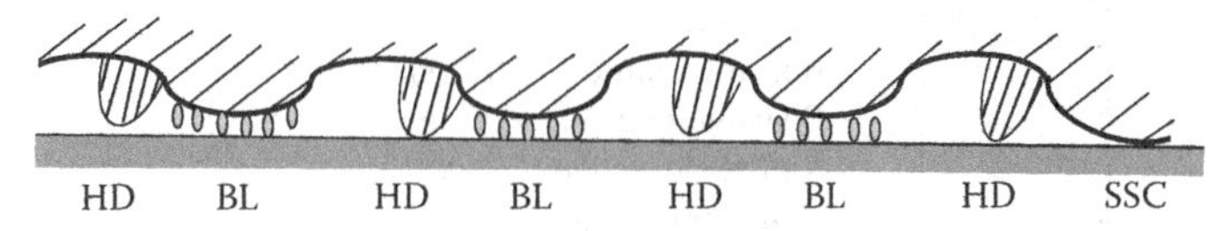

FIGURE 6.4
Visualization of *mixed-film lubrication*. For simplicity, the combined surface roughness of both contacting surfaces is illustrated on only one surface.

equipment or as a retrofit. These small lift pads are to *support the rotor* on a thin oil film at *startup* and subsequently at *shutdown* when the entire rigidly coupled rotor driveline is very slowly rotated for many hours by a motor-driven *turning gear* at around 3–5 rpm, to allow the *rotor to cool down* axisymmetrically, thus avoiding a hard-to-remove thermal bow of the rotor.

As with dry and solid-film lubrication, *mixed-film-bearing lubrication* is neither well modeled nor visualized like full-film-bearing lubrication a la the RLE Equation 1.10 and Figure 1.11. To visualize the mechanics of so-called mixed-film lubrication requires creating a view of the mixed-film phenomenon at the microscopic level. Based on bearing friction coefficient measurements (see Figure 1.1a), mixed-film lubrication is postulated to entail (1) a collection of very small localized pockets of thin-film hydrodynamic action intermingled with (2) a collection of very small localized boundary-lubricated pockets plus (3) probably some localized spots with a "little" solid-to-solid rubbing where the replenishment of organic/synthetic boundary lubricating molecules (compare Figure 1.3 to Figures 1.8 and 1.11) are *sheared off* at a rate faster than their *replenishment* by the chemical reaction between the contacting surfaces and oil additives is able to match. Figure 6.4 is the author's attempt to capture this in a single illustration.

Thin-film lubrication can be thought of as an improved version of mixed-film lubrication where the bearing load is supported almost entirely by hydrodynamic action. Based on the author's own experimental research, achieving thin-film lubrication involves an initial favorable wear-in process in which the *softer* of the two surfaces encounters sufficient improvement of its surface finish from the super-finish machining imparted by the harder surface, allowing the gradual development of a hydrodynamic fluid film that completely separates the two surfaces by a very thin oil film that supports the entire bearing load.

The laboratory-controlled bearing loading fixture of the author's test rig configuration is illustrated in Figure 1.21 (Adams 2017). This test rig was developed to research the load capacity for large steam turbine generator journal-bearing load capacity under slow-roll journal-surface-speed turning-gear operation. These journal bearings employ tin-based babbitt white-metal liners (Figure 5.21). Figure 6.5 shows for two test cases the time plot of bearing friction coefficient from startup under full-bearing load through to the conclusion of the 24-hour test time with constant full load. Test A is consistent with an ensuing wear-in to a hydrodynamic thin film completely supporting the bearing load, whereas Test B is not.

For acceptable cases typified by Case A in Figure 6.5, post-test surface roughness measurements showed that the load-carrying portion of the bearing surface supporting the applied load had its original machined surface finish of 16 μ-inch rms wear-in improved to 2–4 μ-inch rms. The test Case B friction coefficient time plot did not show a wear-in to

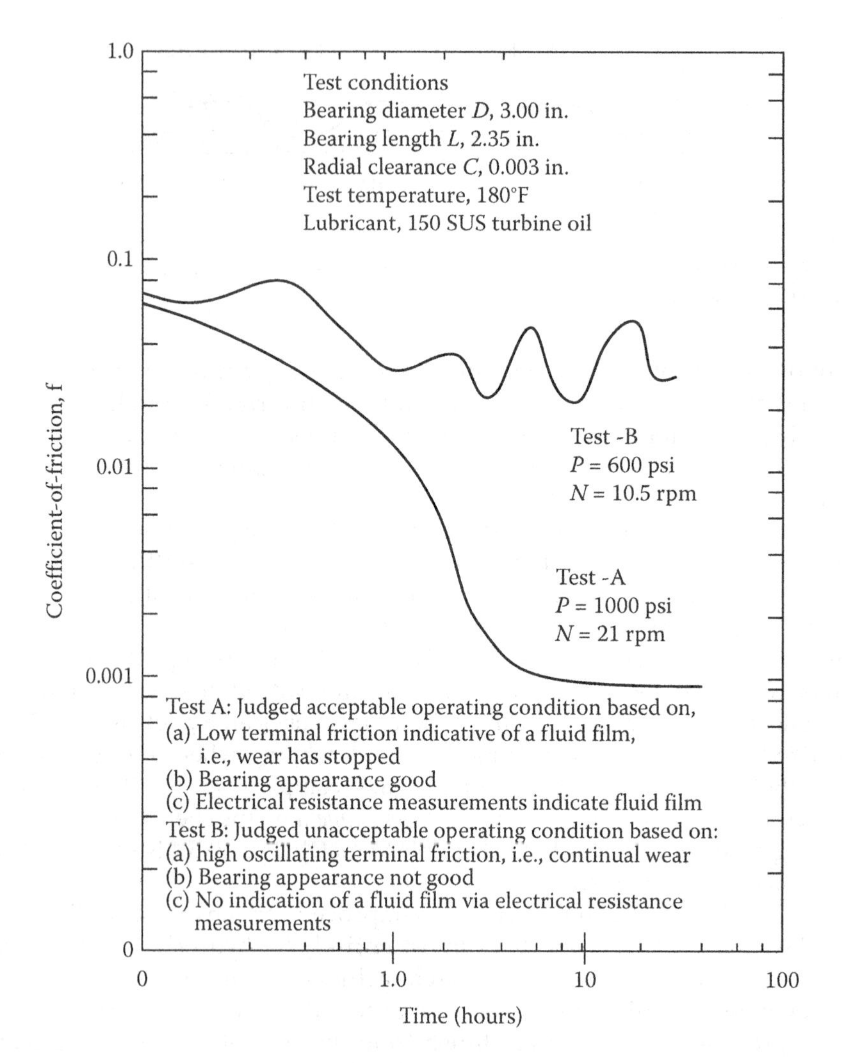

FIGURE 6.5
Measured bearing friction coefficient from several-hour test.

a condition where the friction coefficient transitioned into a steady low value indicative of hydrodynamic lubrication. These two test cases demonstrate a race between two distinct phenomenological possibilities: (1) Case A: the bearing babbitt (softer surface) has its surface finish sufficiently improved quick enough by the wear-in before the bearing babbitt surface gets heavily "whipped" and degraded, or (2) Case B: the initial bearing surface continues to be degraded from wear imposed by the journal and never develops a load-supporting hydrodynamic thin oil film.

To absolutely confirm that the test examples typified by the Figure 6.5 Case A did truly establish a full hydrodynamic oil film separating bearing and journal; (1) electrical resistance across the bearing-journal contact was measured, (2) oil-film pressure within the bearing was measured, (3) after-test bearing surface finish was measured, and (4) appearance of the active portion of the bearing surface was closely examined. For the Case A-type of test friction results, all four complementary criteria confirmed the establishment of a hydrodynamic thin oil film completely separating bearing and journal, as exampled in Figure 6.6 (Adams 2017).

Interestingly, the Case A measured friction coefficient showed a wear-in to a full hydrodynamic oil film separating bearing and journal in under 10 hours, whereas the electrical resistance measurement did not indicate complete electrical separation until about 75 hours. Clearly, the electrical resistance measurement is an even far more demanding criterion for identifying a complete thin film of lubrication separating the two surfaces. Figure 6.7 illustrates the complete test rig. To utilize the electrical resistance method, the test rig was designed with the rotor electrically isolated from everything else except at the bearing–journal interface. Thus, the electrical resistance measurement indicates when the rotor becomes totally isolated electrically, which requires total non-contact between bearing and journal.

6.3 Pivot Bearings

Pivot bearings (PB) are traditionally well known as the bearing type used in escape-mechanism spring-powered watches, as pictured in Figure 6.8. All of the wheels are mounted on shafts supported at both ends by pivot bearing, made of jewels in fine watches. The author's Rolex Oyster Perpetual *self-winding* wrist watch gains one minute a week, doesn't run down, and needs no battery.

(a)

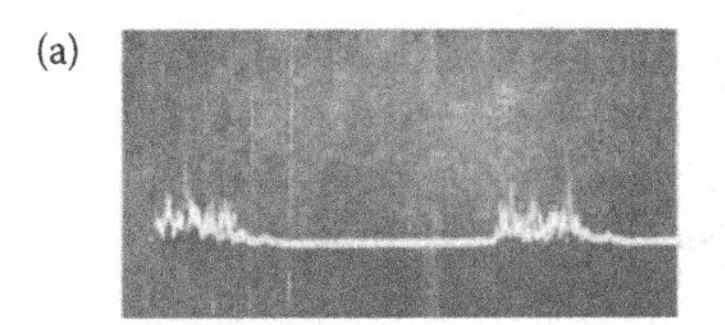

Running time 50.3 hours
Meter reads 0%–30% separation

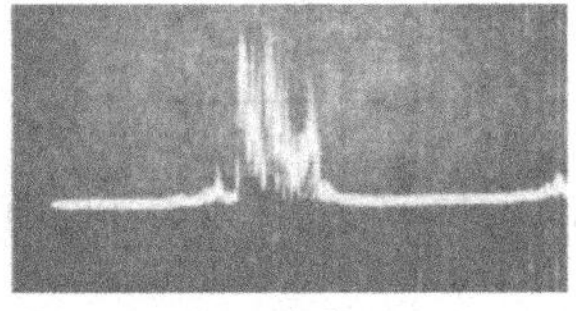

Running time 51 hours
Meter reads 0%–55% separation

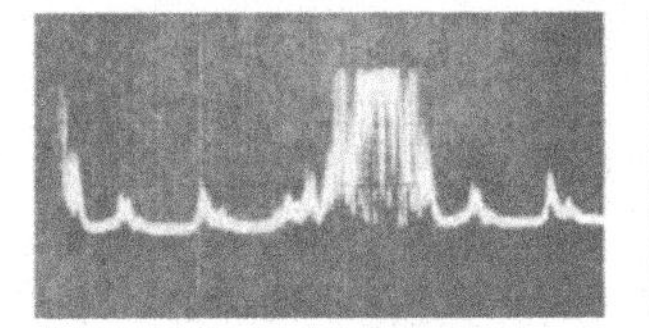

Running time 54.4 hours
Meter reads 5%–100% separation

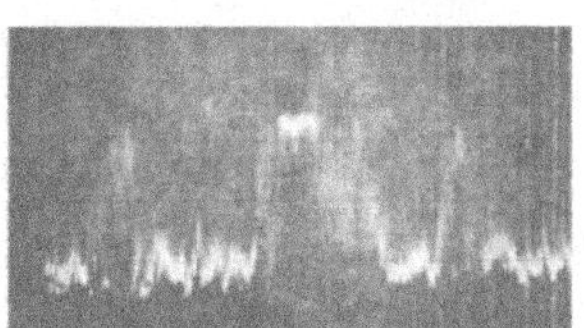

Running time 73.3 hours
Meter reads 60%–100% separation

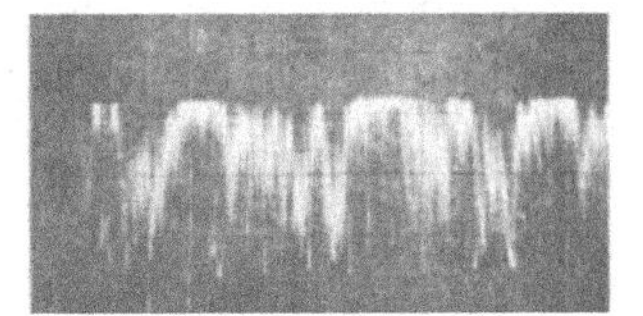

Running time 73.5 hours
Meter reads 90%–100% separation

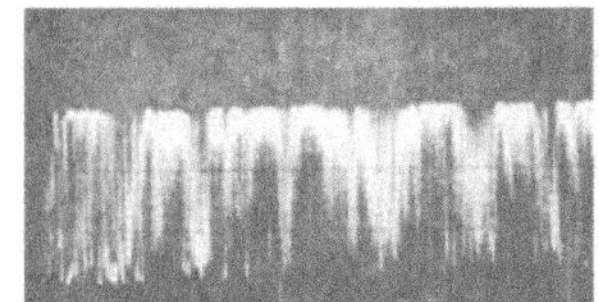

Running time 75.8 hours
Meter reads 100% separation

Oscilloscope picture

FIGURE 6.6
Confirmation of a complete hydrodynamic oil film at slow roll; (a) electrical resistance signal. *(Continued)*

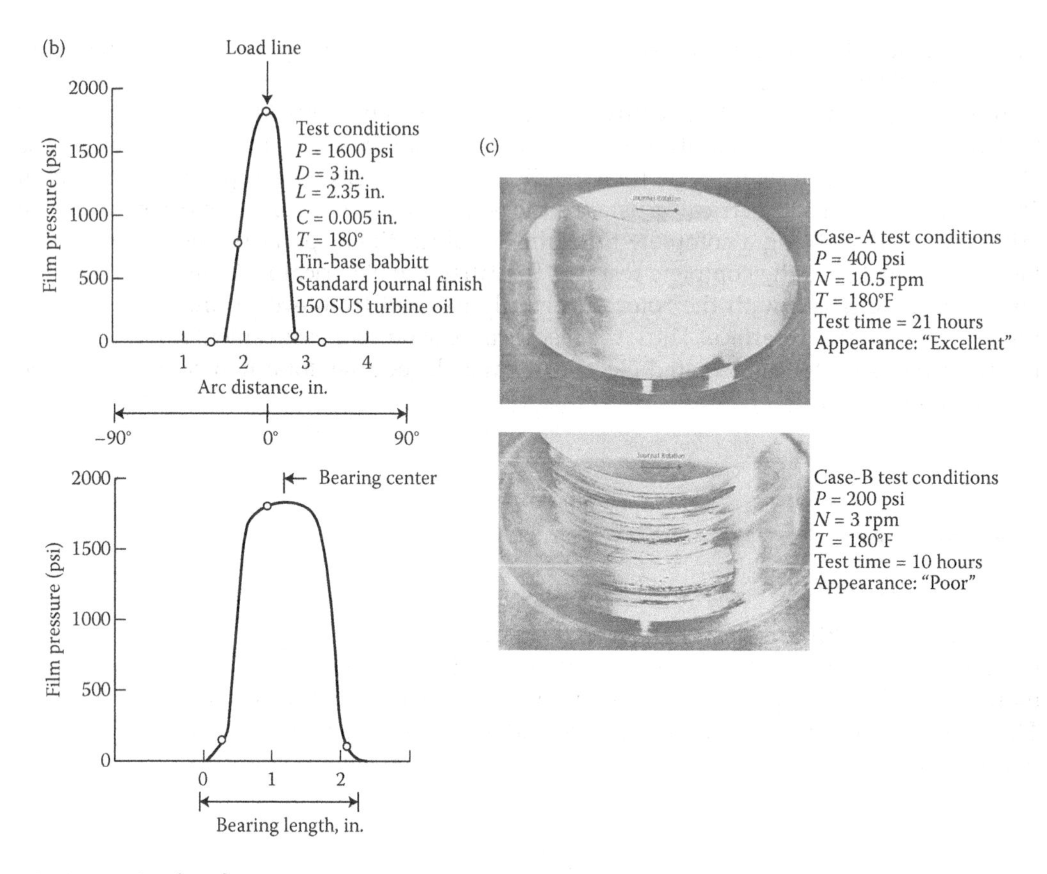

FIGURE 6.6 (Continued)
Confirmation of a complete hydrodynamic oil film at slow roll; (b) oil-film pressure measurement, (c) post-test photo of bearing.

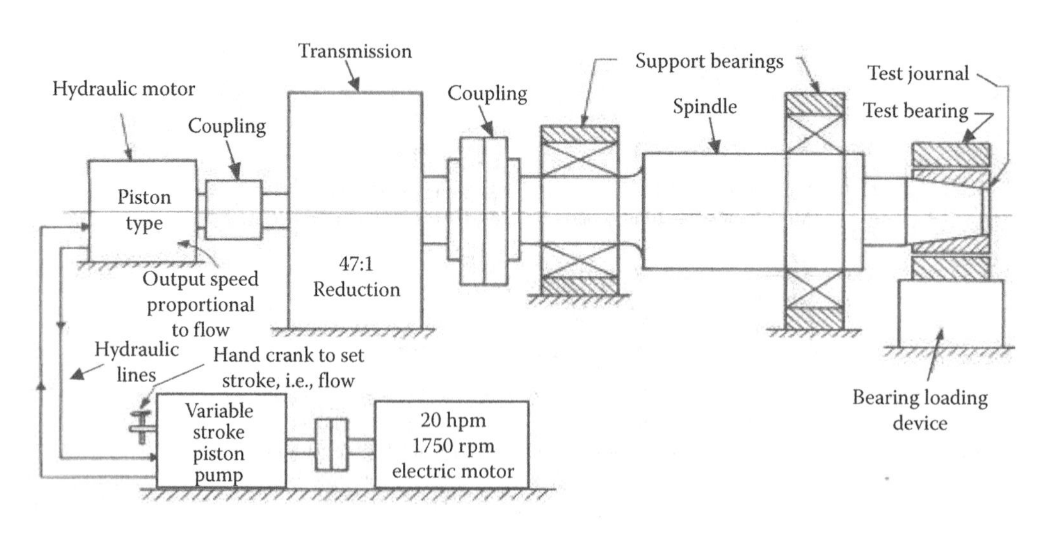

FIGURE 6.7
Complete test rig for slow-roll journal-bearing load-capacity research.

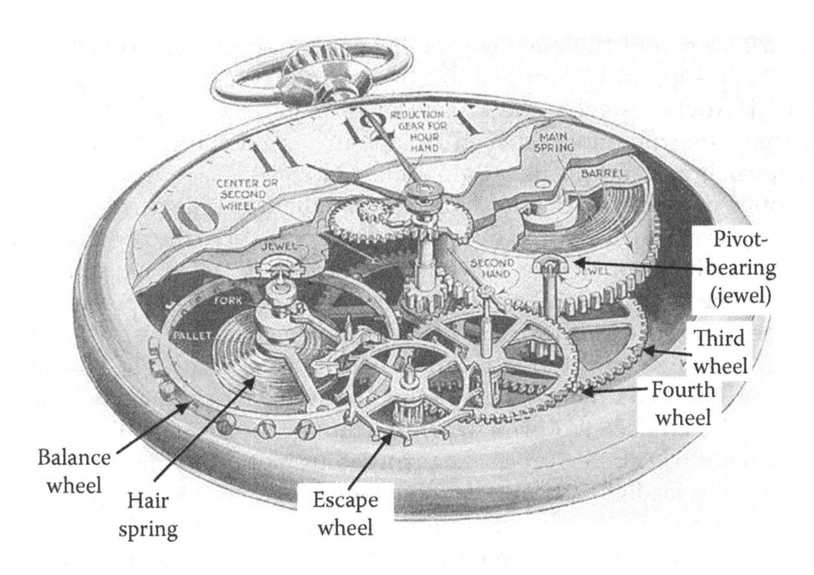

FIGURE 6.8
Escape-mechanism watch.

PBs are of course now utilized on many applications where: (1) the load is small, (2) where low friction is required, and (3) where accurate radial and axial positioning are required. PBs may be touted by their manufacturers as being *frictionless*. This characterization is of course just as erroneous as when it is used to characterize rolling element bearings. As the author points out in Chapter 2, a truly frictionless bearing would lead to a *perpetual-motion machine*, a concept that only "bicycle mechanics" totally ignorant of the fundamental laws of thermodynamics might claim is possible. However, the low-friction character of pivot bearings is clearly an asset.

Figure 6.9 illustrates PBs. The rotational energy losses from contact friction between the pivot point and the bearing are a function of the elastically flattened contact footprint

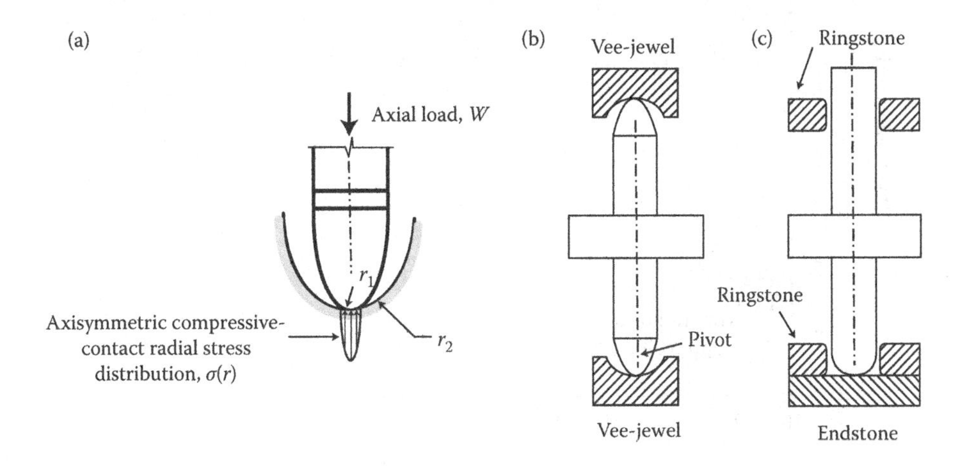

FIGURE 6.9
Pivot bearings; (a) detailed vee-jewel axially loaded with contact stress illustrated, (b) vee-jewel supported, (c) endstone and ringstone supported.

pressure distribution and the effective coefficient of friction. The ringstone support is used when sturdy support is required and the associated higher bearing friction torque is acceptable. Hardened steel is the typical pivot material. Vee-jewels, endstones, and ringstones are made from material such as sapphire, synthetic spinel, borosilcare glass, beryllium copper, brass, and the like (Raimondi et al. 1968).

Pivots are often manufactured with sharp tips that then become permanently blunted in service. But for greater precision and more repeatable operation, the pivot is manufactured to a small radius. The pivot angle is typically around 60° or larger, but smaller values are also employed. The bearing jewel radius is typically 2–3 times the pivot radius, being a compromise between (1) avoiding lateral play between pivot and jewel and (2) minimizing pivot-on-jewel rubbing contact area to minimize friction torque. To minimize friction torque, pivot radius needs the smallest radius that will not produce permanent plastic deformation of the contact footprint when encountering the maximum-applied load. Figure 6.10 (Raimondi et al. 1968) gives the pivot radius plotted against the pivot axial load for the pivot axially loaded against a flat endstone. The line *AB* in Figure 6.10 plots the locus for the minimum pivot radius that can support the applied load without permanent deformation, while providing a safe margin for overload. The straight line *AB* in Figure 6.10 satisfies the following equation.

$$r = 9.39 \times 10^{-3} W^{1/2} \tag{6.11}$$

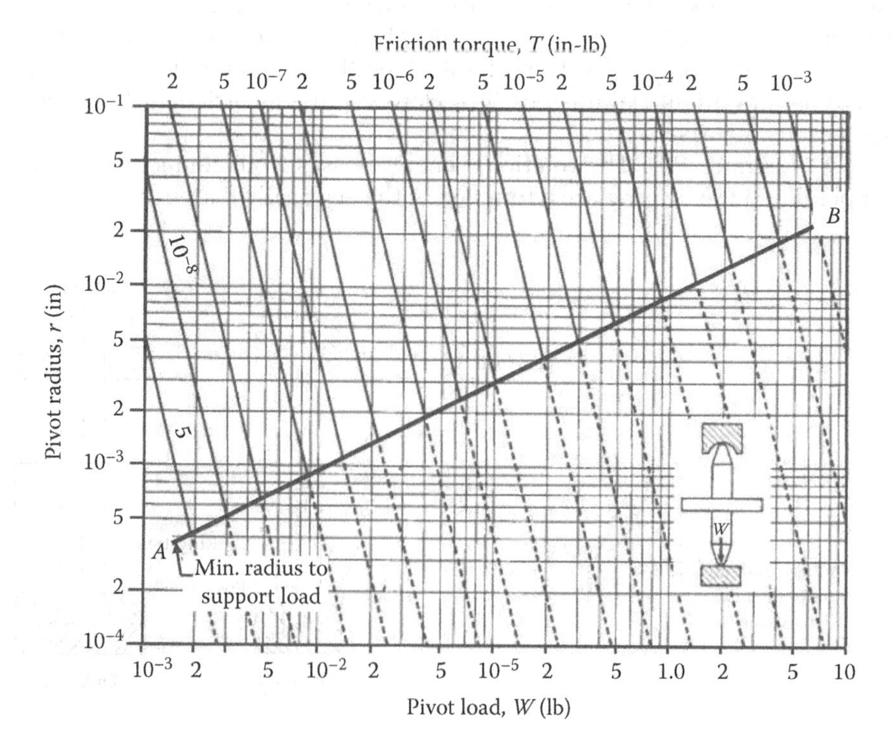

FIGURE 6.10
Pivot radius, pivot load, and friction torque for axially loaded pivots.

The maximum compressive stress in the contact region for a spherically tipped pivot in a spherical jewel is given as follows (Raimondi et al. 1968).

$$\sigma_c = 0.918\sqrt[3]{\frac{W}{\left[(1-\nu_1^2)/E_1+(1-\nu_2^2)/E_2\right]^2}}\left(\frac{n-1}{n}\right)^{2/3}\left(\frac{1}{2r_1}\right)^{2/3} \tag{6.12}$$

where

W = load on pivot (lb)
r_1 = radius of pivot (in.)
r_2 = radius of jewel (in.)
$n = r_2/r_1$
E_1 = pivot material modulus of elasticity (psi)
E_1 (steel) = 30×10^6
E_2 = jewel material modulus of elasticity (psi)
E_2 (sapphire) = 60×10^6
v_1 = pivot material Poisson's ratio
v_2 = jewel material Poisson's ratio

The value of r obtained from Figure 6.10 gives the radius when used with a plane end stone. If the pivot is to be used with a jewel of radius r_2 then a smaller pivot radius r_1 may be employed and still maintain the same maximum contact stress. Thus, with $r_2 = nr_1$ the pivot radius r_1 is given as follows.

$$r_1 = r\left(1-\frac{1}{n}\right) \tag{6.13}$$

The *friction torque* for an axially loaded pivot is based on the classical elasticity solution for the radially varying compression footprint contact stress distribution $\sigma_c(r)$ illustrated in Figure 6.9a, and assuming a constant friction coefficient, f. The equation derived for friction torque follows by integrating the differential torque dT for the differential annulus d, r from zero to R and θ from zero to 2π, as follows.

$$\begin{aligned} T &= \int_0^R dT = 2\pi f\int_0^R \sigma_c(r)r^2dr \\ &= \left(\frac{3}{16}\right)^{4/3}\pi f W^{4/3}\left[\frac{4(1-\nu_1^2)}{E_1}+\frac{4(1-\nu_2^2)}{E_2}\right]^{1/2}\left(\frac{n}{n-1}\right)^{1/3} r_1^{1/3} \end{aligned} \tag{6.14}$$

A sample numerical problem is presented by Raimondi et al. (1968), who state that regarding stress, a pivot radius determined for a given *axial lo*ad will be adequate for the same load in the *radial direction.*

Deriving the equation for the friction torque in a radially loaded PB is much simpler since the friction force is all approximately at radius r_1 as illustrated in Figure 6.11, yielding the following simple equation which is plotted in Figure 6.12 using $f = 0.15$:

$$T = fWr_1 \tag{6.15}$$

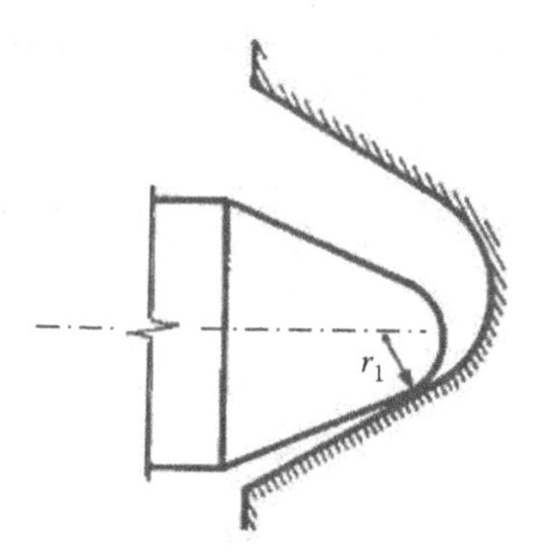

FIGURE 6.11
Radially loaded pivot.

6.4 Flexural Bearings

Bearings that flex or twist in an elastic fixture are useful when only limited translation or twisting motion is involved. Figure 6.13 illustrates the major flexural bearing (FB) configurations. Every one of these has the following features in common.

1. Do not require lubrication
2. Relatively unaffected by hostile factors such as dirt and temperature
3. Have no loose parts
4. Have no wearing parts
5. Withstand rough handling
6. Do not gum up
7. Withstand load reversals
8. Operate with zero clearance

All these flexure configurations need to be designed to tolerate both maximum static stresses as well as to accommodate the time-varying cyclic stresses without encountering a material fatigue failure. Of course, all of the configurations in Figure 6.13 readily lend themselves to finite element analysis (FEA) to achieve the desired design deflection/stiffness properties versus prescribed force/moment inputs. However, for non-large displacements, pre-FEA formulations were long used and still efficiently can be. The two examples illustrated in Figure 6.14 typify the structural deflection-kinematics, stress and stiffness design plots constructed for flexural bearings by fundamental solid mechanics formulations, long before the advent of FEA.

The FB plots presented here are all based on elastic beam theory for small deflections and should provide reasonable approximations for load capacity and angular displacements up to 10°. Commercially available designs are available for displacements up to 30°. To prevent overstressing, stops may be employed to limit displacement. The use of FBs in accurate torque and force measurement systems, shifting of the center of rotation may produce significant measurement error unless the angle of rotation is maintained below 1 or 2°, or unless a special feature is employed to control pivot point shifting under larger rotation angles.

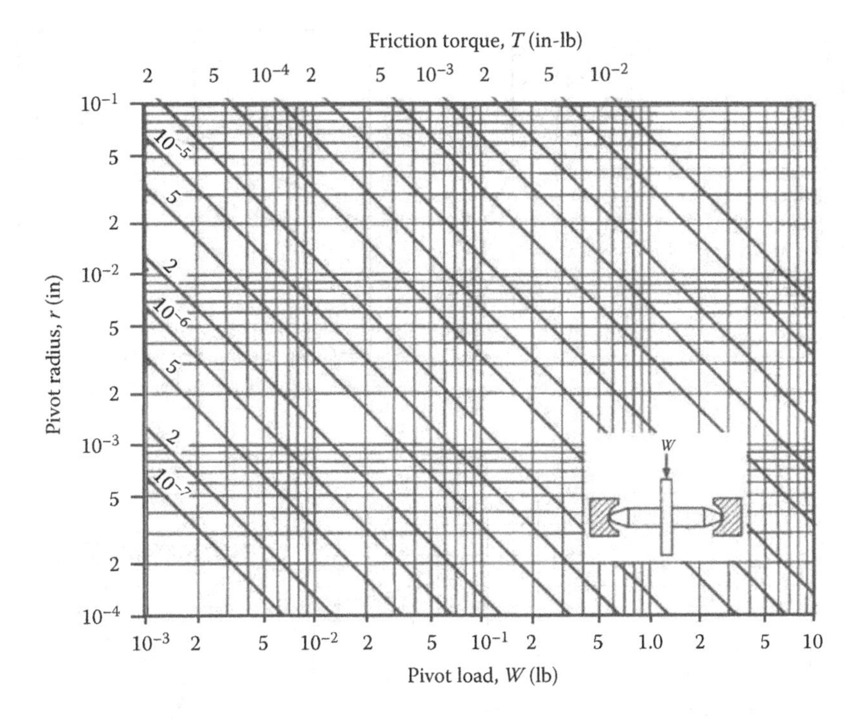

FIGURE 6.12
Pivot radius, load and friction torque for radially loaded pivots.

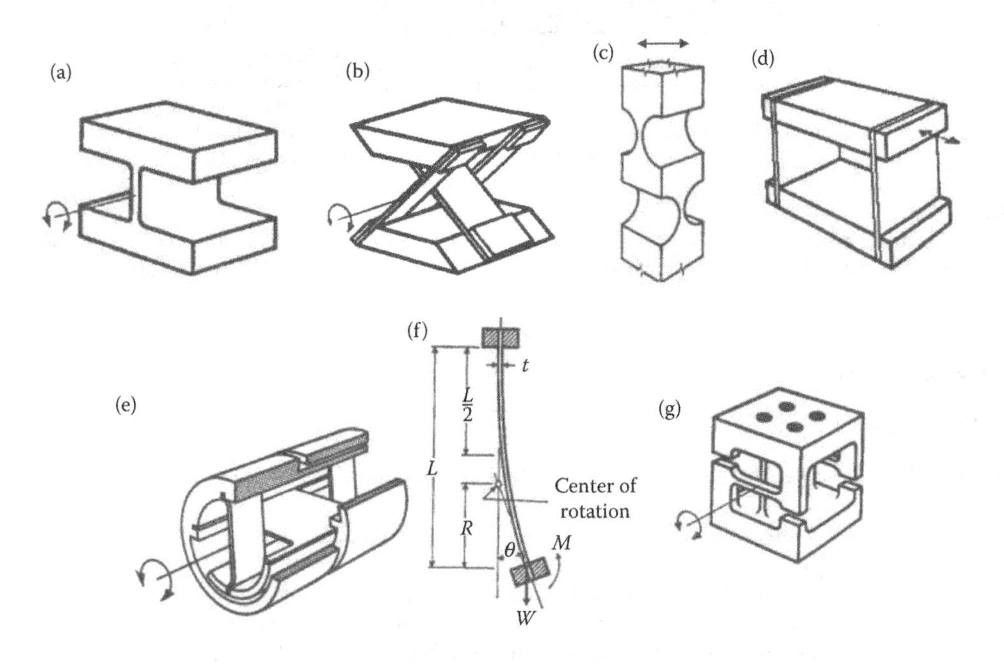

FIGURE 6.13
Prominent flexural bearing examples; (a) simple integral flexure pivot, (b) cross-spring pivot buildup, (c) rotation about two axes, (d) parallel spring support for translation motion, (e) flexure pivot, (f) simple flexure, (g) flexure pivot.

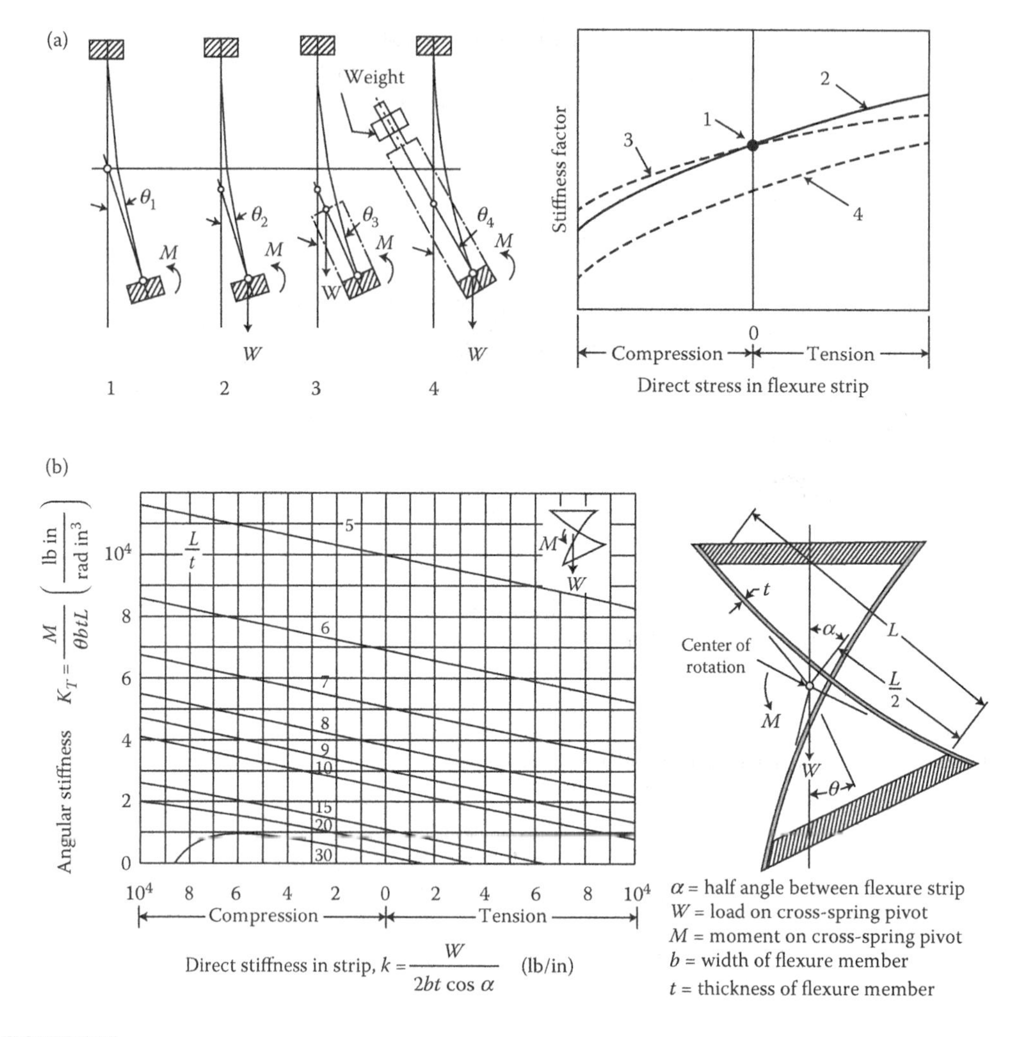

FIGURE 6.14
Flexural bearing design plots from pre-FEA fundamental solid mechanics formulations; (a) simple flexure, (b) cross-spring pivot-buildup. More sophisticated analysis like FEA analysis for larger displacements.

References

Adams, M. L., *Rotating Machinery Research and Development Test Rigs*, CRC Press/Taylor & Francis, Boca Raton, FL, 2017.

Harris, T. A. and Kotzalas, M. N., *Rolling Bearing Analysis*, 5th ed., CRC Press/Taylor & Francis, Boca Raton, FL, 2007.

Larson, C. M. and Larson, R., Lubricant additives, Chapter 14, In: *Standard Handbook of Lubrication Engineering, American Society of Lubrication Engineering (ASLE)*, O'Connor, J. J., Boyd, J. and Avallone, E. A. (Handbook Editors), McGraw-Hill, New York, p. 24, 1968.

Raimondi, A. A., Boyd, J., and Kaufman, Analysis and Design of Sliding Bearings, Chapter 4, In: *Standard Handbook of Lubrication Engineering, American Society of Lubrication Engineering (ASLE)*, O'Connor, J. J., Boyd, J., and Avallone, E. A. (Handbook Editors), McGraw-Hill, New York, p. 118, 1968.

7

Hydrodynamic Journal and Thrust Bearings

The extensive research and development of hydrodynamic bearings started with Reynolds (1886), resulting in the Reynolds lubrication equation (RLE), Equation 1.10, and continues to the present (Adams 2017b). Chapter 1 focuses primarily on the fundamental technology of fluid-film lubrication and bearings, of which *hydrodynamic fluid-film lubrication is the major phenomenon*. Our most personal machine, the *automotive engine,* operates on hydrodynamic lubricated bearings as follows: (1) main bearings, (2) connecting rod bearings, (3) wrist pin bearings, and camshaft bearings, Figure 7.1a. The main bearings keep the crankshaft rotating in place in spite of the piston forces transmitted to the crankshaft by the connecting rods, thus constraining the crankshaft to convert reciprocating motion into rotation. The *main bearings* and *rod bearings* benefit from their journals continuously rotating relative to the bearings at engine rotational speed with an adequate continuous supply of oil. These bearings thus maintain adequate full-film lubrication not only from the (1) *journal continuous rotation* but also from the (2) *squeeze-film effect* energized by the inherently large bearing-load reversals naturally occurring in a piston internal combustion engine.

In contrast, the *wrist-pin bearings* that connect the rods to their respective pistons experience only a slight oscillatory rotational effect as the rods only angulate back and forth a relatively small angle for each revolution of the crankshaft. Thus, wrist pin bearing hydrodynamic effect is mainly from the *squeeze-film effect* that is energized by the large wrist-pin-load reversals. One prominent way employed for wrist-pin bearings to get their oil supply is through a hole in the connecting rod that vents oil from its hydrodynamic oil-film pressure to the wrist-pin bearing clearance, as illustrated in Figure 7.1b. A typical dynamic orbit of the wrist pin within its bearing is illustrated in Figure 7.1c (Adams 2010), showing the squeeze-film action. Related to this, the troubleshooting case study for the refrigerator piston compressor presented in Chapter 14 highlights the interaction of wrist-pin dynamic orbit and oil supply to the wrist-pin bearing.

In the modern world, our other closest technology-based need is *electric power generation,* which requires many machines to simultaneously operate in concert, including: (1) a steam or combustion gas turbine driving, (2) an alternating-current electric generator, necessitating (3) various pumps and (4) fans, all mostly turning on hydrodynamically oil-lubricated fluid-film bearings. Figure 7.2 illustrates schematically the machine at the heart of the modern large electric power plant, the large turbine generator set, and the hydrodynamic full-film bearings that support the end-to-end rigidly coupled multi-bearing turbine generator rotor.

Because there are more than two radial bearings supporting the long driveline of the rigidly coupled rotors of these machines, the journal-bearing loads are statically indeterminate, which thus necessitates accurate elevation setting of the radial-bearing centerlines to (1) properly distribute the weight of the rotor among all the journal bearings, and (2) to avoid excessive shaft cyclic bending stresses that would occur if the journal-bearing centerlines were adjusted to a straight-line horizontal alignment. The long rotor static centerline is basically a catenary (called the *sag line*) with the mid-span rotor centerline approximately

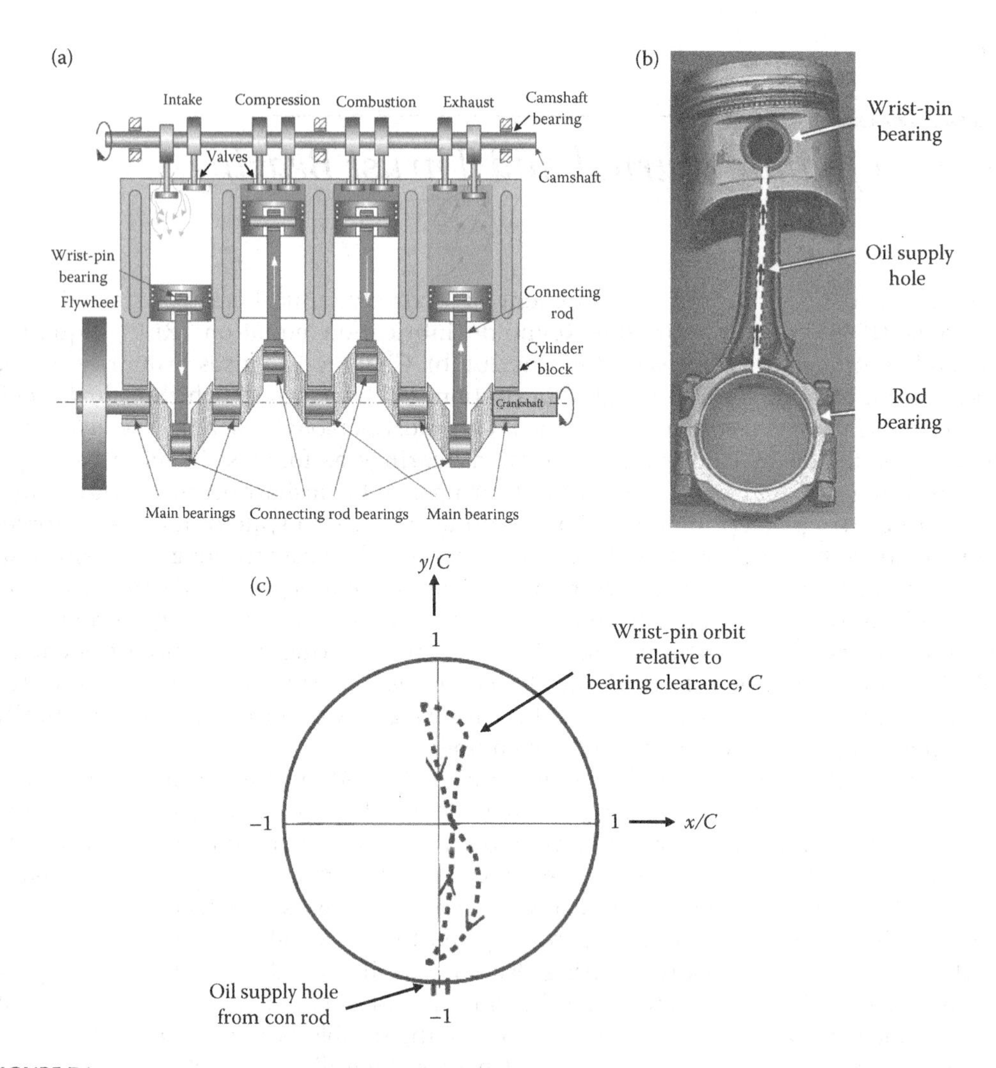

FIGURE 7.1
(a) Automotive four-stroke four-piston engine, (b) wrist-pin bearing oil supply hole, (c) piston compressor wrist pin cyclic orbit within wrist-pin bearing.

2–3 cm lower than a straight line joining the two opposite ends of the quite long rotor on larger sized units. Figure 7.3 compares the relative rotor deflection shapes for (1) an actual bearing elevation sag-line alignment and (2) a straight-line alignment. Clearly, aligning the bearings all at the same elevation would result in (1) a non-optimum distribution of the total rotor weight among the bearings, and (2) needlessly large cyclic shaft bending stresses.

When journal-bearing elevation settings drift out of tolerance over time, usually due to support structure shifting, the sharing of rotor weight by the journal bearings can significantly change from the intended load sharing. That is, the rotor weight does not change but its support distribution among the bearings then does. Such journal-bearing-load distribution shifting can usually be detected (by savvy alert operators) from the likely resulting symptoms of (1) excessive rotor vibration as well as (2) bearing-film temperature shifting (Adams 2010), see Figure 5.21.

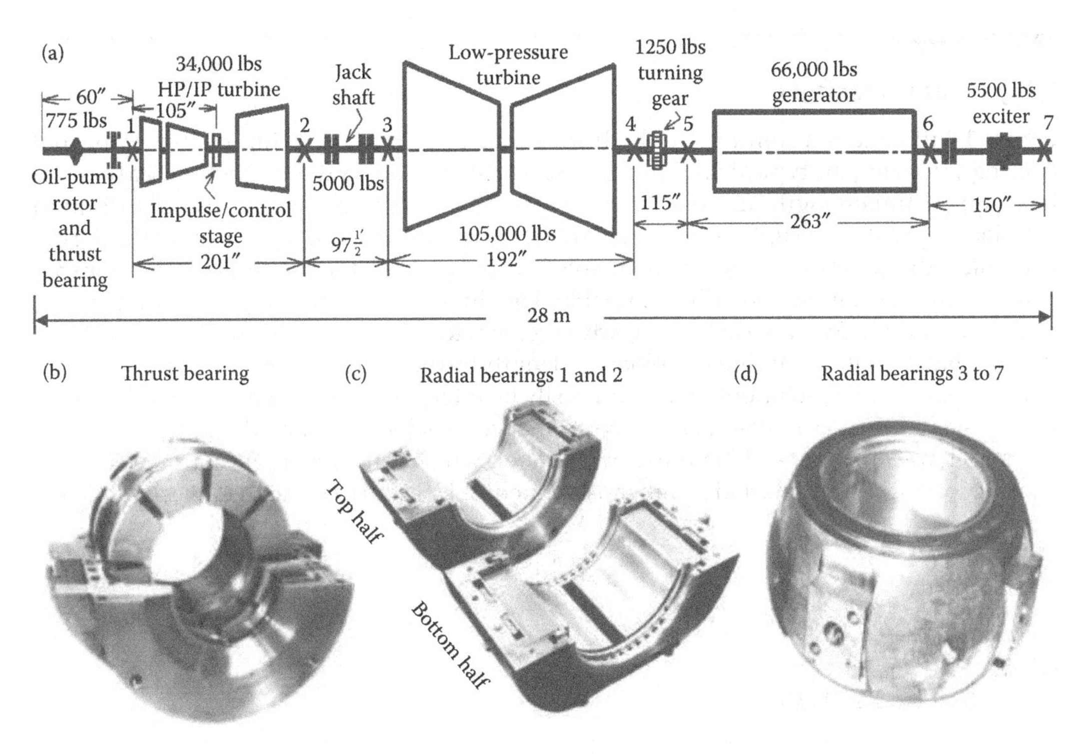

FIGURE 7.2
(a) Schematic of 240 MW steam-turbine generator set (not to scale), (b) double-acting tilting-pad thrust bearing, (c) four-pad tilting-pad journal bearing, (d) cylindrical-sleeve journal bearing.

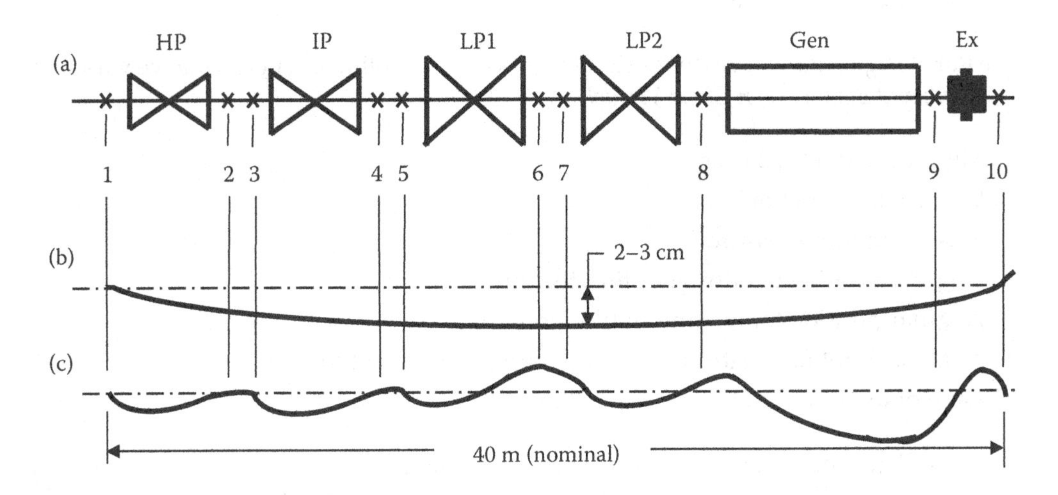

FIGURE 7.3
(a) Large 650 MW steam-turbine generator rotor driveline (radial deflection greatly exaggerated), (b) typical sag-line-bearing alignment, and (c) rotor static deflection with straight-line bearing alignment; HP ~ high-pressure turbine, IP ~ intermediate-pressure (reheat-cycle) turbine, LP ~ low-pressure turbine, GEN ~ generator, EX ~ exciter (illustration not to scale).

7.1 Journal Bearings

Figure 1.8 illustrates a summary of hydrodynamic lubrication of cylindrical-bore journal bearings, showing the typical fluid-film pressure distribution for the 360° cylindrical journal bearing; (1) finite-length, (2) long-bearing, and (3) short-bearing pressure distribution solutions, (4) with film rupture, and (5) without film rupture. Until the advent of the digital computer, the pressure-computational solution of the RLE for the finite-length journal bearing was not computationally obtainable. The three-part landmark paper by Raimondi and Boyd (RaB) (1958) provided a great advancement to designers of journal bearings. Their design charts contain families of plots for length-to-diameter (L/D) ratios of 0.25, 0.5, 1, and ∞ (long-bearing solution) which can easily be interpolated for other intermediate L/D ratios. The RaB dimensionless-bearing performance variables are plotted as functions of the bearing characteristic speed (Sommerfeld number) which is the inverse of a dimensionless-bearing load. The Sommerfeld number as defined in U.S. journal design circles as follows.

$$S = \left(\frac{R}{C}\right)^2 \frac{\mu N}{P} \tag{7.1}$$

where

P = unit load $\equiv W/DL$,
W = load,
N = journal rotational speed (rev/sec),
μ = viscosity,
D = bearing and journal nominal diameter,
R = $D/2$,
L = bearing axial length,
C = bearing radial clearance.

The RaB design charts have dimensionless plots for the following performance variables.
Raimondi and Boyd Design Chart Variables

1. Minimum film thickness
2. Coefficient of friction
3. Maximum film pressure
4. Angular position of minimum film thickness
5. Angular position for maximum film pressure
6. Total lubricant flow naturally sheared into the lubricant film
7. Ratio of total-to-axial side leakage flows

These RaB journal-bearing design charts for the cylindrical 360° journal bearing are replicated in all undergraduate machine design textbooks. As relayed here in the Acknowledgments, it was the author's good fortune to work under Dr. Al Raimondi during his employment at the Westinghouse Corporate R & D Center near Pittsburgh (1971–1977), prior to becoming a professor.

In the 10 years following the RaB (1958) journal-bearing publication, the speed and availability of digital computers advanced considerably, including the development of software compilers like Fortran that allowed engineers to forego programing in *machine language*. That gave birth to the development of commercially available general-purpose computer codes to create RaB-type dimensionless design charts for virtually any journal-bearing configuration such as those illustrated in Figure 1.10. The main developers of these early generation fluid-film computer codes were at the Franklin Institute Research Laboratories (Philadelphia, PA), where the author was employed (1967–1971). Castelli and Shapiro (1967) developed a non-iterative 2D finite-difference solution to the Reynolds lubrication equation (RLE) that was near universally adopted by all software developers working in the fluid-film lubrication field.

There is now a plethora of such fluid-film computer codes commercially available (Adams 2010), such as the prominent COJOUR code developed by Mechanical Technology Inc. (MTI), originally as a mainframe code sponsored by the Electric Power Research Institute (EPRI) and subsequently marketed by MTI for PC use (Pinkus and Wilcock 1985). It is worth noting that the use of such software to develop RaB-type design charts for specific journal-bearing configurations *follows the reverse procedure* as when the design charts are subsequently utilized for bearing design analyses. As demonstrated here in the subsequent design calculation examples (Section 7.1.6), the designer specifies the bearing-load *magnitude* and *angular direc*tion (Figure 7.4) and uses the design charts to determine the minimum film-thickness magnitude and angular position as well as the other design-chart outputs; (a) required lube flow to bearing, (b) friction coefficient, (c) bearing-film operating temperature, and so on. However, when developing the design charts, the minimum film-thickness distribution (i.e., magnitude and direction of journal eccentricity) must be specified and the resulting output is the load magnitude and radial direction. That is, to solve the RLE for a data point, the film-thickness distribution must be specified, that is, the magnitude (dimensionless) and angular direction with respect to bearing geometry like an axial feed groove or pocket. That process is summarized as follows (Adams 2010).

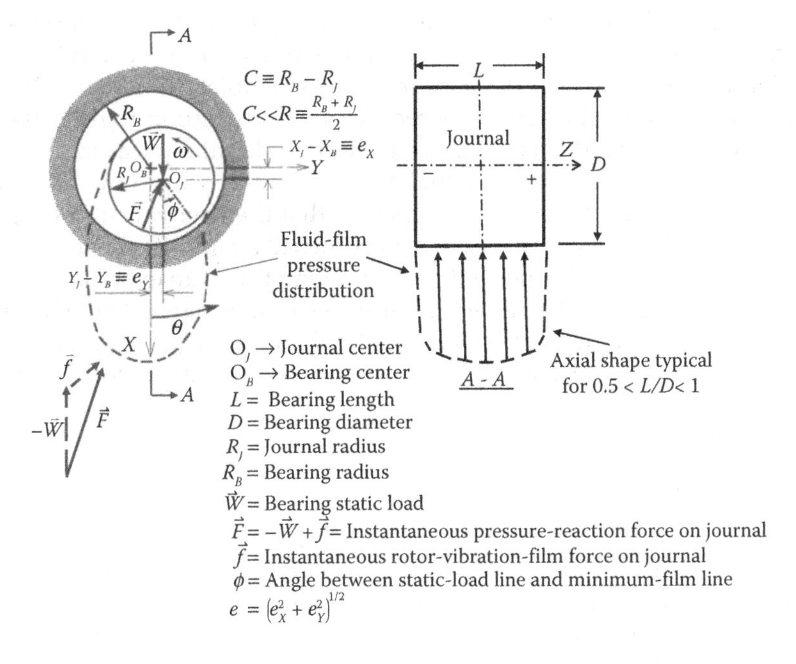

FIGURE 7.4
Journal-bearing configuration and nomenclature.

For a single RLE journal-bearing solution point:

1. Specify $e \equiv \sqrt{(e_x^2 + e_y^2)}$, $\phi = \arctan(e_y/e_x)$, $e_x = x_J - x_B$, $e_y = y_J - y_B$
2. With journal-to-bearing axial alignment, $h = C - e_x\cos(\theta) - e_y\sin(\theta)$, because $e \ll R$, bearing radial clearance, $\tau = R\theta$ giving,
$dh/d\tau = (e_x/R)\sin(\tau/R) - (e_y/R)\cos(\tau/R) = (e_x/R)\sin\theta - (e_y/R)\cos\theta$, and
$dh/dt = -\dot{e}_x\cos(\tau/R) - \dot{e}_y\sin(\tau/R) = -\dot{e}_x\cos\theta - \dot{e}_y\sin\theta$, $dh/dt = 0$ for charts
3. Solve the RLE for the pressure distribution $p = p(\tau, z)$
4. Integrate $p(\tau, z)$ over the journal cylindrical surface to get x and y forces on the journal.

$$\begin{aligned} W_x &= -\int_{-L/2}^{L/2}\int_0^{2\pi R} p(\tau, z)\cos\theta\, d\tau\, dz \\ W_y &= -\int_{-L/2}^{L/2}\int_0^{2\pi R} p(\tau, z)\sin\theta\, d\tau\, dz \end{aligned} \tag{7.2}$$

5. Calculate resultant radial load and its angle.

$$W = \sqrt{W_x^2 + W_y^2}, \quad \theta_W = \arctan(W_y/W_x) \tag{7.3}$$

By performing the above steps, 1 through 5, over a suitable range of values for $0 \le e/C < 1$ and ϕ, enough solution points are generated to construct design curves as the same as those of Raimondi and Boyd. As stated earlier, the sequence of computations in design analyses is the reverse of the above sequence. That is, one starts by specifying the bearing load, W, and its angle θ_W and uses design curves preassembled from many RLE solutions to determine the corresponding journal eccentricity, e, and attitude angle, ϕ. Because CPU running-time with modern PCs is not a cost factor as it definitely was in the era of using computer main frames, some RLE journal-bearing PC software now have the option which essentially iterates on steps 1 through 5 so that the user can simply specify load magnitude and direction to obtain resulting eccentricity e and its angle ϕ, for example, COJOUR PC code (Pinkus and Wilcock 1985). Although this is an expedient short cut that bypasses the significant task of generating RaB-type design charts, it deprives the user of visualizing the overall characteristics of the particular journal-bearing-type as functions of design parameters to be selected like L/D ratio as well as bearing-load magnitude and direction.

7.1.1 Cylindrical 360°

Unlike rolling element bearings (REB), which must be selected and purchased from an REB manufacturer's catalog, the machine designer employing fluid-film journal bearings (FFJB) has the option to either (1) design and manufacture their own bearings or (2) select and purchase from a journal-bearing manufacturer's product line. In mature rotating-machinery types employing FFJBs, the journal-bearing design particulars have evolved over time from early configurations and sizes of the machine-type as the product evolves and operating feedback from the field is factored into product improvements. That product evolution scenario of course varies from product to product and therefore cannot be completely treated in a general textbook. This subsection presents the RaB dimensionless design charts, Figures 7.5–7.9, covering the following journal-bearing performance variables of (1) minimum film thickness, (2) friction coefficient, and (3) lubricant supply flow. Sample design calculations utilizing those RaB design charts and those for the other journal-bearing types are subsequently covered in Section 7.1.6.

7.1.2 High-Load Cylindrical 360°

Solutions of the full two-dimensional RLE Equation 1.10 for the RaB design charts in Figures 7.5–7.9 were computed assuming *no axial misalignment* between journal and bearing. That is, the journal and bearing centerlines are tacitly assumed to be parallel. Most cylindrical journal bearings are sized so that when operating under load the nominal design eccentricity ratio is $0.5 < \varepsilon = e/C < 0.75$, reflecting the compromise between (1) maximizing the minimum film thickness and (2) avoiding large-amplitude self-excited sub-synchronous rotor vibration, a result of having over-sized too-lightly-loaded journal bearings (Adams 2010). That range of eccentricity ratio ($0.5 < \varepsilon < 0.75$) combined with adequate *bearing-to-journal assembly alignment* justifies the engineering assumption of zero misalignment for which Figures 7.7–7.9 were developed.

Furthermore, since the bearing liner material should be significantly softer (e.g., babbitt) than the typical steel journal, a tolerable amount of *bearing edge loading* of a new bearing will quickly be relieved by wearing-in of the initial bearing edge rubbing by the journal. Prior to modern automobile engine precision manufacturing, engine bearings (see Figure 7.1) were manufactured with a 10 mil (thousands of an inch) babbitt liner to accommodate new-engine wear-in, given all the manufacturing tolerances inherent throughout the engine at that time.

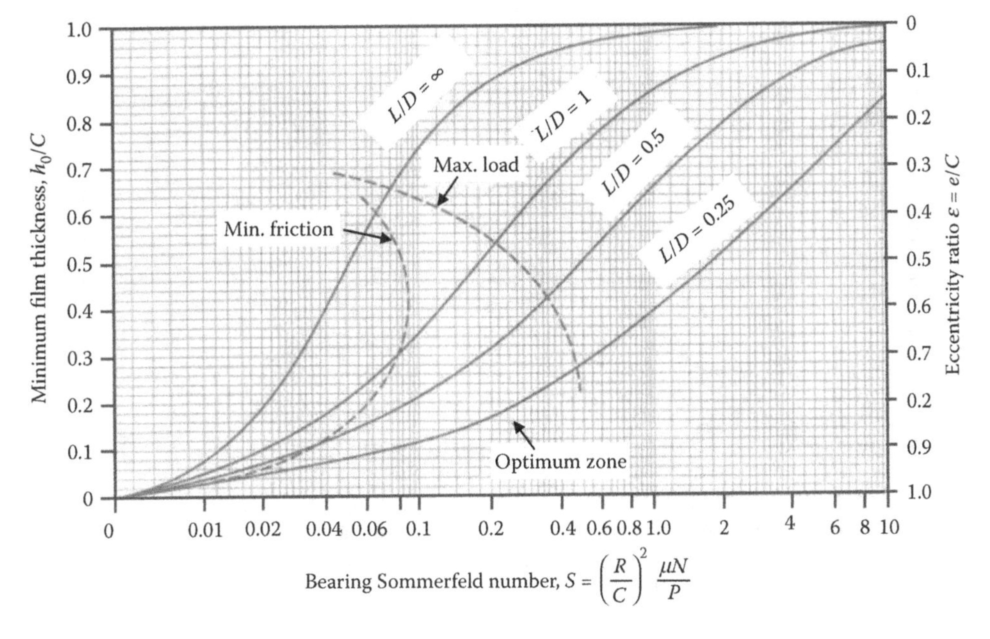

FIGURE 7.5
Minimum film thickness.

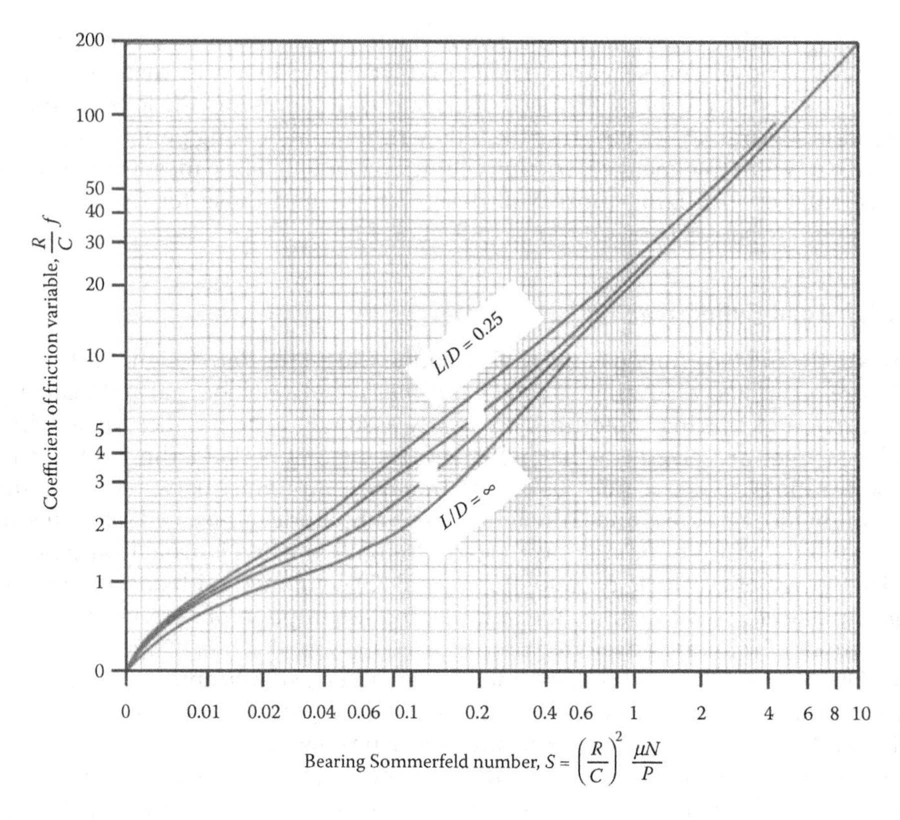

FIGURE 7.6
Friction coefficient.

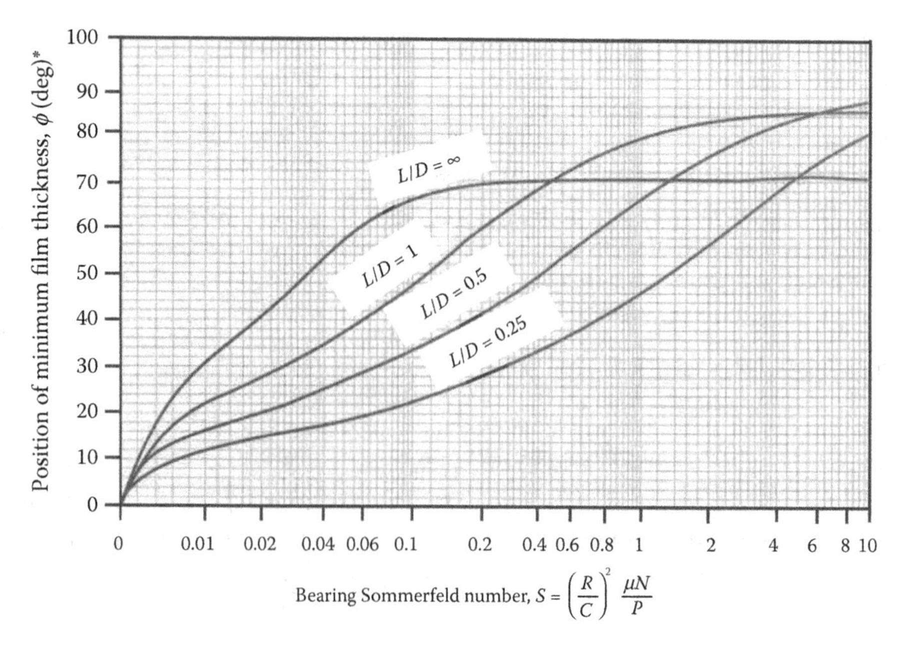

FIGURE 7.7
Angular position of minimum film thickness.

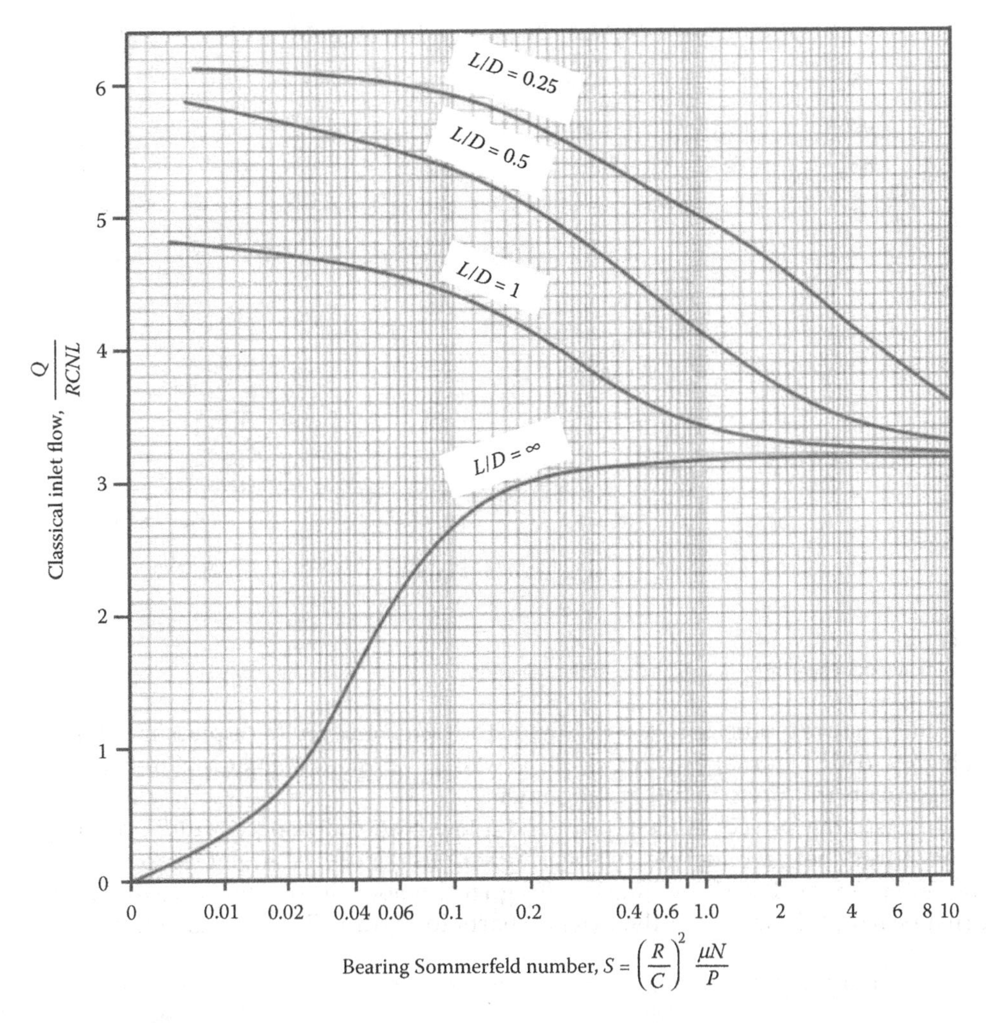

FIGURE 7.8
Lubricant inlet flow naturally sheared into the film.

However, for very *high-load high-eccentricity* ($\varepsilon > 0.9$) journal-bearing applications, misalignment should be eliminated by employing a *self-aligning feature*. This was determined by the author in 1971 when consulting for a positive-displacement pump manufacturer of high-pressure sliding-vane hydraulic pumps. The client's just released new sliding-vane pump product was suffering over a 20% journal-bearing failure rate within the first 10 hours of operation. They needed a quick fix before their new sliding-vane pump product irretrievably lost total credibility. Having then fortunately sold only about 200 pumps, they stopped production until their bearing failure problem was solved. Having just read a then-recent machine design magazine article by Adams (1970) on axial-grooved journal bearings, they retained the author who was then employed by the Franklin Institute Research Laboratories (FIRL). The author followed up by employing a new fluid-film-bearing software package that he helped develop at FIRL. That software was a RLE solver that embodied a *variable grid-spacing option* (in both circumferential and axial directions) in its finite-difference formulation. The variable-sized grid spacing feature made it feasible to accurately accommodate the very

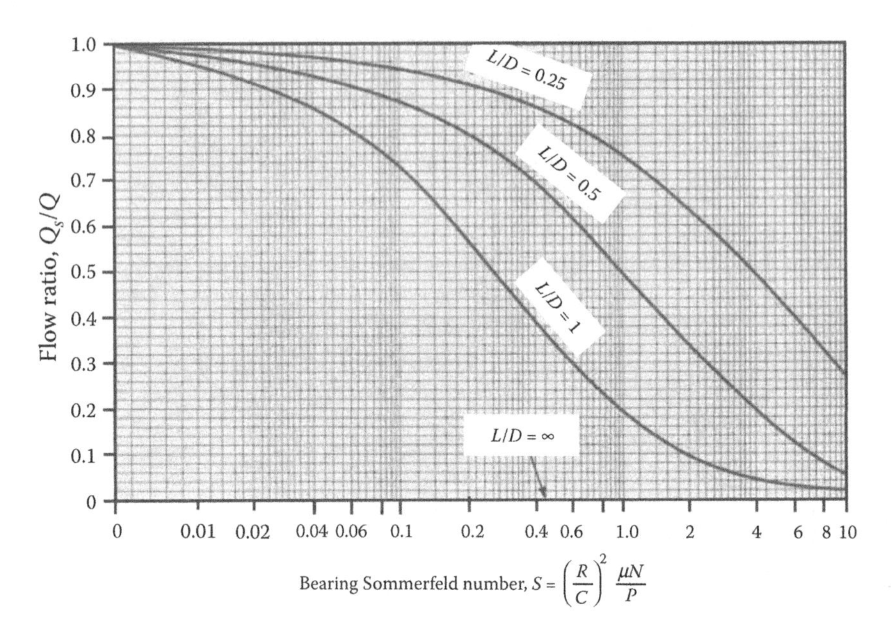

FIGURE 7.9
Ratio of flow out of bearing two axial sides to the inlet flow.

high-film pressure gradients in the vicinity of the localized very-high-pressure film portion of a highly loaded journal bearing, while as well accommodating journal-to-bearing axial misalignments from zero to full misalignment.

That consulting project quickly provided a low-cost, quite-successful fix of the pump manufacturer's new-product journal-bearing serious deficiency. And it led to a subsequent publication (Adams 1971) providing the design technology needed for successful high-load journal bearings, from which the design charts in Figures 7.10 through 7.15 for $L/D = 1$

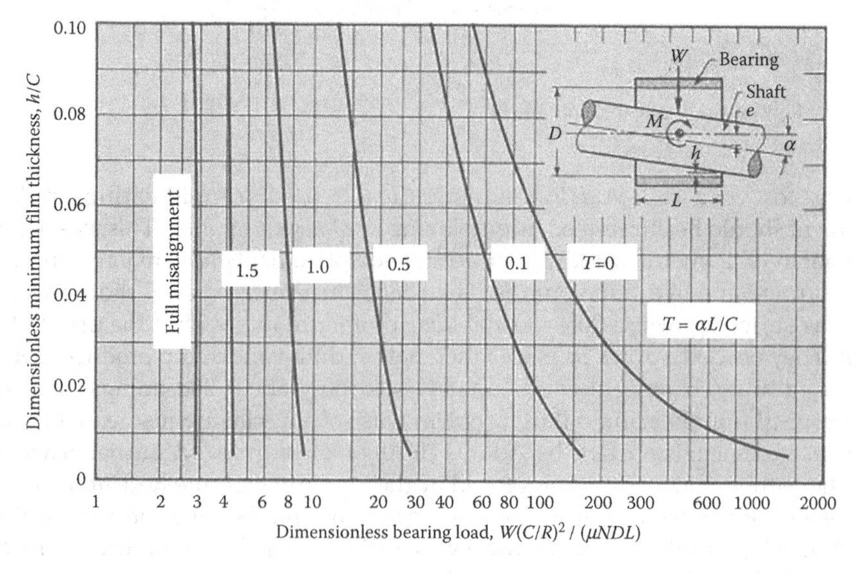

FIGURE 7.10
Minimum film thickness versus load with misalignment.

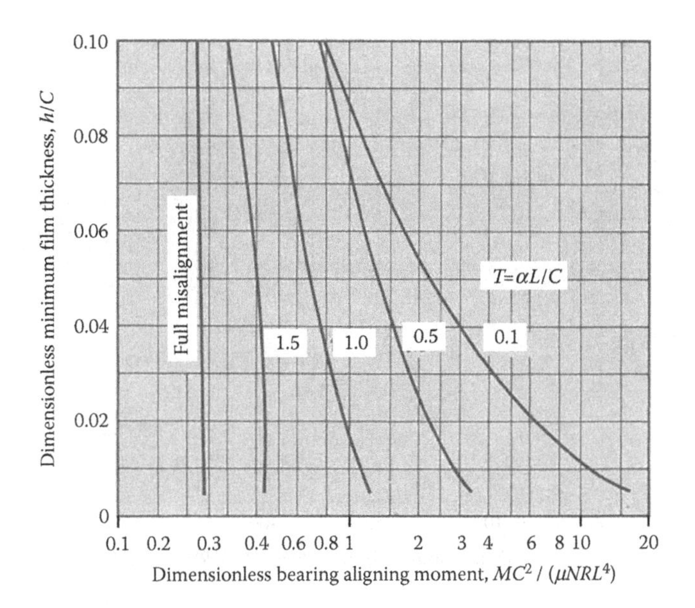

FIGURE 7.11
Minimum film thickness versus bearing aligning moment.

are replicated. On a much larger machine (320 MW steam turbine generator unit), Clavis et al. (1977) used the DC signal portions from sets of x-y non-contacting displacement measurement proximity probes (Figure 5.12c) at each axial side of the journal-bearings, and additionally journal bearing oil-film pressure measurements, to diagnosis that the journal-bearings in large turbogenerators could encounter significant axial misalignment during some normal operating transients.

7.1.3 Two-Axial-Groove Cylindrical

Cylindrical journal bearings with two axial equally spaced oil-supply grooves 180° apart have been a commonly used configuration for many decades. One application researched by the

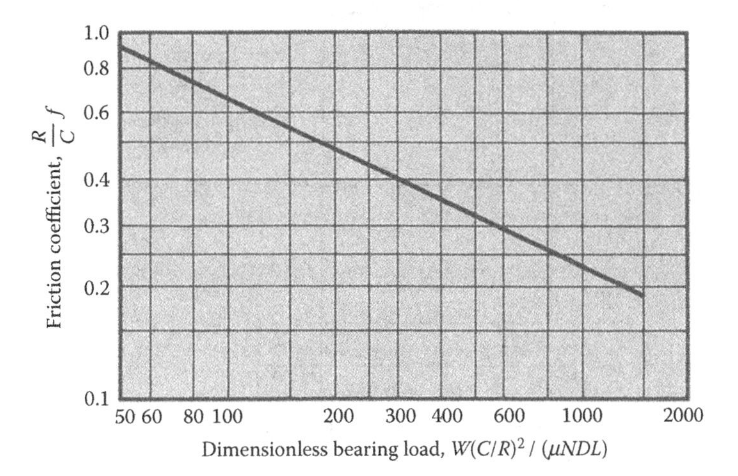

FIGURE 7.12
Compute bearing-film friction coefficient versus load.

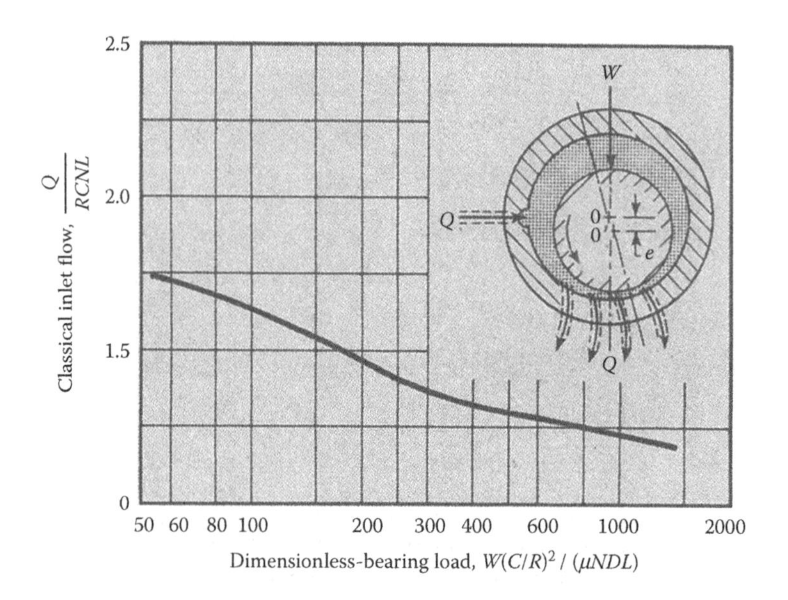

FIGURE 7.13
Film inlet flow versus load.

author was for main propeller geared-transmission drivelines in large ocean-going vessels. Large ocean-going ships are typically powered either by *gas turbines* or powered by a single direct-connected massive *diesel* piston engine. The choice of whether to employ a direct-connected diesel engine as opposed to a group of two or more combustion gas turbines is dictated by where on the globe the ship is expected to mainly operate. The choice is based on the relative cost of various fuels available in the ship's main operating global region.

Adams and Rippel (1969) produced a comprehensive design manual for a major manufacturer of geared-driveline transmissions for large ocean ships that employ gas

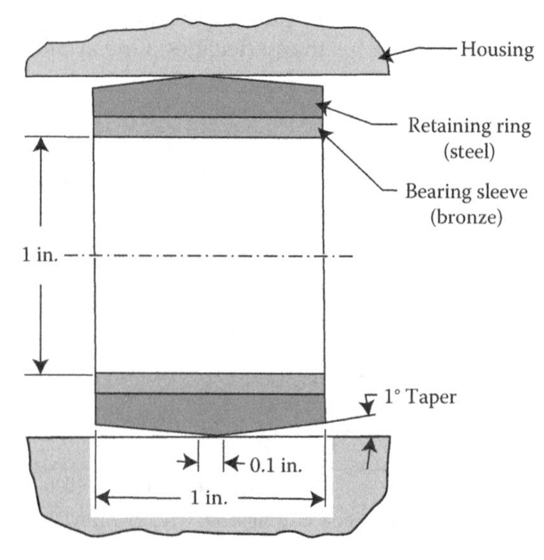

FIGURE 7.14
Self-aligning journal bearing with varying static loads up to 3000 lb.

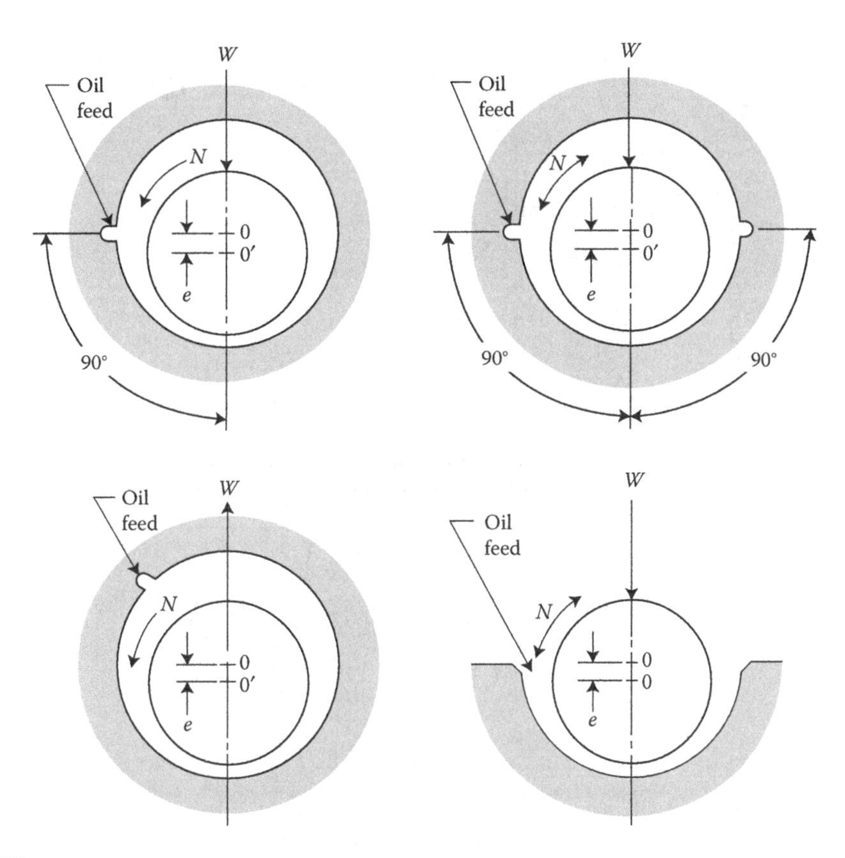

FIGURE 7.15
Feed-groove configurations for high-eccentricity (high-load) journal bearings; configurations (b) and (d) can be used for both rotational directions.

turbines as the propulsion power source. That manufacturer had apparently experienced some occasional journal-bearing distresses in their propeller driveline transmissions, prompting them to retain the Franklin Institute Research Laboratories (FIRL) to research their 2-axial-groove-bearing configuration.

The 2-groove hydrodynamic journal bearing has (1) film friction and (2) required lubricant supply magnitudes nearly the same as provided by the Raimondi and Boyd (1958) charts for the 360° journal bearing (Figures 7.7–7.9). However, Adams and Rippel (1969) discovered that the *load-carrying characteristic* of the equally spaced 2-groove bearing is fraught with a previously unrecognized inherent *alarming characteristic* regarding the *angular direction* of the applied-bearing radial load relative to the two axial groove locations. That discovery emerged from the design-chart field maps that Adams and Rippel developed for their client on the equally spaced 2-groove journal bearing. They utilized FIRL's then newly developed fluid-film-bearing software to create design charts for the 2-groove bearing. Figure 7.16 (angles γ, θ and ϕ specified in Figure 7.17) replicate the unique and most important Adams–Rippel bearing performance design charts that clearly show the *alarming characteristic* of bearing-radial-load direction relative to the oil supply grooves. *Disclaimer*: The now-outdated choices for the dimensional units used in defining the bearing-load factor A were not this author's choice, that is, it's like a dimensionless load but it's clearly not dimensionless.

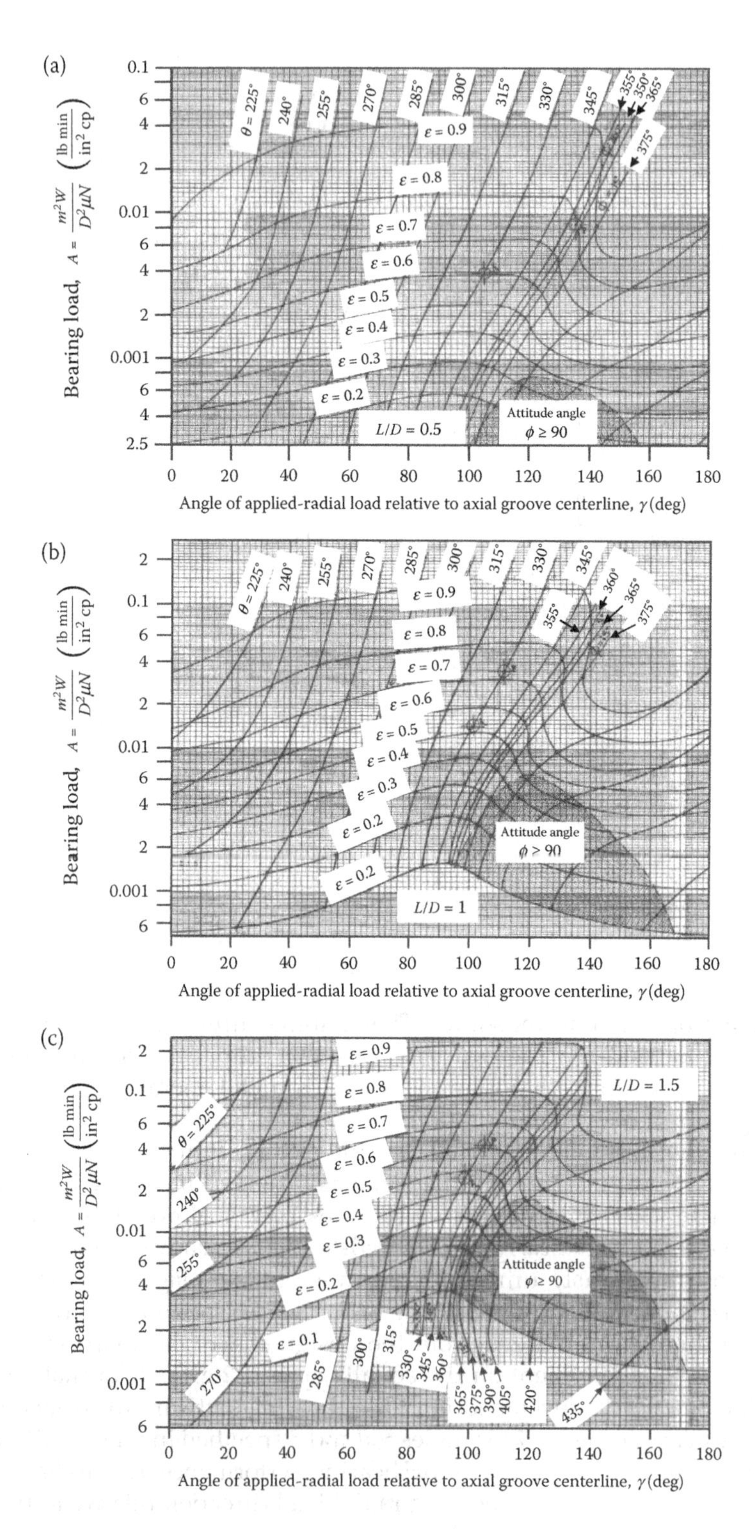

FIGURE 7.16
Load characteristics for the two-groove hydrodynamic journal bearing; (a) $L/D = 0.5$, (b) $L/D = 1$, (c) $L/D = 1.5$, $m \equiv 2C \times 10^3/D$.

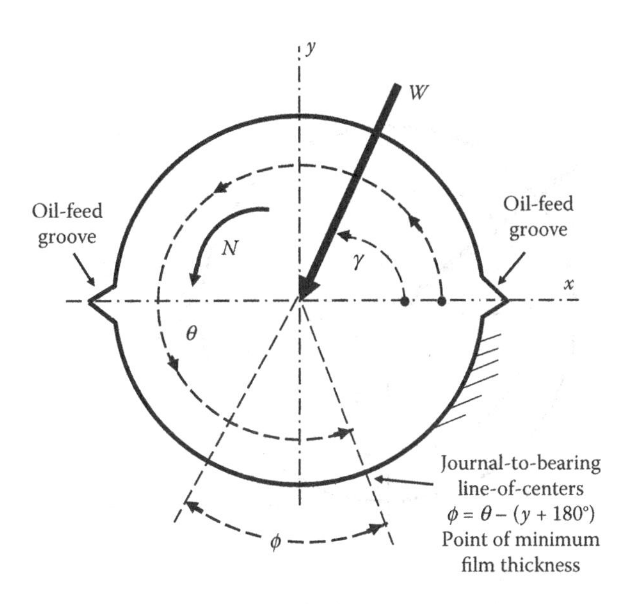

FIGURE 7.17
Angles specified in Figure 7.16.

With ocean ship main-propeller transmissions, the journal-bearing loads are the vector sum of (1) the constant *dead weight* of the shaft-with-gear components and (2) the helical-gear radial-plane component of *tooth loading* at the tooth-pressure angle. The gear tooth loading portion of total journal-bearing load is *constant in direction* but *quite variable in magnitude.* This is because the torque transmitted to the propeller varies considerably over its full wide operating range of speed and transmitted power. Thus, the bearing total radial-load vector encounters significant changes in direction over its full range of operation, making the *load-direction characteristic* information clear from Figure 7.16 a *warning* for such applications of the equally spaced two-groove bearing. A potentially better configuration for applications with significant load-angle variations in operation is a *single axial feed groove* located at an angle significantly away from where the minimum film thickness could ever locate within the full operating range of the bearing.

This can be fully appreciated from an examination of Figure 7.16 focused on the load-magnitude characteristic at load-direction $\gamma \simeq 140°$ for high load ($\varepsilon = 0.9$), for any of the three ratios, $L/D = 0.5, 1, 1.5$. The Figure 7.18 illustration is based on the Figure 7.17b chart for $L/D = 1$, and clearly shows why the alarming load-direction sensitivity occurs with this bearing configuration. Basically, in satisfying static force equilibrium between the applied load W and the equilibrating hydrodynamic-film pressure distribution, the minimum film thickness position occurs at the downstream feed groove. The film pressure at the minimum film thickness must be at the feed-groove pressure to satisfy the groove-pressure boundary condition as illustrated in Figure 7.18. For the example illustrated in Figure 7.18, as the load angle γ transitions from 140° to 150° the load factor A changes from 0.13 to 0.018, Figure 7.17b. That is a reduction in A of 0.13–0.018 = 0.112 which yields $(0.112/0.13) \times 100\% = 86\%$ *reduction in load capacity* at $\varepsilon = 0.9$. Under the applied load corresponding to $A = 0.13$, that would result in a considerable static eccentricity increase beyond $\varepsilon = 0.9$ in the transition of γ from 140° to 150°. And that would surely put the bearing operating point well into the *very-high-load category*, thus definitely requiring a bearing *self-aligning feature* as explained in Section 7.2.2. The three-axial groove, four-axial-groove cylindrical bearing, and so on

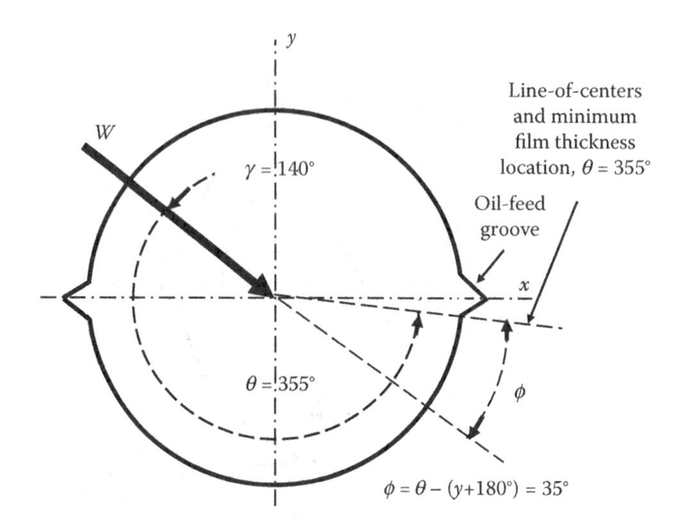

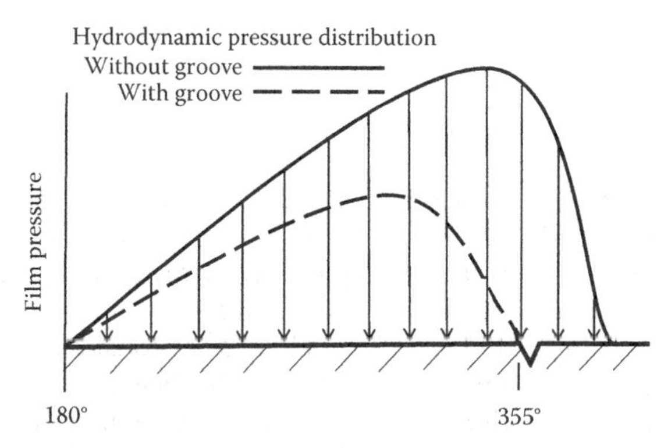

FIGURE 7.18
Angular locations with a 140° load angle for $L/D = 1$ with $\varepsilon = 0.9$, and the resulting effect of groove location on hydrodynamic pressure distribution.

should certainly also not be employed for applications where the load direction experiences large operating swings in angular direction that would potentially put bearing operation in the mode exampled by Figure 7.18. That is, where the load direction causes the minimum film thickness to be at a lubricant supply-pressure-groove.

7.1.4 Partial-Arc and Preloaded Multi-Arc (Lobe)

In large bearing diameter applications where the applied journal-bearing radial-load direction is well confined to a nearly single direction, for example, dominated by the supported dead weight of the rotor, the partial-arc configuration is most often applied to reduce the power dissipated by the bearing. A prominent example of this is in large two-pole steam turbine AC generator sets for electric power generation, where each bearing can dissipate hundreds of horsepower under normal operation. The upper portion of such large partial-arc journal bearings generally has two very narrow concentric cylindrical surface extensions of the partial-arc-bearing cylindrical surface that axially boarder a deeper cylindrical pocket, at lubricant supply pressure, to provide full circular radial-clearance journal containment in case of an emergency such as from the loss of large turbine blades at full speed (Adams 2010).

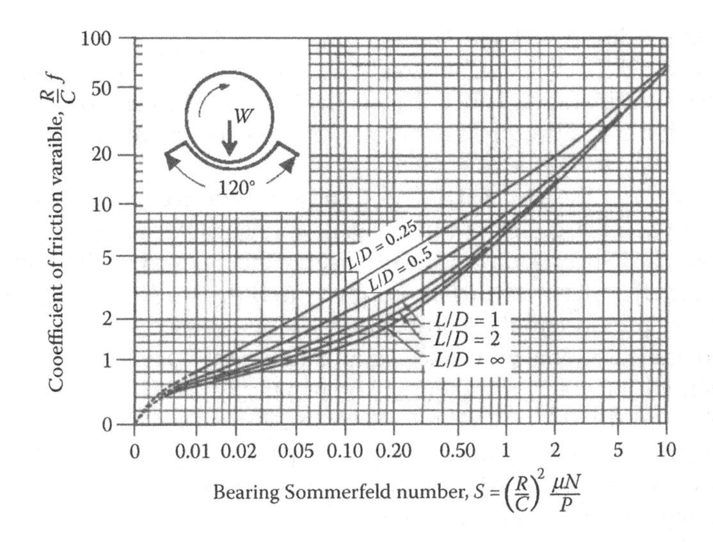

FIGURE 7.19
Friction coefficient for 120° partial-arc journal bearing.

The power dissipated in a partial-arc journal bearing is significantly less than that of a comparable sized full 360° cylindrical configuration at the same load and journal speed.

Raimondi et al. (1968) provide a complete set of design charts for the centrally loaded 120° partial-arc journal bearing of the same combination as Figures 7.5–7.9 for the 360° cylindrical bearing. The specific one of those charts here replicated (Figure 7.19) is for the friction coefficient, shown here to illustrate the relative reduction in power loss compared to the 360° bearing. Utilizing both Figures 7.6 and 7.19 for $f(R/C)$ plotted values, plus a little physical insight as subsequently explained, Table 7.1 was produced to show an example comparison in friction coefficient between the 120° and 360° cylindrical journal-bearing configurations for $L/D = 0.5$. This comparison is comparable to a comparison of bearing-dissipated power.

Recalling that the bearing Sommerfeld number S (a dimensionless speed) is the inverse of a dimensionless load, as used for Figures 7.10–7.13, the trend in the Table 7.1 right column makes physical sense. That is, as load is progressively increased, a progressively larger portion of viscous power dissipation naturally occurs in the localized higher-pressure smaller-film-thickness distribution region. In contrast, as load is progressively decreased, the viscous power dissipated progressively becomes more evenly distributed throughout the film, of which the 120° bearing has 240° less of than the 360° bearing. Accordingly, as load is postulated to approach zero, $W \to 0$, $S \to \infty$, the values of the three design-chart

TABLE 7.1
Journal-Bearing $f(R/C)$ Friction Value Comparison Points for $L/D = 0.5$

S	f(R/C): 120°	f(R/C): 360°	$[f(R/C)_{360°}] \div [f(R/C)_{120°}]$	
0	————	————	1.00 ←	Based on physical insight
0.1	2.2	3.4	1.55	From f (R/C) values picked off Figures 7.6 and 7.19
0.2	3.2	6.0	1.88	
1.0	9.0	24	2.67	
∞	————	————	3.00 ←	Based on physical insight

points in the right column of Table 7.1 should approach 360/120 = 3, and that is how the right column in Table 7.1 trends. Conversely, as load is progressively increased, $W \to \infty$, $S \to 0$, the power dissipated is concentrated in the local film region where pressure is relatively quite high and film thickness distribution is relatively quite small, so the right column will approach the value of 1.

Figure 7.20 illustrates the 160° partial-arc configuration used extensively by one of the major manufacturers of large steam turbine generator sets. It is referred to as the *viscosity-pump bearing* because it incorporates a clever offset of the relieved top half oil-supply pocket so that with a nearly tangential surface entrance into the load-carrying bottom 160° arc, the inlet boundary pressure to the load-bearing 160° arc is boosted a bit above the oil inlet pressure to provide additional assurance against lubricant starvation to the partial arc.

Preloaded multi-arc journal bearings (PMAB), usually called *lobe bearings*, are illustrated for the two-lobe and three-lobe configurations in Figure 1.10. The individual arcs of the PMAB are designed with either *pitched* or *unpitched* partial lobes as illustrated in Figure 7.21. General performance charts for PMABs, comparable to the those in Figure 7.5–7.9 for the 360° cylindrical bearings, are not compatible to such a compact form as those for the 360° bearing. That is because there are four more parameter variables; (1) number of equally spaced lobes, (2) the preload radial offset of the lobes, (3) pitch angle of lobes (zero or pitch value), and of

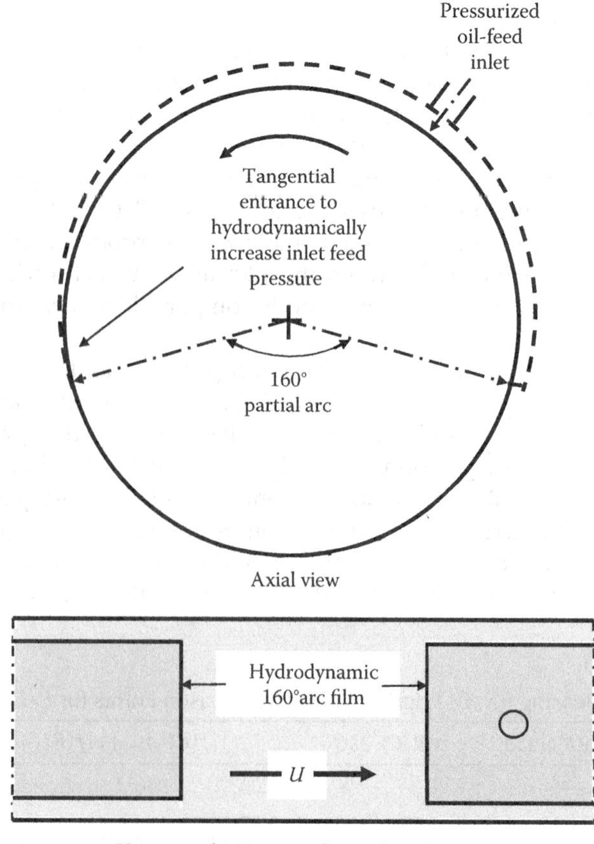

FIGURE 7.20
Viscosity-pump 160° partial-arc journal bearing.

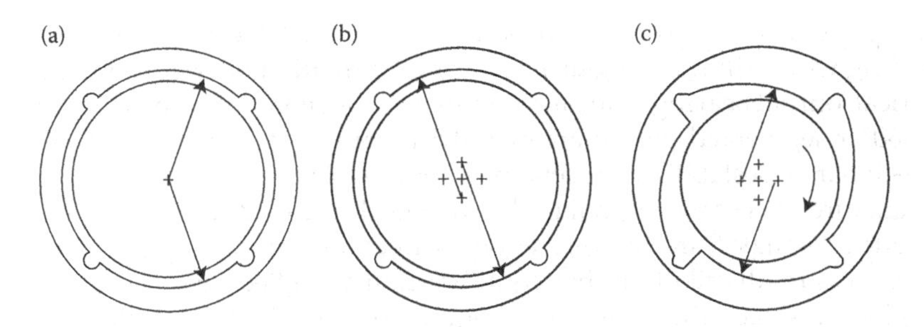

FIGURE 7.21
Four-arc bearing; (a) cylindrical with grooves, (b) unpitched lobes, (c) pitched lobes.

course (4) the radial direction of the bearing applied load relative to one of the feed grooves separating the individual lobes. Thus, to analyze specific PMAB configurations, one must employ fluid-film-bearing software to compute the bearing performance parameters for the specific input parameter combinations, thereby conducting the parametric studies necessary to optimize the specific bearing. The unpitched configuration performs equally well in both directions of rotation, whereas the pitched version is only good for one rotational direction as illustrated in Figure 7.21.

The PMAB is a lower-cost competitor to the tilting-pad journal bearing (TPJB) in that PMABs also can prevent the large-amplitude self-excited subsynchronous rotor vibration known as *oil-whip* that can occur when cylindrical journal bearings become too lightly loaded under some operating condition(s) in an application. On the other hand, a factor weighing in favor of the TPJB is that it can be designed for the same application with less power dissipated than the comparable PMAB, and better adapts to load direction variations. The lower power dissipation of the TPJB is a significant monetary trade-off in large-bearing applications such as large steam turbine generator sets, that is, bearing first cost versus bearing power dissipation cost over time.

The two major U.S. traditional large steam turbine generator manufacturers (GE and Westinghouse) differed on this. That is, for the high-pressure (HP) turbine section of the unit driveline (see Figures 7.2 and 7.3), Westinghouse employed four-pad TPJBs while GE employed four-lobe pitched PMABs. These HP-turbine-bearing choices were to prevent *oil whip* rotor vibration of the HP turbine during the lessening of net downward loads on the HP turbine journal bearings caused by the added upward rotor force from the impulse-turbine (first-stage) during nozzle partial-emission, used to control turbine output reduction at unit part-load operation (Adams 2010). The use of impulse turbine partial emission to regulate unit power output is because U.S. fossil-fueled electric power plants employ constant-pressure boilers. Whereas, European fossil-fueled electric power plants employ variable pressure boilers to regulate power output, so they do not need an impulse-stage as the first-stage of the HP turbine, and thus do not encounter unloading of the HP turbine bearings under part-load operation.

7.1.5 Tilting-Pad

The tilting-pad journal bearing (TTJB) evolved in the second half of the twentieth century as one approach to combat self-excited rotor-vibration phenomena such as (1) bearing *oil whip*, (2) HP turbine *steam whirl*, (3) other fluid dynamic mechanisms such as in *centrifugal pumps* and *compressors*, and (4) other rotor-dynamic destabilizing mechanisms like *rotor-spline sliding friction* (Adams 2010, 2017a).

Illustrating the 360° cylindrical journal bearing, Figure 7.4 provides the first clue to understanding why a tilting-pad journal bearing (1) should not cause oil whip like fixed-arc cylindrical journal bearings can, and (2) can provide more net rotor-vibration damping than say 360°or segmented partial-arc cylindrical journal bearings to help suppress the other rotor-dynamic destabilizing mechanisms. Figure 7.4 clearly illustrates that the radial-displacement direction of the journal under-load leads the load-direction line by the angle ϕ (called "attitude angle") in the co-rotational direction, a necessary positioning to yield a fluid-film pressure distribution that equilibrates the applied bearing load. Amplifying that Figure 7.4 "message," the load-vector diagram in Figure 7.22 shows that the cylindrical journal-bearing fluid-film force has (1) a component opposing the journal eccentricity displacement and (2) a component in the tangential co-rotational direction perpendicular to the eccentricity direction. With a dynamic perturbation, the tangential component of the fluid-film force provides an energy source to potentially excite co-rotational self-excited rotor vibration orbiting if sufficient damping to suppress that "negative damping" effect is not present. This is explained with more mathematical rigor in Adams (1987, 2010). In contrast, the force produced by the hydrodynamic pressure distribution on each TPJB pad aligns with the journal displacement component into the pad-pivot point since in the static-equilibrium state the moment about the pivot point must be zero, as Figure 7.23 illustrates.

Figure 7.24 further illustrates the unique rotor dynamic characteristics of the TPJB with emphasis on how its rotor dynamic characteristics react to bearing static-load direction. Specifically, the fluid film of each pad can be conceptualized as a radial spring. In addition, Figure 7.24 correspondingly illustrates how the bearing load will automatically put the journal eccentricity between two pads. Referring to the TPJB model illustrated in Figure 5.7b, the Figure 5.11 nonlinear time-transient simulated unbalance journal orbital vibration displays chaotic motion without bearing preload.

Two of the author's actual industrial troubleshooting experiences are also embodied in Figure 7.24. One trouble-shooting example involved a European turbine generator manufacturer's rotor-vibration problem caused by having most of the bearing load carried by one pad. The author's corrective fix was to have the bearing inner assembly rotated by 25° in the direction of rotation, more evenly distributing the bearing load between two pads. Also, the second example shown had a likewise successful troubleshooting result on a large boiler feed pump with five-pad TPJBs as illustrated in Figure 7.24 (Adams 1981). When bearing radial-load angular-direction ranges over a wide portion of the 360° possibility, then preloading the TPJB is strongly preferred so that all the pads retain some significant loading under all operating conditions, including when the load is directed right into one of the pad pivot points, as exampled in Figure 7.25 for the three-pad TPJB,

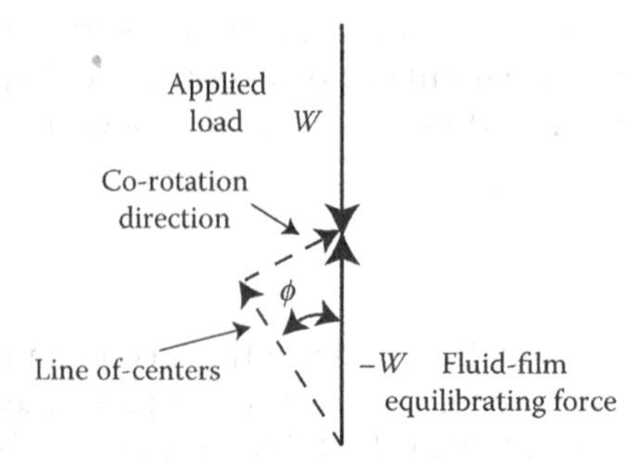

FIGURE 7.22
Cylindrical journal-bearing static–force–equilibrium vector diagram.

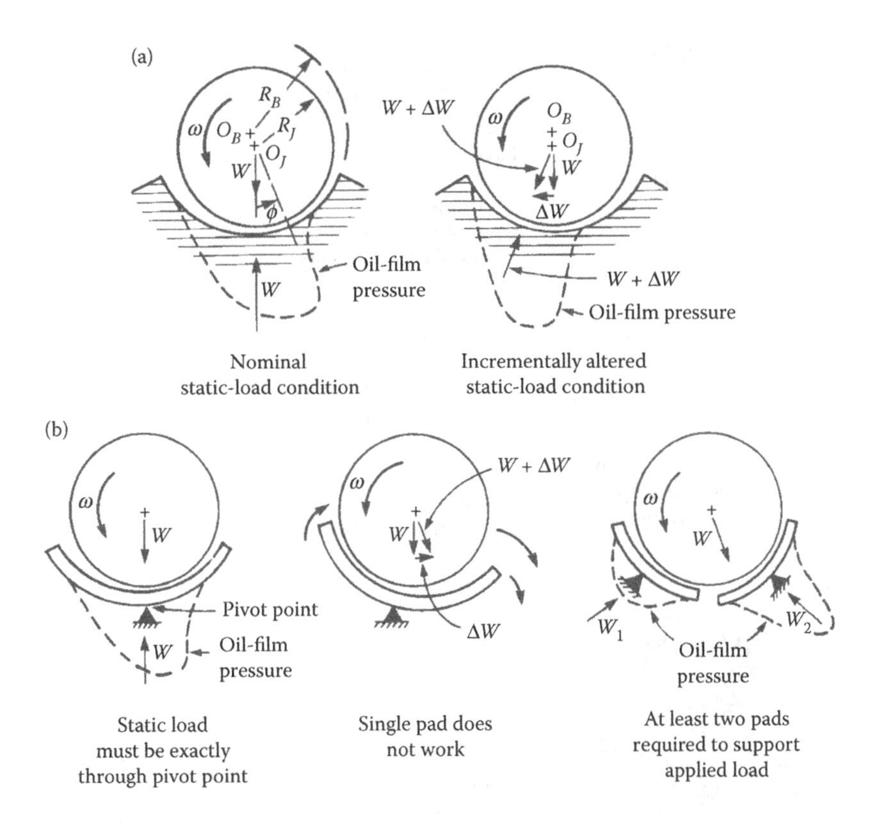

FIGURE 7.23
Comparison between (a) cylindrical and (b) tilting-pad bearings.

both for load into the pivot direction and load between pivots. Unloaded TPJB pads are vulnerable to "sprigging," a phenomenon where the unloaded pad does not have a stable static equilibrium position and consequently experiences self-excited pad flutter that can readily lead to fatigue damage to the leading edge of the pad, as detailed in the Chapter 12 troubleshooting case study.

Since three points define a circle, the three pivot points of a three-pad TPJB automatically define a circle, which is not the case with the number of pads >3. Thus, preloading the three-pad TPJB does not require precision adjustment of pivot points to define a circle if the preload is applied through a preloaded flexural component. The preload can then be held essentially constant by employing a preload flexural component with a compliant stiffness (k_{PL}) preload deflected (δ_{PL}) by an order-of-magnitude greater distance than the bearing clearance. As illustrated for the three-pad bearing in Figure 7.25, with spring preloading the applied-bearing load W can change without the preload changing. The six-pad TPJB devised and built by Turbo Research Inc. (TRI), described in the Chapter 16 troubleshooting case study, does not have its six pivot points adjusted to a single pivot circle. Adams and Payandeh (1983) showed that unloaded tilting pads can exhibit subsynchronous self-excited vibration (*pad flutter*) leading to Babbitt bearing liner fatigue damage on the leading-edge portion of the pad. It is a basic shortcoming of TPJBs that unloaded pads are likely to *incur pad flutter* if the unloaded pad(s) cannot find a stable static equilibrium position.

A unique application utilizing TPJBs is the electric canned-motor driven centrifugal pump operating in very high ambient pressure environments, like pressurized water reactor (PWR) main coolant pumps. The canned-motor pump was invented at the Westinghouse

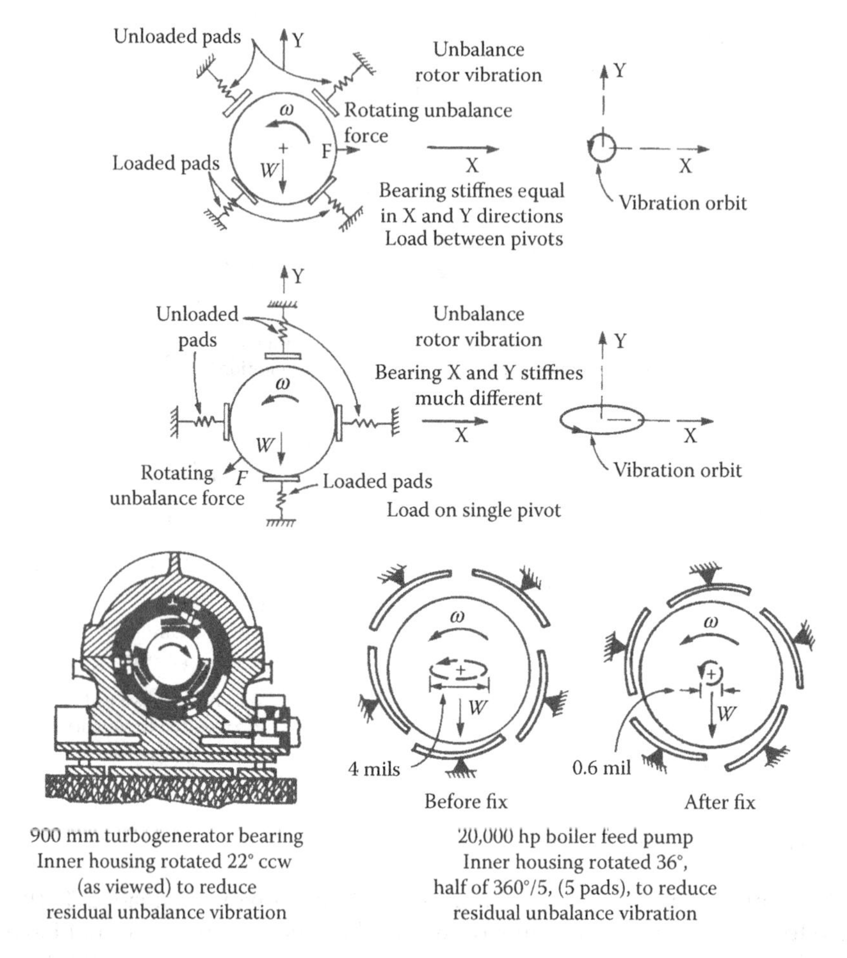

FIGURE 7.24
TPJB load-direction influence on residual rotor unbalance vibration.

Corporate Research Laboratories, Pittsburgh, circa early 1950s. A schematic of such a pump is illustrated in Figure 7.26. The major manufacturer of these pumps was originally Westinghouse EMD, now owned and operated by the Curtiss-Wright Co.

With the canned-motor approach, the entire pump rotor–bearing system is statically sealed within the PWR primary loop (pressure nominally 150 bars), the major benefit being the elimination of the need for a dynamic shaft seal. With this arrangement, what would normally be the *air gap* between motor rotor and stator is instead a *water gap* annulus filled with primary-loop water at the nominal 150 bars pressure. At that pressure, film rupture (cavitation) within the bearing films as well as within the motor can-annulus water gap does not occur. The motor water gap has a radial clearance nominally larger than 10 times the bearing radial clearances, to minimize viscous losses in the motor water gap while maintaining the needed electromechanical interaction between motor rotor and stator. Because of the approximately 10% energy efficiency loss in the can-annulus motor water gap, commercial PWR nuclear power plants transitioned to shaft-sealed PWR main coolant-pump configurations circa early 1960s, but canned-motor configurations are still employed in nuclear power ocean vessels. Since the motor is submerged in water, its stator and rotor are each manufactured with a tightly fitted hermetically sealing thin-walled

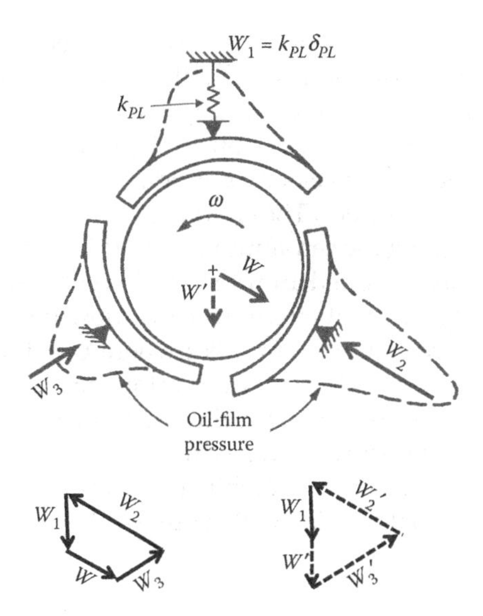

FIGURE 7.25
Schematic of preloaded three-pad TPJB demonstrating constant preload.

metal can to statically seal the motor electricals from the water. With the motor annular water-gap roughly 10 times greater than the bearing-radial-clearance magnitude, the rotational Reynolds number within the motor water-gap annulus is high enough that the gap circumferential flow regime and pressure distribution is governed by the convective fluid inertia effect, not the viscous effect that governs the hydrodynamic, albeit turbulent, bearing water films.

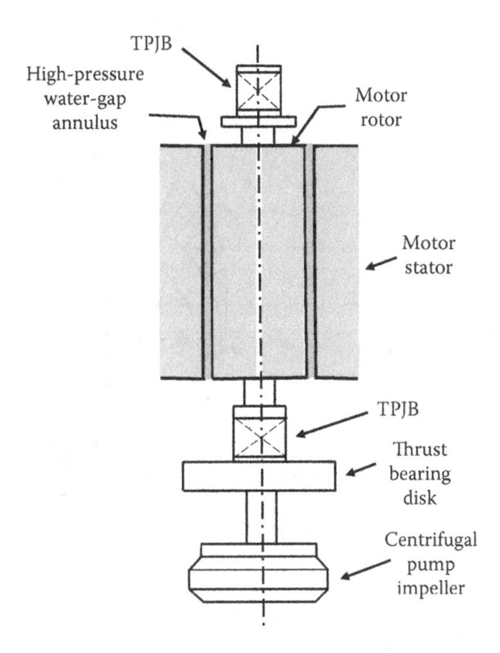

FIGURE 7.26
Canned-motor centrifugal pump rotor and motor stator.

Because the can annulus circumferential flow is governed by fluid convective inertia, its circumferential pressure distribution is virtually the opposite of that for a 360° journal bearing as illustrated in Figure 7.27c and d. Conceptually, the core of the annulus has much flatter circumferential velocity profile than Couette flow (Figure 1.11), with relatively very thin boundary layers that satisfy the velocity boundary conditions, zero on the stator-can surface, and $R\omega$ on the rotor-can surface. The circumferential pressure distribution is closely approximated by conceptualizing the can-annulus film-thickness circumferential variation as a connected nozzle-diffuser wrapped around itself in 360°. So employing the elementary Bernoulli equation assuming constant circumferential flow per unit of annulus-length by using the assumed average velocity $Q = (R\omega/2)C$; where C is the can-annulus radial gap radial clearance, R = can-annulus radius, and ω = rotor speed. As correctly surmised from Figure 7.26c and d, the can annulus acts as a quite strong negative radial stiffness, unlike a journal bearing which is like a positive radial stiffness.

Consequently, when the can annulus is combined with two TPJBs, the rotor will find the statically stable equilibrium radial position with a quite *high bearing load* with *high-eccentricity*

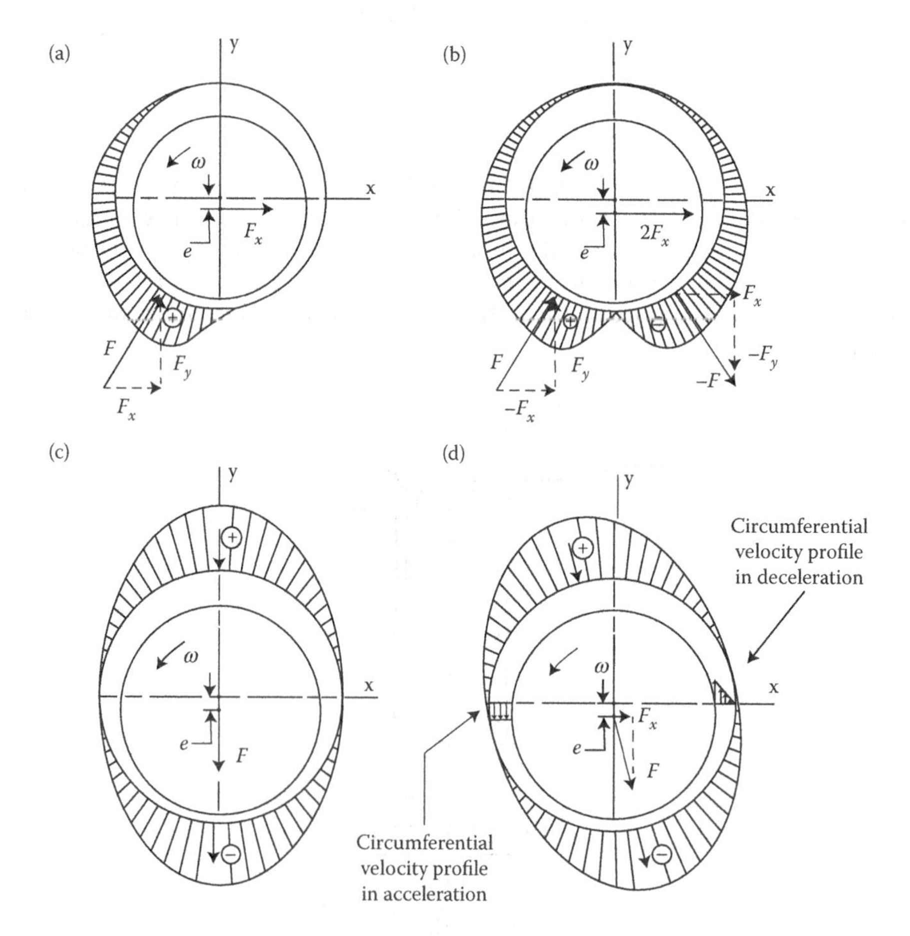

FIGURE 7.27
Circumferential pressure distributions relative to ambient; (a) 360° journal bearing operating with atmospheric ambient pressure (cavitation); (b) 360° journal bearing operating with very high ambient pressure (no cavitation); (c) and (d) high Reynolds number fluid annulus with high ambient pressure. (From Adams, M. L., *Rotating Machinery Vibration—From Analysis to Troubleshooting*, 2nd ed., CRC Press/Taylor & Francis, Boca Raton, FL, 2010.)

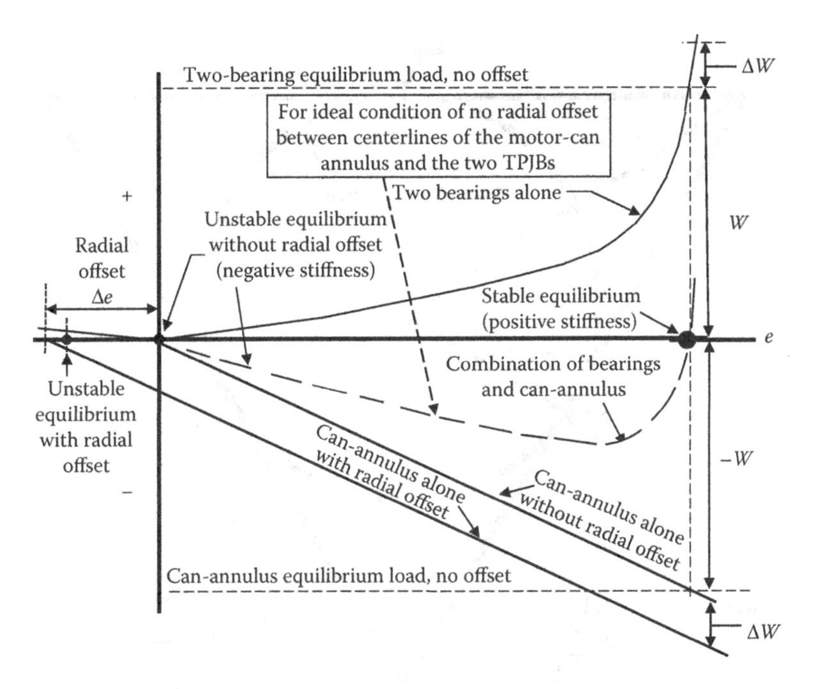

FIGURE 7.28
Canned-motor pump-bearing load and eccentricity at equilibrium.

directly between two bearing pads of the two aligned TPJBs, as illustrated in Figure 7.28. Clearly, the rotor will not equilibrium locate where the net-radial stiffness is negative. Of course, due to manufacturing tolerances, the can-annulus centerline will not be exactly coincident with the axis line joining the two TPJB center points. Figure 7.28 illustrates the journal-bearing-can-annulus radial force equilibrium with and without centerline-radial offset. Clearly, with the offset the TPJB load will be correspondingly larger by the shown ΔW than the load W for the ideal case with no radial offset between the can annulus and the bearing centerlines.

In practice, the computation for static equilibrium state must be iterative because the bearing force–displacement characteristic is quite a stiffening nonlinearity, as illustrated in Figure 7.28, and because the can-annulus is like a third bearing, making the bearing and balancing can-annulus loads statically indeterminant. The can-annulus force-displacement characteristic is essentially linear within the journal-bearing clearance restriction (as illustrated) due to its radial clearance being more than 10 times that of the TPJB-radial clearance. So, while the eccentricity ratio of the TPJB-loaded two pads on each bearing are approximately $\varepsilon = 0.95$, that radial displacement yields a can-annulus eccentricity ratio of no more than 0.1. The author has determined that a can-annulus is close to linear out to an eccentricity of about $\varepsilon = 0.3$, roughly three times the bearing clearance. To be as accurate as possible in the iterative computation to determine the bearing-can static-equilibrium position and loading state, rotor static deflection under the bearing-can equilibrating forces should also be considered, which adds an outer iteration loop to the static equilibrium analysis.

Figure 7.29 shows that e_{max}, the maximum geometrically possible journal eccentricity for a three-pad TPJB is twice the pivot clearance. That is, if the journal were placed in contact with any two adjacent pads, it would be radially eccentric *twice the pivot clearance.* Given the dominating strength of the can-annulus decentering force, this is an obvious shortcoming when compared to TPJBs with the number of pads >3. The illustration in Figure 7.29 provides e_{max}/C_P for the 3-, 4-, 5-, and 6-pad TPJB, an indication of how the

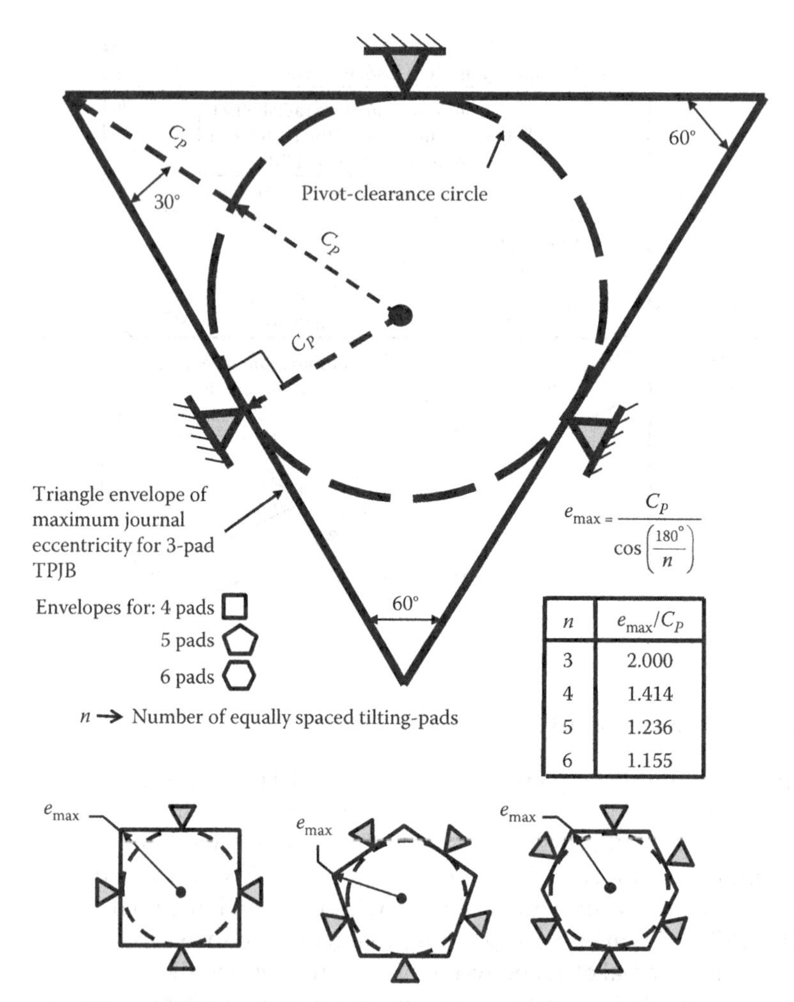

n	e_{max}/C_P
3	2.000
4	1.414
5	1.236
6	1.155

FIGURE 7.29
Influence of TPJB number-of-pads on the maximum eccentricity.

can-annulus decentering force is reduced as the TPJP pad number is increased. Because when used in conjunction with a canned motor, the TPJB e_{max}/C_P value is approximately the relative maximum expected bearing load with a specific can-annulus. For example, in simply changing from a three-pad TPJB to a four-pad configuration, the expected reduction in bearing loads is approximately $[(2 - 1.414)/2] \times 100\% = 30\%$, utilizing the tabulation in Figure 7.29. For a five-pad configuration that would be 38% load reduction and for a six-pad configuration 42% load reduction.

Figures 7.30–7.32 show a quite novel three-pad inside-out TPJB configuration, developed by Adams and Laurich (2005) and detailed in Adams (2017b), for next-generation centerless grinders. The new bearing configuration is an inside-out version of the conventional three-pad TPJB. Two of the three pivoting pads employ fixed-support pivot points while the pivot point of the third pad is supported by a hydraulically actuated radial-position preloading

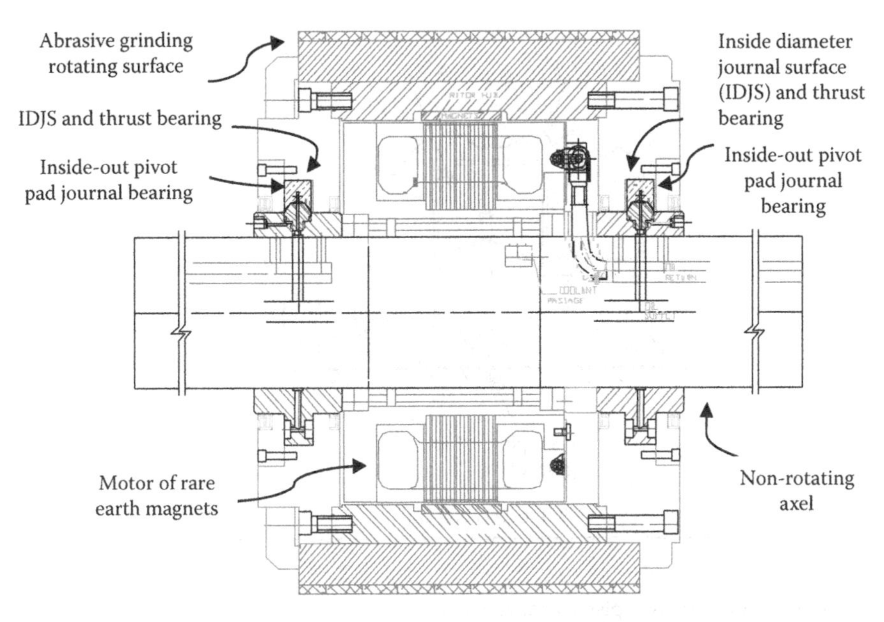

FIGURE 7.30
Centerless grinder spindle with internal 50 hp motor.

(a) Journal rotation

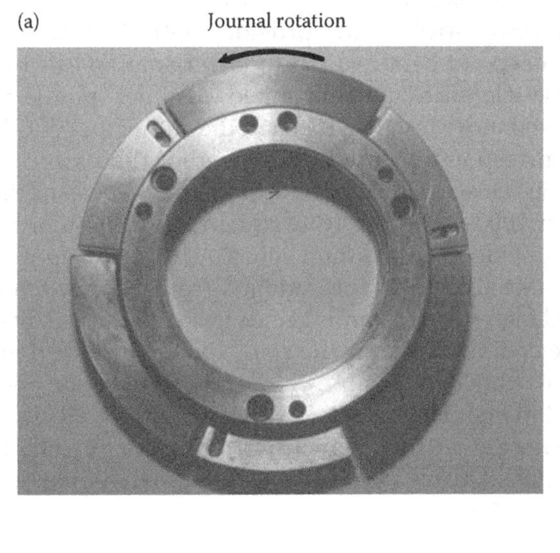

(b)

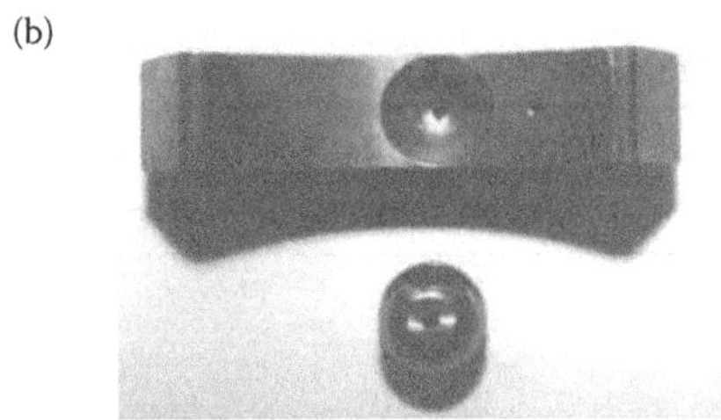

FIGURE 7.31
Bearing prototype tested; (a) complete, (b) single pad.

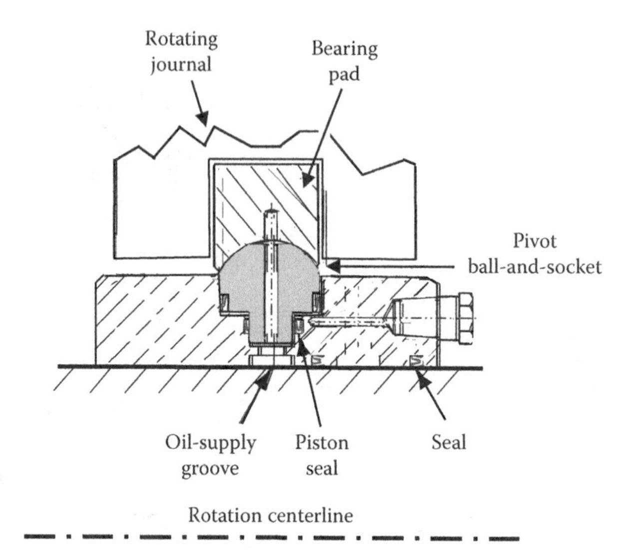

FIGURE 7.32
Preloading-pad hydraulic actuation piston arrangement.

piston. This radial-pad actuation provides a real-time controllable preload to the other two pads and thereby controls in real-time the overall radial stiffness of the bearing. This spindle-bearing stiffness control feature provides optimization of a grinding cut as to whether it's an early deep cut better suited to lower spindle-bearing stiffness or a light shallow finishing close-tolerance cut better suited to higher spindle-bearing stiffness.

Extensive-bearing laboratory test data taken in the author's CWRU laboratory compared favorably with the corresponding predicted theoretical-bearing performance (Adams 2017b). The comparisons between measured and predicted-bearing redial stiffness results shown in Table 7.2 for 2000–3000 rpm were quite reassuring that the new unique TPJB configuration could provide the maximum spindle stiffness values up to 1 million lb/in. (175 million N/m) predicted for 7000 rpm, and judged necessary to fully exploit the then-recently demonstrated high-speed ceramic grinding compliance (Kovach and Malkin 1998).

The most extensive compendium for the static-load and rotor-dynamic characteristics of all the commonly used hydrodynamic fluid-film journal bearings, including TPJBs, is given by Someya et al. (1988). That text of 323 pages contains 460 figures and 115 tables. It also contains extensive comparisons between computationally obtained journal-bearing performance charts and extensive experimental results from industrial test rigs. That text

TABLE 7.2
Predicted and Measured Bearing Stiffness, 2000 to 3000 rpm

		Stiffness		
Speed (RPM)	**Actuator Force (lb)**	**Measured (lb/in.)**	**Predicted (lb/in.)**	**Error (%)**
2000	233.3	223,077	277,792	19.7
2250	207.4	242,857	230,059	5.6
2500	219.7	275,610	239,888	14.9
2750	213.3	303,846	305,781	0.6
3000	281.9	329,167	321,592	2.4

was developed as a major project of the Japan Society of Mechanical Engineers (JSME) through a concerted team effort of several Japanese industry and academia-bearing specialists, from June 1979 through May 1982. The CD accompanying this book contains a compendium of journal-bearing tabulated data from Someya et al., as well as from other therein cited sources.

7.1.6 Journal-Bearing Design Computation Examples

So that journal-bearing design information charts are readily applicable to any combination of bearing size and the other dimensional variables, the charts are of course developed as fully dimensionless (or pseudo dimensionless like the Figure 7.16 *load factor* outdated units-combination).

The first example here utilizes the Raimondi and Boyd (RaB) design charts, Figures 7.5–7.9. This example is a typical one of many to be found in any of the many undergraduate *machine design texts* of which all have replicated the RaB charts. *But an important point of clarification, not revealed in any of the undergraduate machine design books, needs first to be explained as follows.*

Although the RaB charts do not explicitly display a lubricant-supply axial groove, the 2D finite-difference Reynolds lubrication equation (RLE) as model needs two axial pressure boundary conditions (BC) because the RLE is an *elliptic partial differential equation,* thus needing BC values (pressure) specified around the entire unwrapped rectangular boundary of the film domain. For the RaB chart computer computations, an ambient-pressure axial-groove-like BC was placed at the angular position of the *maximum film thickness,* that is, 180° from the minimum film-thickness position. The way the development of the RaB charts handled this constitutes a *flaw* since in an actual 360° journal bearing, the position of the oil supply axial groove is fixed in place, so it naturally does not circumferentially relocate as changing-load magnitude also changes the attitude angle ϕ, that is, position of the maximum film thickness. An alternative later devised modeling approach is to employ the so-called *joined-boundary condition* that forces the pressure and pressure-gradient to have the same axial distributions at both circumferential ends of the 2D finite-difference grid. However, that approach is also flawed since an actual 360° journal bearing has to have an oil-feed groove. But for most applications where the bearing-load direction is primarily close around one direction, like rotor weight, this flaw in the RaB charts is but a minor inconsistency. But it is clearly shown by Figure 7.16 as exampled in Figure 7.18, there is a dramatic-load-capacity downside if the applied-load direction places the minimum film thickness at an oil groove.

Proceeding with design computation examples, a necessary ingredient is given in Figure 7.33, a viscosity-temperature chart for a range of commonly employed oils. It clearly shows, as does Figure 1.1c, the strong influence of temperature on viscosity.

Cylindrical 360° Journal Bearing Computation

Bearing load, $W = 1000$ lb (4448 N)

Diameter, $D = 3$ in. (7.2 cm)

Length, $L = 3$ in. (7.2 cm), so $L/D = 1$ and $P = 1000/(3 \times 3) = 111$ psi

Radial clearance, $C = 0.006$ in. (0.114 mm)

Speed, 1800 rpm = 30 rps

Lubricant, SAE 30 W, $\gamma =$ Weight density of lubricant = 0.03 lb/cu-in.

Inlet oil temperature, $t_i = 90$°F (32°C)

$c_p =$ Specific heat of lubricant = 0.4 Btu/lb-deg F

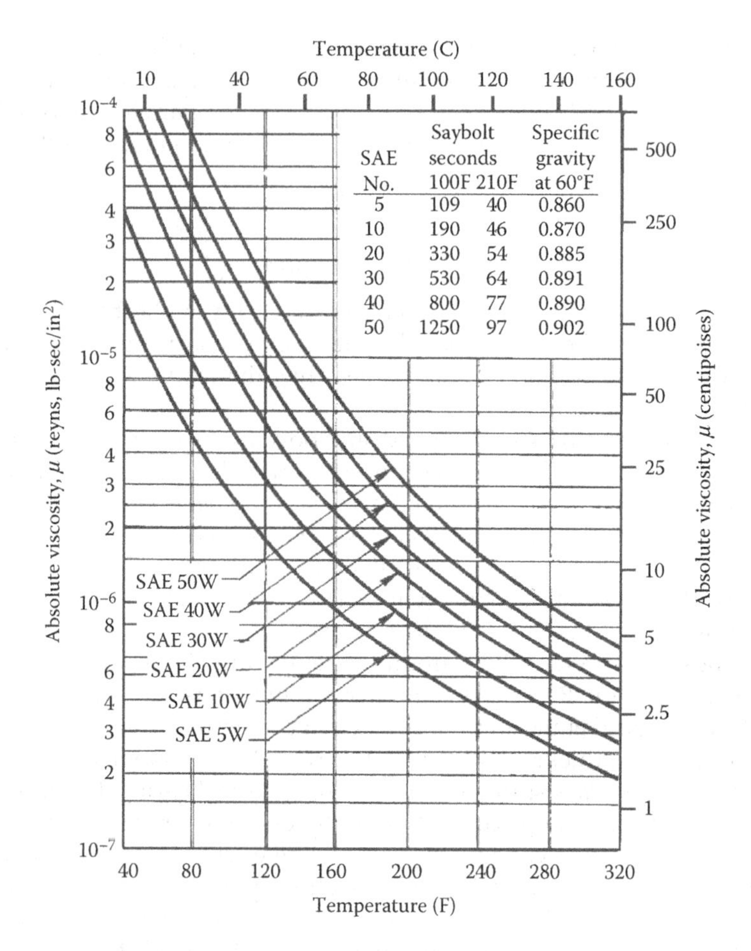

FIGURE 7.33
Oil-viscosity chart.

Many machine design texts show sample problems where the average film temperature t_{avg} is assumed a priori, but that is not indicative if an actual design computation procedure because that circumvents the fact that there is an elementary heat-balance phenomenon involved. That is, the amount of energy dissipated in the oil film must be equal to the amount of heat carried away from the bearing to achieve a steady state t_{avg} film temperature. The computation for that involves an iteration as this sample problem demonstrates. One additional usually neglected fact is that the oil-film temperature is not uniformly at t_{avg}, but that assumption becomes significant primarily on large higher surface-speed journal bearings such as on large 3600 rpm steam turbine generator sets (Figure 7.2). Proper analysis of such large bearings requires coupling the RLE to the energy equation over the finite-difference-analysis 2D grid, and also accounting for heat conducted away by the bearing structure. This sample problem assumes that both the (1) film variable-viscosity affect and the (2) the bearing heat-away conduction affect are negligible. That is, all carried-away heat is by the lubricant leaving the oil film. Uniform film temperature-viscosity assumption is implicit not only in the (1) RaB design charts for the 360° bearing, but also for the (2) high-load high-eccentricity charts (Figures 7.10–7.13), as well as the two-axial-groove bearing charts (Figure 7.16).

Before the important design calculation for journal eccentricity and minimum film thickness, the operating temperature of the oil film must first be determined so that the operating oil

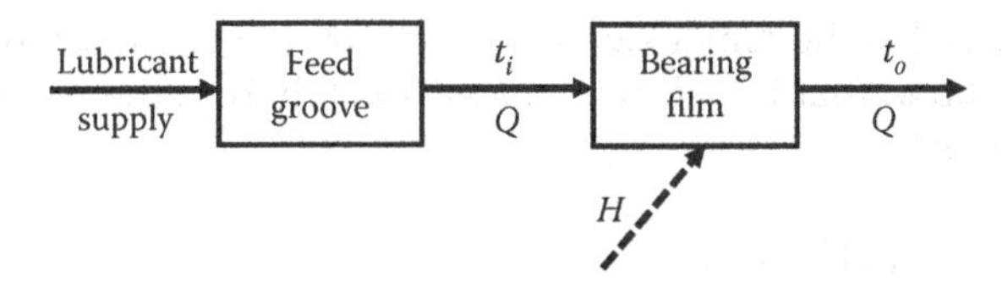

FIGURE 7.34
Simple heat-balance schematic for hydrodynamic-bearing film.

viscosity is determined and thus the operating Sommerfeld number determined. So, the design analysis starts with the heat balance iterative process (Figure 7.34) as demonstrated in the following five-step iteration.

The RaB journal-bearing design charts, a pioneering development in their time (1958), are notorious for their odd-ball and difficult-to read scales. Therefore, contained on the CD accompanying this book is a folder with Figures 7.5–7.9 in PDF files. That way the charts can be enlarged on the PC monitor with vertical and horizontal construction lines superimposed when picking off values. The Figure 7.33 oil-viscosity-temperature chart is likewise contained for user convenience in the same folder of the CD accompanying this book.

Step 1: Estimate value for discharge drain out temperature $t_o = 120°F$, to compute first iteration for $t_{avg} = (t_i + t_o)/2 = (90° + 120°)/2 = 105°$. From Figure 7.33 for 30 W oil at 110°F, viscosity $\mu = 7.775 \times 10^{-6}$ reyns (lb sec/in^2).

Step 2: Compute Sommerfeld number S.

$$S = \left(\frac{R}{C}\right)^2 \frac{\mu N}{P} = \left(\frac{1.5}{0.006}\right)^2 \left(7.75 \times 10^{-6} \times \frac{30}{111}\right) = 0.131$$

Step 3: Compute viscous power loss; Figure 7.6 for $S = 0.131$, and $L/D = 1, f(R/C) = 4.4$. Then $f = 4.4(0.006/1.5) = 0.0176$:

$$H = 2\pi fWRN = 2\pi \times 0.0176 \times 1000 \times 1.5 \times 30 = 5033 \text{ in lb/sec}$$

Step 4: Compute bearing film flow. From Figure 7.8 for S = 0.131 and $L/D = 1$, Q/(RCNL) = 4.3: Q = 4.3 × 1.5 × 0.006 × 30 × 3 = 3.48 cu-in/sec.

Step 5: Compute new value for t_{avg} based on Steps 1 through 4 and a simple heat balance: $t_o = t_i + H/(Q\gamma c_p J) = 90 + 5033/(3.48 \times 0.03 \times 0.4 \times 9{,}336) = 102.9°F$ Repeating Steps 1 through 5 twice more starting with $t_o = 102.9°F$, yields convergence to about $t_{avg} = 93°$ to $95°F$, $\mu = 10.8 \times 10^{-6}$ reyns (lb sec/in^2) giving $S = 0.182$, which from Figure 7.5 yields $e/C = 0.5$ to give minimum film thickness 0.003 in. and $\phi = 58°$. This is too lightly loaded and thus could potentially lead to oil-whip self-excited vibration. It would be advisable to reduce the bearing width, say to 1.5 in., $L/D = 0.5$ and repeat the complete design analysis here exampled to achieve e/C near 0.7, interpolating between the L/D family of curves if necessary.

The next design analysis example pertains to a *high-load journal-bearing design* using charts in Figures 7.10–7.13. For user convenience, these design charts are likewise also contained on the CD accompanying this book. The high-load journal-bearing design analysis example here closely parallels the previous example utilizing the RaB charts. The only basic difference is that the dimensionless charts are plotted as a function of the inverse of a Sommerfeld number, that is, a *dimensionless load*, whereas the Sommerfeld number is like a *dimensionless speed*.

Cylindrical high-load ($e/C > 0.9$)
Bearing load, $W = 3000$ lb (13,344 N)
Diameter, $D = 1$ in. (2.4 cm)
Length, $L = 1$ in. (2.4 cm), so $L/D = 1$ and $P = 3000/(1\times1) = 3000$ psi
Radial clearance, $C = 0.001$ in. (0.0024 mm)
Speed, 1800 rpm = 30 rps Misalignment $T = 0$
Lubricant, SAE 20 W, $\gamma =$ Weight density of lubricant = 0.03 lb/cu-in.
Inlet oil temperature, $t_i = 120°F$ (49°C)
$c_p =$ Specific heat of lubricant = 0.4 Btu/lb-deg F

Step 1: Estimate value for film average temperature $t_{avg} = 200°F$. From the figure for 20 W oil at 200°F, $\mu = 1.17\times10^{-6}$ reyns (lb-sec/in^2)

Step 2: Compute the dimensionless load.

$$\bar{W} = \frac{W(C/R)^2}{(\mu NDL)} = \frac{3000(0.001/0.5)^2}{(1.17\times10^{-6}\times30\times1\times1)} = 342$$

Step 3: Compute the viscous power loss. From Figure 7.12 for $\bar{W} = 342$, $f(R/C) = 0.37$, yielding $f = 0.37(0.001/0.5) = 0.00074$.

$$H = 2\pi fWRN = 2\pi\times0.00074\times3000\times0.5\times30 = 209 \text{ in lb/sec}$$

Step 4: Compute bearing film flow. From Figure 7.13 for $\bar{W} = 342$, $\bar{Q} = 1.35$,

$$Q = \bar{Q}RCNL = 1.35\times0.5\times0.001\times30\times1 = 0.0202 \text{ in}^3/\text{sec}$$

Step 5: Compute new value for average film temperature t_{avg} based on Steps 1 through 4, $t_o = t_i + H/(Q\gamma c_p J) = 120 + 209/(0.0202\times0.03\times0.4\times9336) = 212°F$ and a simple heat balance. Repeat Steps 1 through 5 until computed film temperature value equals assumed value, which gives 210°F with giving $\mu = 1.013\times10^{-6}$ reyns (lb-sec/in^2) and $\bar{W} = 387$ lb.

Step 6: Compute the minimum film thickness from Figure 7.10 for $\bar{W} = 387$ and misalignment $T = 0$ postulated on the use of a self-aligning feature such as illustrated in Figure 7.14, $h/C = 0.017$ ($e/C = 1 - 0.017 = 0.983$), so $h = 17\ \mu$ in.

Step 7: Using Figure 7.11 with $h/C = 0.017$ and the smallest misalignment curve $T = 0.1$ gives $\alpha = 0.0001$ rad where $\bar{M} \equiv MC^2/(\mu NRL^4)$ yields the following computation. $M = 7.0\times1.035\times10^{-6}\times30\times0.5\times(1)^4/(0.001)^3 = 109$ lb-in. The alignment stiffness is thus computed as $k = 109/0.0001 = 1{,}090{,}000$ lb-in./rad = 19,000 lb-in./deg

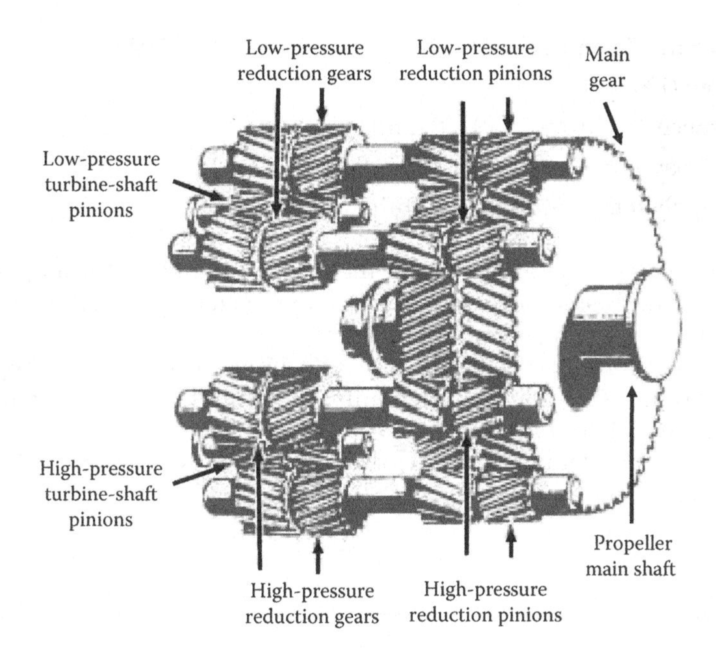

FIGURE 7.35
Large ocean ship turbine-to-propeller transmission driveline.

The calculated minimum film thickness $h = 17\ \mu$ in. is substantially smaller than what will actually occur because of the very-high-film pressure peak (approximately 25,000 psi) that such a small minimum film thickness would produce. The bearing will locally deform a minute amount in such a way as to reduce the peak pressure, that is, spread the load more evenly across the bearing, Similarly, the predicted oil-film pressure distribution aligning-reaction/stiffness to the selected misalignment is astoundingly large, and would most likely be considerably lessened in the actual bearing due to the aforementioned slight localized deformation of the bearing. The aforementioned sliding-vane-pump bearing-failure problem was completely eliminated by the fix illustrated in Figure 7.14.

The next design analysis example is for the *two-axial-groove journal bearing* that supports the turbine pinion shafts in an ocean ship propulsion transmission like that illustrated in Figure 7.35, taken from an actual application ocean ship propulsion transmission (Adams and Rippel 1969). Such transmissions provide a two-stage speed reduction from high-pressure and low-pressure drive turbine speeds to propeller shaft speed. Two journal bearings (one on either side of each pinion gear) equally share the pinion loads. The gear-load magnitude varies with N^2 and has a constant 22.5° pressure angle. The dead weight load (350 lb) is of course constant in direction and magnitude. At full forward speed, 6000 rpm, the gear-load-per-bearing is 7500 lb. The maximum reverse speed is 3000 rpm with a gear load of 5250 lb.

Equally Spaced two-Axial-Groove Cylindrical

Bearing load, $W = 7350$ lb (4448 N)

Load direction, $\theta = 140°$, Figure 7.16b, worst-case scenario

Diameter, $D = 6$ in. (14.4 cm)

Length, $L = 6$ in. (7.2 cm), so $L/D = 1$ and $P = 7350/(6{\times}6) = 204$ psi

$N = 6000$ rpm (100 rps)

Radial clearance, $C = 0.007$ in. (0.168 mm)

Lubricant: 53 Saybolt seconds at 210°F, SAE 20 W

Lubricant supply temperature. $t_i = 120$°F

Employing the same steps as shown for the prior 360° and high-load journal-bearing design analysis examples, successive iterations on predicted average film temperature t_{avg} converged with the following iteration summary.

Iterations number for t_{avg}: (1) 170°F, (2) 141°F, (3) 150°F, (4) 149.3°F, yielding a predicted eccentricity ratio $e/C = 0.76$, and predicted minimum film thickness 0.0016 in. (0.038 mm), by utilizing Figure 7.16b for $L/D = 1$.

Design analysis computations for TPJB follow similar steps as exampled for all the prior fixed-arc configurations.

7.2 Thrust Bearings

The term *thrust bearing* is synonymous with its two functions: (1) to carry the cumulative sources of rotor *axial loads* (e.g., from gears, turbomachinery stages, electric-motor magnetics, rotor dead weight on vertical rotors) and (2) to hold the rotor in its designed *axial position* with respect to the non-rotating portion of the machine. The simplest and most basic form of a hydrodynamic bearing is the fixed-pad slider bearing, illustrated in Figure 1.12 with its fluid-film pressure distribution and theoretical foundation. Like all hydrodynamic fluid film bearings, the slider-bearing functions by lubricant being drawn into a converging wedge-shaped space, producing fluid pressure that counteracts the applied bearing load while preventing contact between the parts in relative sliding motion. Virtually any converging film-thickness-distribution shape can so produce a load-supporting fluid film to separate sliding surfaces. For the thrust bearing employing fixed pads (as opposed to tilting pads), configurations having a uniform converging slope starting from the inlet is illustrated in Figure 7.36.

A practical option long employed is to have the sloped portion terminate in a flattened area which better accommodates the sliding-bearing surfaces when the rotor is not rotating, and the two mating surfaces are in contact. Additionally, that option also enhances the fluid-film *load capacity* in operation since the minimum film thickness is not at a near-ambient boundary-condition outlet pressure, thus avoiding the same less-than-optimum situation as illustrated for the two-axial-groove journal bearing in Figure 7.18 based on the load-capacity field maps in Figure 7.16. For that fixed-pad load-capacity enhancer, the *optimum value for* λ illustrated in Figure 7.36 is approximately 0.15. For a given minimum film thickness h_2, the load-capacity optimum is $h_1/h_2 \cong 2$. The fixed-pad slider bearing will of course enjoy $h_1/h_2 \cong 2$ only near one value of load, otherwise smaller than 2 for lighter loads and larger than 2 for heavier loads. That fact helped give rise to the tilting-pad slider-bearing invention since during overload variations it will automatically seek the same h_1/h_2 optimum value to maintain the same film pressure distribution shape to place the centroid of the pressure distribution force directed through the pad pivot point to maintain zero-moment static equilibrium of the tilting pad (see Figure 7.43).

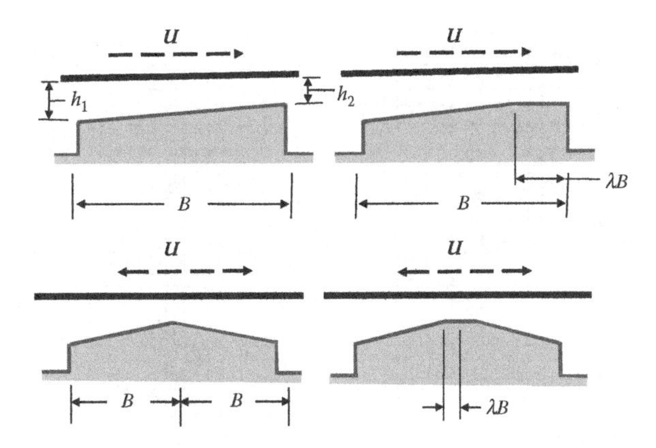

FIGURE 7.36
Typical slider-bearing fixed-pad configurations.

7.2.1 Fixed Pad

A thrust-bearing utilizing fixed slider-bearing pad is illustrated in Figure 7.37 and pictured in Figure 1.12a. Design charts by Raimondi et al. (1968) for the single fixed-pad slider bearing are duplicated in Figures 7.38–7.42. They also were generated using rectangular finite-difference-grid numerical solutions, simplifying the individual fixed pad to be a rectangular area as defined in Figure 1.14.

Application of these design charts is virtually the same as the design computation examples presented in Section 7.1.6 for journal bearings. The following is an example.

Load-per-pad $W = 3000$ lb, $U = 1200$ in/sec, $\mu = 6\times10^{-6}$ reyns, $B = L = 3$ in.,

$$\frac{(h_1 - h_2)}{B} = 0.001$$

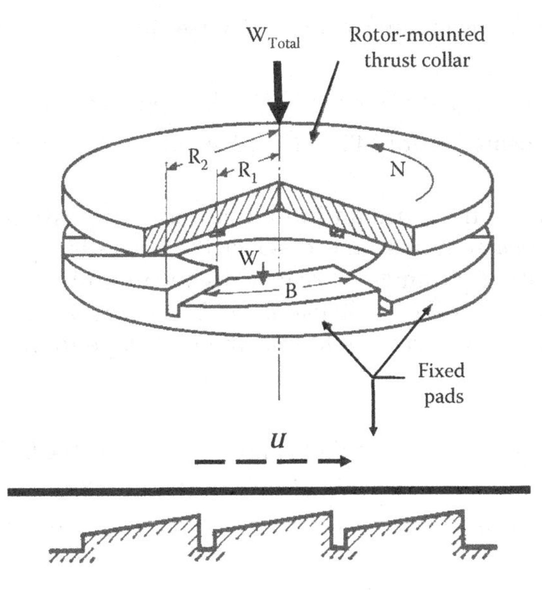

FIGURE 7.37
Thrust-bearing utilizing fixed-slider pads.

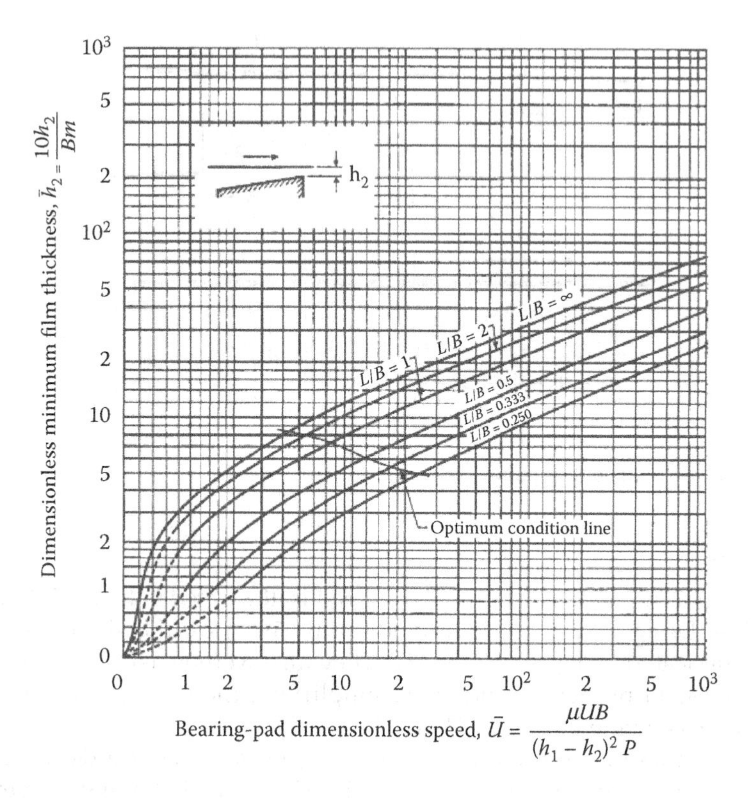

FIGURE 7.38
Chart for determining minimum film thickness.

1. *Minimum film thickness*: Using Figure 7.38 yields $\bar{h}_2 = 6.0$ and $h_2 = 0.002$ in.
2. *Friction power loss*: Using Figure 7.39 yields $\bar{f} = 6.4$ and $f = 0.0064$ yielding power loss $H = 4.19$ hp.
3. *Lubricant flow*: Using Figure 7.40, $\bar{Q} = 0.65$ yielding $Q = 7.02$ cu-in./sec.
4. *Temperature rise*: using Figure 7.42, $\Delta\bar{t} = 12$ yielding $\Delta t = 42.8$°F.

In designing a bearing there are an infinite number of possibilities, so determining an *optimum configuration* requires one to select the operating parameter, or weighted combination of operating parameters, one seeks to minimize or maximize. Here are the fixed parameters. $P = W/BL = 450$ psi, $U = 1200$ in/sec, $\mu = 3\times10^{-6}$ reyns, $B = 3$ in., L = 6 in. (so $L/B = 2$). The objective is to find the pad slope to maximize the minimum film thickness h_2.

1. *Minimum film-thickness parameter*: From Figure 7.38 where the $L/B = 2$ line intersects the "optimum condition line" yields, $\bar{U} = 5.5$ and $\bar{h}_2 = 8.1$
2. *Slope* and *maximum minimum film thickness* are then determined as follows

$$(h_2)_{\max} = 0.0017 \text{ in. and slope } \frac{h_1 - h_2}{B} = \sqrt{\frac{\mu U}{PB\bar{U}}} = 0.0007.$$

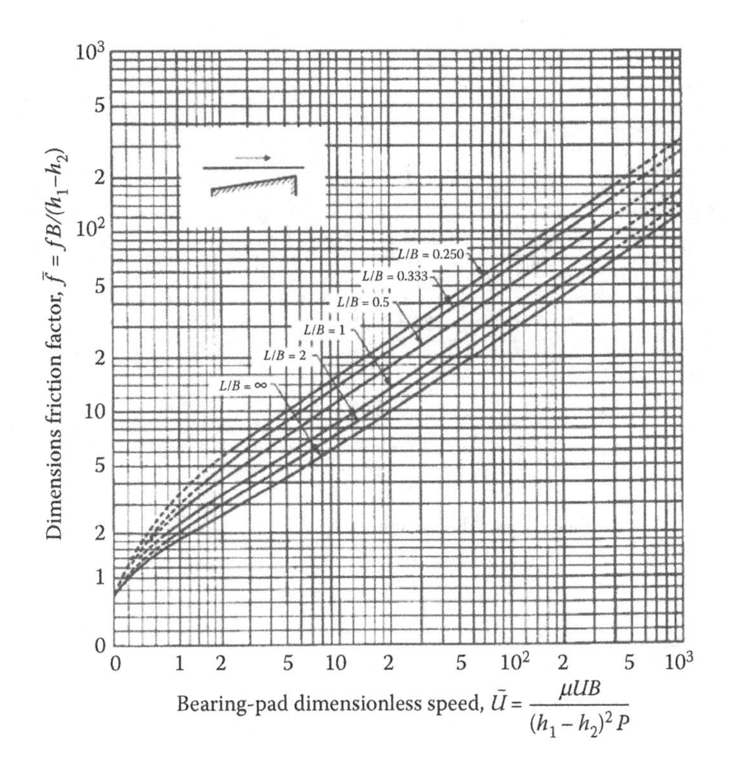

FIGURE 7.39
Chart for determining friction coefficient.

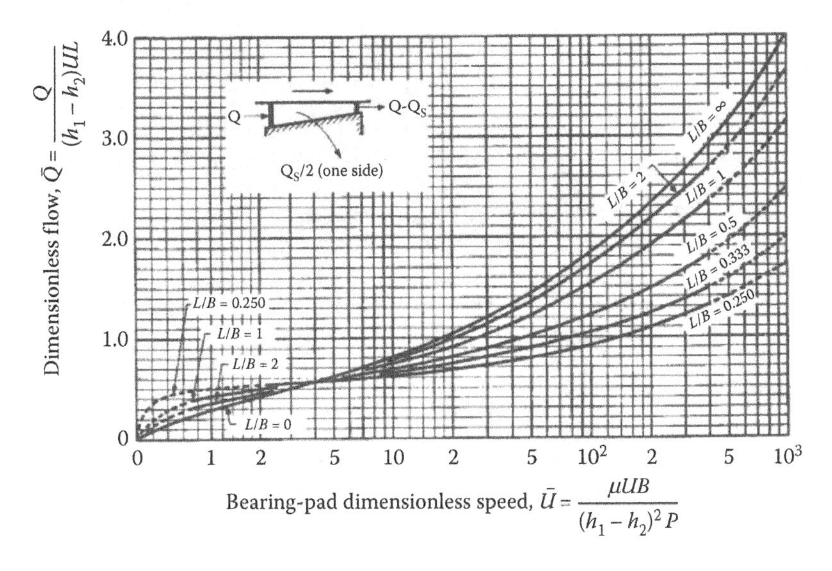

FIGURE 7.40
Chart for determining lubricant flow.

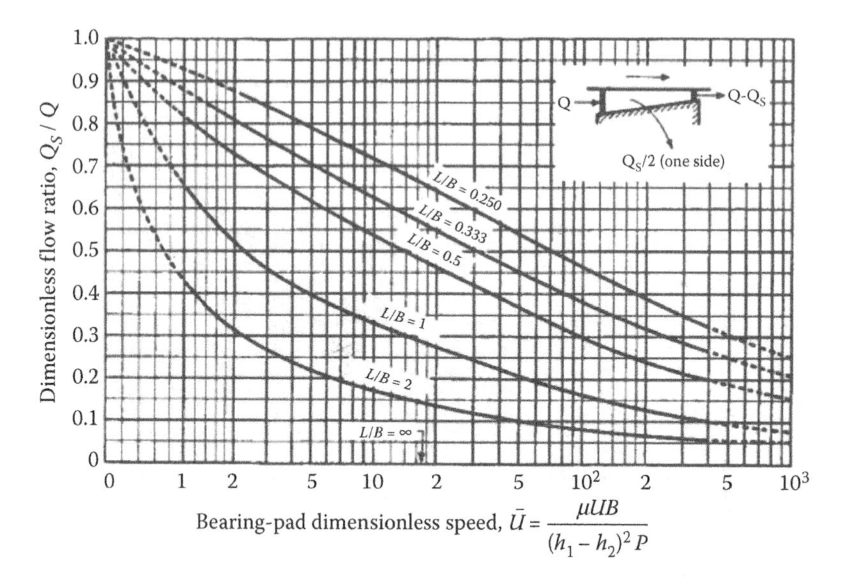

FIGURE 7.41
Chart for determining side-flow-out pad.

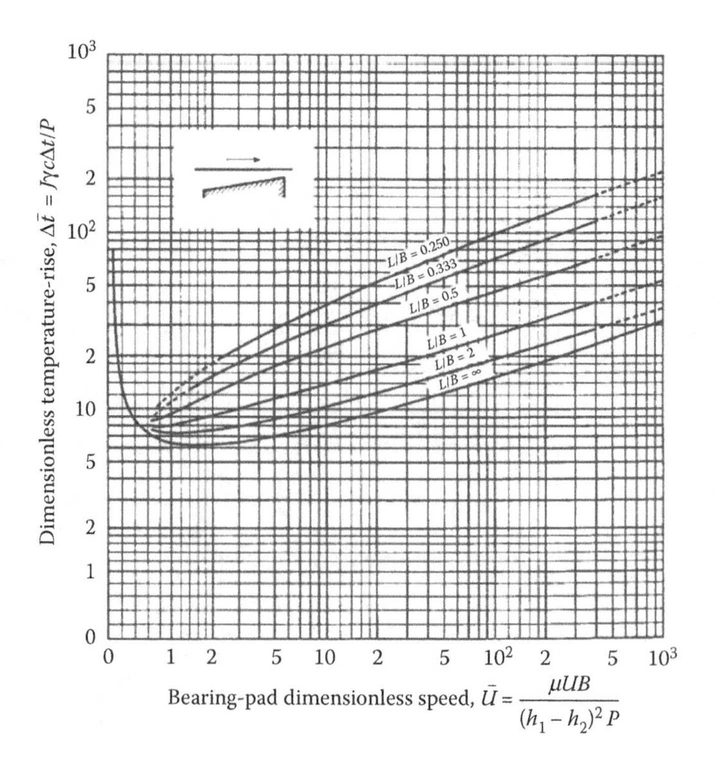

FIGURE 7.42
Chart for determining lubricant temperature rise from inlet to outlet.

7.2.2 Tilting Pad

Tilting-pad bearings, often referred to as pivoted-pad bearings, made their early appearance in the 1950s to optimize the load-carrying capacity of hydrodynamic thrust bearings. That is reported in the early pioneering development work by Abramowitz (1955, 1956) and Raimondi (1960). Prominent machinery types quick to employ tilting-pad thrust bearings (TPTB) were (1) vertical canned-motor PWR nuclear reactor coolant pumps for submarines and early-generation commercial PWR nuclear power plants (see Figures 7.26–7.29), and (2) large power plant steam turbine generator sets (see Figure 7.2). Subsequently, TPTBs are now extensively employed on many other types of rotating machinery. A TPTP is illustrated in Figure 7.43.

Application of these design charts is virtually the same as the design computation example presented in Section 7.2.1 for the fixed-pad bearing.

Step 1: Select optimum $\bar{x}/B$ value from Figure 7.44 for L/B value.

Step 2: Use Figures 7.45–7.48, iterating to converge to film temperature.

Step 3: Solve for minimum film thickness-bearing flow and power dissipation.

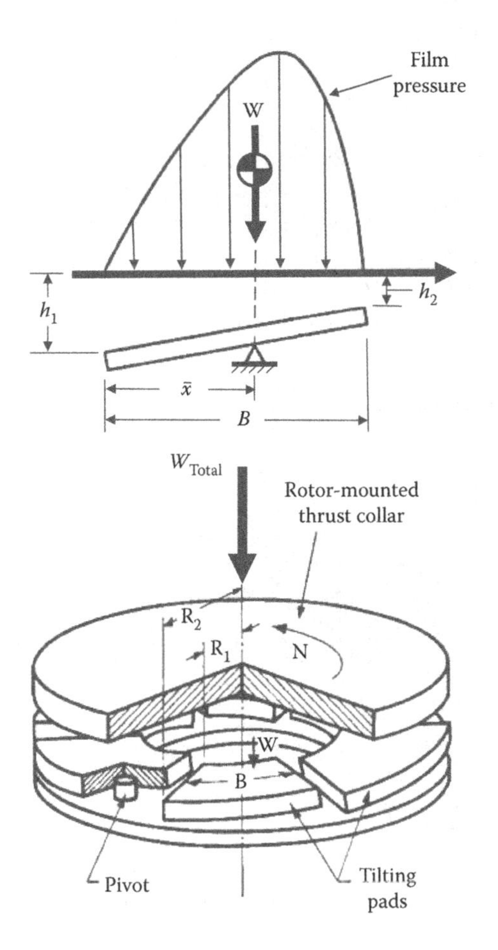

FIGURE 7.43
Thrust-bearing utilizing tilting pads.

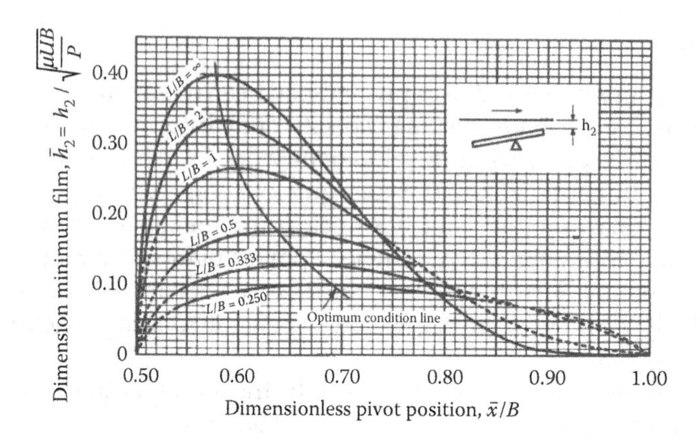

FIGURE 7.44
Chart for determining minimum film thickness.

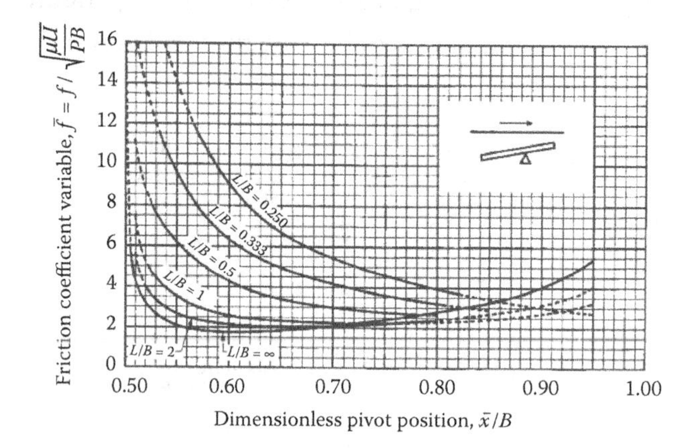

FIGURE 7.45
Chart for determining friction coefficient.

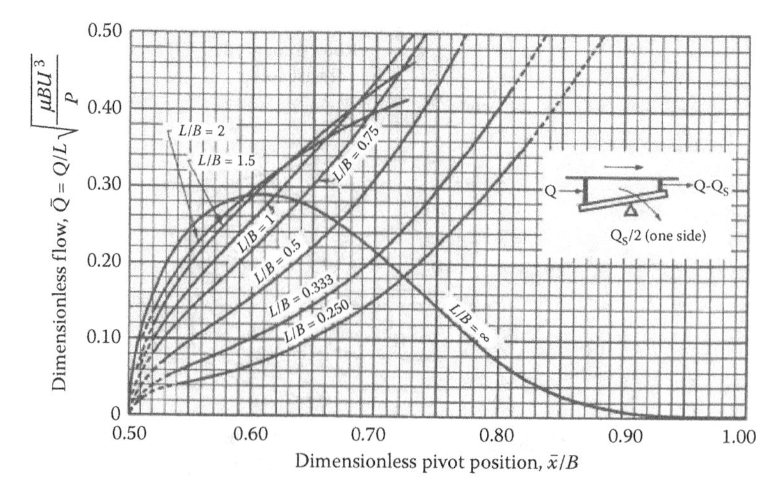

FIGURE 7.46
Chart for determining lubricant flow.

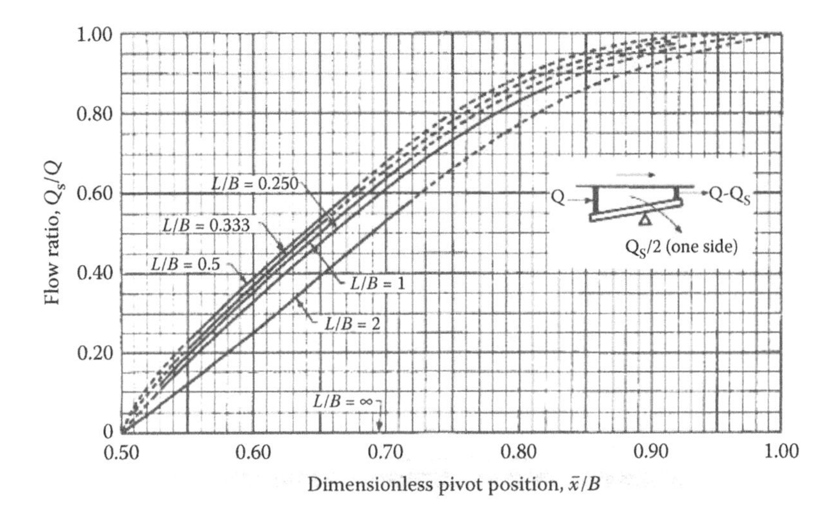

FIGURE 7.47
Chart for determining side-flow out-of-pad.

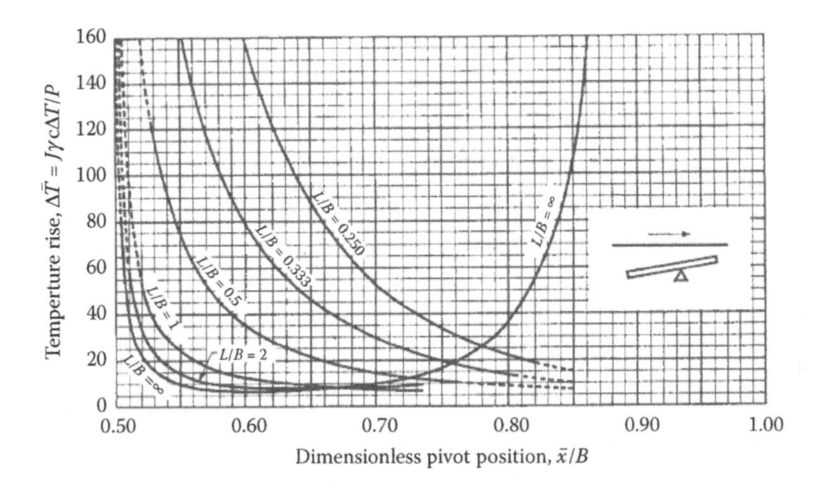

FIGURE 7.48
Chart for determining lubricant temperature rise from inlet to outlet.

The Figure 7.44 design chart indicates that the pivot point must be located downstream of the pad midpoint, that is, $\bar{x}/B > 0.5$, based on a uniform viscosity distribution within the lubricant film. Because in that case with a centrally located pivot, $\bar{x}/B = 0.5$, any pressure distribution developed for an initial $h_1 > h_2$ would have its momentary center of force (see Figure 7.43) downstream of the pivot point and thus not permitting a static equilibrium-of-moment about the pivot point. Furthermore, in some applications flat tilting pads in TPTBs have exhibited difficulty on machine startup from zero rotational speed because there is no automatic converging startup film thickness like there is for the journal touching the journal bearing at startup (see Figure 1.1b). There are two options often employed to remove this potential startup deficiency of tilting flat pads: (1) use of hydrostatic lifts, or (2) a crowned-pad surface, Figure 7.49. In applications where rotation can be in either direction, a centrally located pad pivot, $\bar{x}/B = 0.5$, must be used and then similar provision must be employed

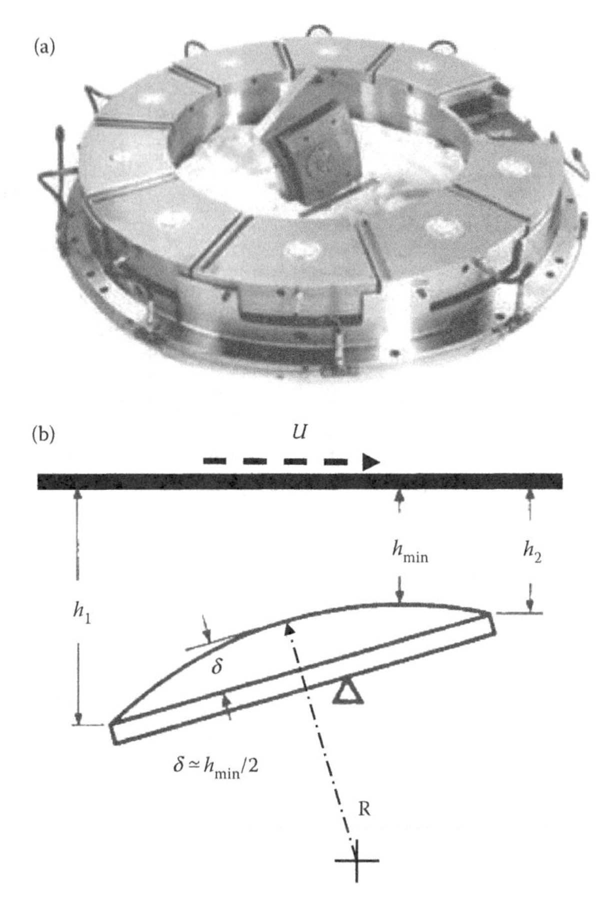

FIGURE 7.49
(a) Tilting-pad thrust bearing with leading edge oil feed grooves and hydrostatic oil-lift pockets, (b) convex crowned pad.

like crowned pads. The use of pad crowning must be carefully prescribed, with the crown height δ approximately half the minimum film thickness (Raimondi et al. 1968). A crowning affect may also result from differential thermal expansion of a tilting pad from the oil-film surface being at a higher temperature than the back side of the pad.

Summary

A portion of the material in this chapter has been in the published literature for decades starting in the 1950s, but by now possibly unfamiliar to the new generation of machine builder engineers. As mentioned early in this chapter, there is a major practical difference between fluid-film bearings (FFB) and rolling element bearings (REB). That is, FFBs (1) can either be engineered and built in-house by the machine builder or (2) simply purchased from an FFB supplier. Whereas, REBs must be purchased from a REB supplier. Over the years, the trend with machine builders using FFBs has gravitated more to the second option, purchasing

FFBs rather than designing and constructing them in-house. This chapter provides a consolidated broad presentation of FFB design information for machine builders who may decide to utilize the option of *designing and building their own* FFBs instead of purchasing them externally. At the same time, even for machine builders who prefer purchasing their FFBs, the design information in this chapter will equip those machine builders with a solid background in FFB design, to sharpen their efforts when purchasing FFBs.

References

Abramowitz, S., Theory for a Slider Bearing with a Convex Pad Surface; Slide Flow Neglected, *Journal of the Franklin Institute*, Vol. 259(3), p. 99. 221–233, March 1955.

Abramowitz, S., Turbulence in a Tilting Pad Thrust Bearing, *Trans. ASME*, Vol. 78, pp. 7–11, 1956.

Adams, M. L., *Axial-Groove Journal Bearings*, Machine Design Magazine, Penton Publishing Co, Cleveland, OH, April 16, pp. 120–123, 1970.

Adams, M. L., *Designing High-Load Journal Bearings*, Machine Design, Penton Publishing, Cleveland, OH, January 7, pp. 100–104, 1971.

Adams, M. L., How to Apply Pivoted-Pad Journal Bearings, *Power Magazine*, McGraw-Hill, New York, October 1981.

Adams, M. L., Insights Into Linearized Rotor Dynamics, Part 2, *Journal of Sound & Vibration*, 112(1), pp. 97–110, Academic Press, London, 1987.

Adams, M. L., *Rotating Machinery Vibration—From Analysis to Troubleshooting*, 2nd ed., CRC Press/Taylor & Francis, Boca Raton, FL, 2010.

Adams, M. L., *Power Plant Centrifugal Pumps—Problem Analysis and Troubleshooting*, CRC Press/Taylor & Francis, Boca Raton, FL, 2017a

Adams, M. L., *Rotating Machinery Research and Development Test Rigs*, CRC Press/Taylor & Francis, Boca Raton, FL, 2017b.

Adams, M. L. and Laurich, M. A., Design, Analysis and Testing of an Inside-Out Pivoted-Pad Journal Bearing with Real-Time Controllable Preload Stiffness, *Proceedings, ASME World Tribology Conference*, Washington, DC, September 2005.

Adams, M. L. and Payandeh, S., Self-Excited Vibration of Statically Unloaded Pads in Tilting-Pad Journal Bearings, *ASME Journal of Lubrication Technology*, Vol. 105, pp. 377–384, 1983.

Adams, M. L. and Rippel, H. C., *Design Manual for 2-Axial Groove Hydrodynamic Journal Bearings*, Final Report F-C2491, Franklin Institute Research Laboratories, Philadelphia, October 1969.

Castelli, V. and Shapiro, W., Improved Method for Numerical Solutions of the General Incompressible Fluid Film Lubrication Problem, *ASME Journal of Lubrication Technology*, Vol. 89(2), pp. 211–218, 1967.

Clavis, A., Lapini, G. and Rossini, T., Diagnosis in Operation of Bearing Misalignments in Turbogenerators, ASME Paper No. 77-DET-14, *Design Engineering Technical Conference*, Chicago, IL, September 1977.

Kovach, J. A. and Malkin, S. (1998), *High-Speed, Low-Damage Grinding of Advanced Ceramics: Phaes-2, Final Report*, Oak Ridge National Laboratory, 1998.

Pinkus, O. and Wilcock, D., *COJOUR User's Manual Guide: Dynamic Coefficients for Fluid-Film Journal Bearings, Electric Power Research Institute, EPRI Research Project 1648–1, Report CS-4093-CCM*, Mechanical Technology Inc., Latham, NY, 1985.

Raimondi, A. A., The Influence of Longitudinal and Transverse Profile on the Load-Carrying Capacity of Pivoted Pad Bearings, *Trans. American Society of Lubrication Engineers (ASLE)*, Vol. 3(2), pp. 265–276, 1960.

Raimondi, A. A. and Boyd, J., A Solution for the Finite Journal Bearing and its Application to Analysis and Design, Parts 1, 2 and 3, *Trans. ASLE*, Vol. 1(1) pp. 159–209, 1958.

Raimondi, A. A., Boyd, J. and Kaufman, H. N., Analysis and Design of Sliding Bearings, *Standard Handbook of Lubrication Engineering,* American Society of Lubrication Engineering (ASLE), Handbook Editors: O'Connor, J. J., Boyd, J. and Avallone, E. A., McGraw-Hill, New York, 1968.

Reynolds, O., *On the Theory of Lubrication and its Application to Mr. Tower's Experiments,* Philosophical Transactions of the Royal Society, London, England, p. 177, Part 1, 1886.

Someya, T., Mitsui, J., Esaki, J., Saito, S., Kanemitsu, Y., Iwatsubo, T., Tanaka, M., Hisa, T., Fujikawa, T. and Kanki, H., *JSME Journal Bearing Databook,* Springer-Verlog, Tokyo, English translation from original text in Japanese, p. 323, 1988.

8

Hydrostatic Bearings

The "bible" on hydrostatic bearings (HB) is the design manual by Rippel (1965). That manual contains a wealth of dimensionless design charts and lucid explanations on how HBs work. That is, how being supplied by only a fluid flow source at constant pressure, HBs can be designed to have a relatively thick load-supporting fluid film with high stiffness and substantial damping. Section 1.5 covers the theoretical foundation and functioning of HBs, in particular Figure 1.18 which employs a DC electric analog circuit, devised by the author for his machine design course at Case. This chapter provides major design charts and their use from Rippel's design manual. Supplementing the application examples presented in Chapter 1, presented in this chapter is a recent development on how HBs can be employed to nearly eliminate the high-decibel high-frequency acoustic noise emissions inherent with high-pitch-line gear sets.

8.1 Advantages, Operation, and Limitations

Hydrostatic bearings (HB) are applicable to nearly any bearing function, with a combination of remarkable unique properties. However, HBs are relatively expensive since a high-pressure continuous lubricant supply flow is required, necessitating (1) a closed-circuit collection of bearing-pad exiting flow, (2) filtration, and (3) cooling, to provide the delivery of clean lubricant at the proper viscosity. Of course, interruption of the lubricant supply system (see Figure 1.19) immediately results in total dysfunction of a hydrostatic bearing. In spite of the relatively expensive cost and complex supply system, HBs have advantages not found all together in any other type of bearing, as follows.

1. High-load-carrying capacity at all sliding speeds, including zero speed.
2. Zero-starting breakaway friction and extremely low running friction.
3. No contact between film-separated surfaces at any speed or load, thus life is dictated by lubricant supply system life.
4. Predictable and adjustable bearing performance with regard to displacement–load (stiffness) characteristic, frictional drag, and temperature rise.

The major advantage of HBs is their ability to precisely position very heavy loads at slow sliding with a minimum of driving force. Consequently, HBs have been successfully applied in machine tools, radio telescopes, large radar antennas, precision rotating machinery test rigs (e.g., Figure 1.21a) and other heavily loaded and slow-moving equipment.

8.2 Flat-Pad Design Coefficients

As illustrated in Figure 1.18, the elementary HB pad consists of a relatively deep recess that is surrounded by a relatively thin lubricant film. Supplementing Figure 1.18, Figure 8.1 illustrates the manner in which an HB pad provides a positive stiffness to load changes. As Figure 1.18 shows, the film is formed by the *sill* or *land* that surrounds the recess or pocket. Since the lubricant-film thickness is much thinner than the recess depth, the pressure drop from recess pressure to pad-ambient pressure essentially occurs completely across the *land*. So as Figure 1.18 illustrates, the upstream flow restrictor and the thin film surrounding the recess are two fluid-flow resistors in series. Therefore, with a fixed pressure drop across the entire HB pad, as a load increase reduces film thickness and thereby its flow resistance, the recess pressure increases, and vice versa for a load decrease.

Before presenting the formulation to quantify operating parameters such as (1) load capacity, (2) supply flow required, and (3) pad-pump power required, pad coefficient charts for load, flow, and pump-power are presented, in Figures 8.2 and 8.3. These are the most prominent charts from Rippel (1965) that contain a number of other pad coefficient charts for other pad geometries. The equations for *load capacity, flow,* and *power* employing the coefficients in Figures 8.2 and 8.3 are as follows, where A_p = pad area.

$$\text{Load capacity of pad,} \quad W = a_f A_p p_r \tag{8.1}$$

$$\text{Flow required by pad,} \quad Q = q_f \left(\frac{W}{A_p} \right) \frac{h^3}{\mu} \tag{8.2}$$

$$\text{Power, required by pad,} \quad H_B = p_r Q = H_f \left(\frac{W}{A_p} \right)^2 \frac{h^3}{\mu} \tag{8.3}$$

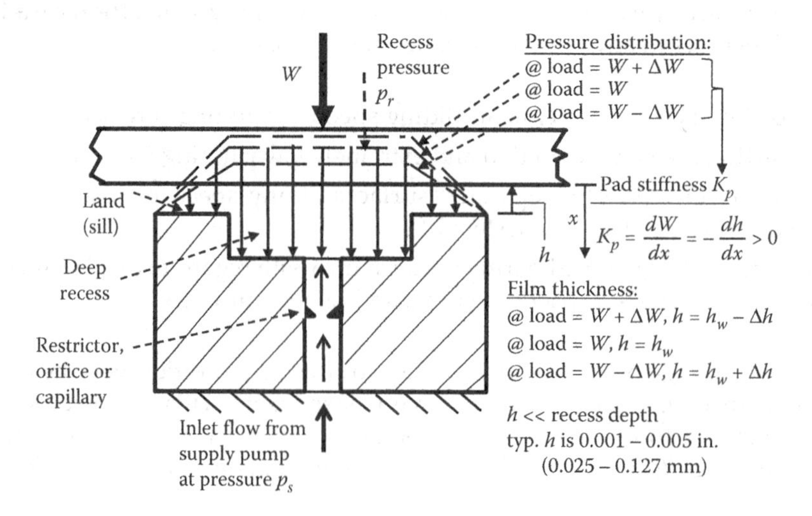

FIGURE 8.1
Hydrostatic pad-film stiffness through flow control compensation.

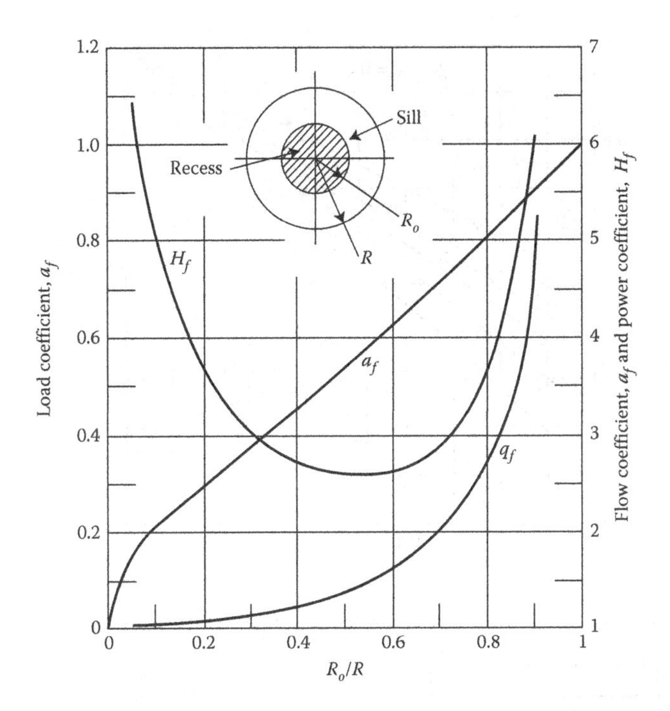

FIGURE 8.2
Concentric-recess circular-pad *load*, *flow*, and *power* coefficients.

$H_f \equiv q_f/a_f$, where it has a minimum value as a function of the ratio of recess-length-to-pad-length, that is, a best-efficiency pad proportion. All pad configurations exhibit a minimum value for H_f as a function of the ratio of *recess-length-to-pad-length*, approximately from 0.4 to 0.6.

For the concentric-recess circular pad (Figure 8.2), the load, flow, and pad pump-power coefficients are determined from a closed form solution of the Reynolds lubrication equation (RLE) for zero motion with a unity pressure in the inner radial boundary (R_o) and recess, and a zero pressure at the outer boundary (R). Without either sliding or squeeze-film velocity, the RLE reduces to the well-known Laplace equation, Equation 1.12. Since for the concentric-recess circular pad there is only one independent special coordinate (radius, r), pad coefficients can be determined from a closed-form solution of Equation 1.12, yielding the following solution equations on which the Figure 8.2 charts are based.

$$\text{Pad-load coefficient,} \quad a_f = \frac{1}{2}\left[\frac{1-(R_o/R)^2}{\log_e(R/R_o)}\right] \tag{8.4}$$

$$\text{Pad-flow coefficient,} \quad q_f = \frac{\pi}{3}\left[\frac{1}{1-(R/R_o)^2}\right] \tag{8.5}$$

$$\text{Pad-power coefficient,} \quad H_f = \frac{2\pi\log_e(R/R_o)}{3[1-(R_o/R)^2]^2} \tag{8.6}$$

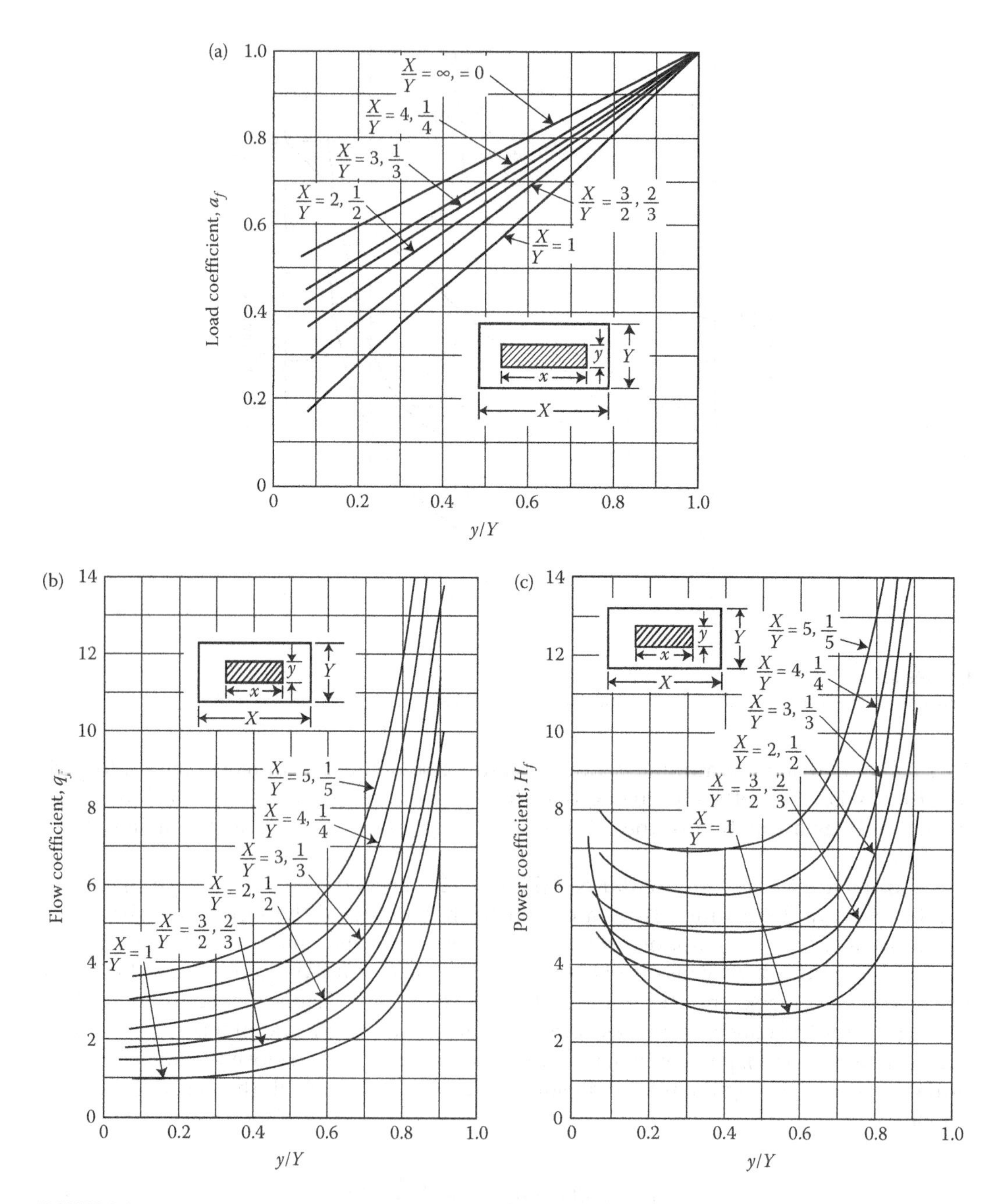

FIGURE 8.3
Rectangular-pad coefficients, (a) load, (b) flow, and (c) power.

Note that consistent with physical insight, as the pad inner radius dimension approaches the outer radius, that is, $R_o \rightarrow R$, the load coefficient $a_f \rightarrow 1$ while both the flow and power coefficients $(q_f, H_f) \rightarrow \infty$. That is, the entire pad area is at recess pressure but the resistance to flow across the concentric land is zero. Of course, $R_o \rightarrow R$ is not a realistic pad geometry because of the absence of the necessary fluid-film-flow variable resistance illustrated in Figure 1.18. The pad area for the concentric-recess circular pad is of course $A_p = \pi R^2$. For the circular pad with a concentric recess bounded by inner and outer sills (Figure 8.4) the pad coefficients are also determined from a closed form solution to Equation 1.12, as follows.

$$\text{Pad-load coefficient,} \quad a_f = \frac{1}{2(R_4^2 - R_1^2)} \left[\frac{R_4^2 - R_3^2}{\log_e R_4/R_3} - \frac{R_2^2 - R_1^2}{\log_e R_2/R_1} \right] \tag{8.7}$$

$$\text{Pad-flow coefficient,} \quad q_f = \frac{\pi}{6a_f} \left[\frac{1}{\log_e R_4/R_3} - \frac{1}{\log_e R_2/R_1} \right] \tag{8.8}$$

$$\text{Pad-power coefficient,} \quad H_f = \frac{q_f}{a_f} \tag{8.9}$$

$$\text{Pad area,} \quad A_p = \pi(R_4^2 - R_1^2) \tag{8.10}$$

For the rectangular flat pad design coefficient charts (Figure 8.3), a closed form solution of the Laplace equation does not exist, thus requiring other solution means. Rippel's charts for the rectangular pad were developed prior to the availability of software employing the finite-difference numerical approach. Rippel used electromagnetic field analog testing as explained in Chapter 1 in conjunction with Figure 1.20.

Example Use of Pad Coefficients

Pad configuration: Concentric-recess circular pad maximum diameter of 10 in, $W = 10{,}000$ lb, $h > 0.002$ in., with viscosity $\mu = 4.4 \times 10^{-6}$ reyns (lb-sec/in^2).

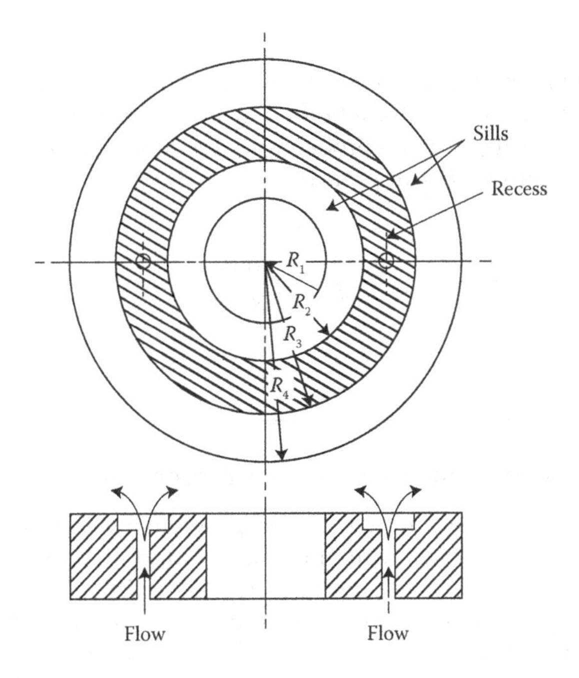

FIGURE 8.4
Circular pad with recess between two concentric sills.

Determine the (1) required recess pressure, (2) flow, (3) power, and (4) lift-off pressure. The pad coefficients are graphed in Figure 8.2 as well as given by Equations 8.1 through 8.3.

Step 1: To minimize pressure and flow requirements, hence power, the largest pad permitted ($R = 5$ in.) is selected and a relative recess size which yields a minimum value of power coefficient $H_f = 0.5$, therefore, $R_o = 2.5$ in. For this geometry pad area $A_p = \pi R^2 = 25\pi$ in^2

Step 2: From Figure 8.2 at $R_o/R = 0.5$, $a_f = 0.5$, $q_f = 1.40$, $H_f = 2.60$.

Step 3: Required recess pressure for a 10,000 lb load from Equation 8.1 is

$$p_r = \frac{W}{a_f A_p} = \frac{10,000}{0.54 \times 25\pi} = 236 \text{ psi}$$

Step 4: Required flow from Equation 8.2 for minimum film thickness 0.002 in.

$$Q = q_f \left(\frac{W}{A_p}\right)\frac{h^3}{\mu} = 1.40\left(\frac{10,000}{25\pi}\right)\frac{(0.002)^3}{4.4 \times 10^{-6}} = 0.325 \text{ in}^3/\text{sec}$$

Greater flow will result in thicker film thickness and vice versa. For example, if a constant-displacement pump of 0.1 gpm (0.385 in^3/sec) is selected, the film thickness from Equation 8.2 computes as follows.

$$h - \left(\frac{QA_p\mu}{q_f W}\right)^{1/3} - \left[\frac{0.385 \times 25\pi \times 4.4 \times 10^{-6}}{1.40 \times 10,000}\right]^{1/3} = 0.0021 \text{ in}$$

Step 5: For computation of liftoff pressure, p_{LO}, the most conservative assumption is that the load surface and pad are both so near to perfectly flat that no pressure diffuses across the sills, which is probably not the case. The following computation is based on that conservative approach.

$$W = p_{LO} A_R; \quad p_{LO} = \frac{10,000}{\pi(2.5)^2} = 510 \text{psi}$$

8.3 Flow-Control Restrictor Influence on Performance

As just shown by the sample problem in the previous section, a single hydrostatic-bearing pad with single recess can function alone simply with a positive displacement constant-flow lubricant pump supplying it. However, hydrostatic bearings are rarely if ever a single-pad, single-recess system, for example, see Figures 1.16, 1.17, and 1.21a. So, to employ the typical multi-pad system supplied by a single supply pressure source, each pad recess needs a flow control restrictor to properly function as illustrated in Figures 1.18 and 8.1. Clearly, compared to more common bearing-types, hydrostatic bearings are relatively complex systems, composed of pads, a pump, and compensating restrictors. The compensating restrictor has as much influence on the bearing capability as the other system components. The compensating element: (1) orifice, (2) capillary, or flow-control valve, must be chosen carefully. This section deals with the design methodology for pad-flow-compensating types.

The most frequently employed flow-control compensating-type elements are the (1) sharp-edged orifice and the (2) capillary tube. Flow of an incompressible fluid through a *sharp-edged orifice* is expressed as follows.

$$Q_o = C_d \frac{\pi d_o^2}{4} \left(\frac{2}{\rho} \right)^{1/2} (p_s - p_r)^{1/2} \tag{8.11}$$

where the orifice discharge coefficient is a function of the Reynolds number as $C_d = 0.2(N_R)^{1/2}$, d_o = orifice diameter, and ρ = mass density of the lubricant.

A *capillary tube* is a long tube of relatively small diameter ($l_c > 20\, d_c$). Laminar flow through the capillary tube, neglecting entrance and exit effects and viscosity changes due to temperature effects is expressed as follows.

$$Q_c = \frac{\pi d_c^4 (p_s - p_r)}{128\, l_c\, \mu} \tag{8.12}$$

where d_c = capillary tube diameter and l_c = capillary tube-length.

Compensation by a flow-control valve (Figure 8.5) is relatively expensive but offers certain advantages, with flow expressed simply as follows.

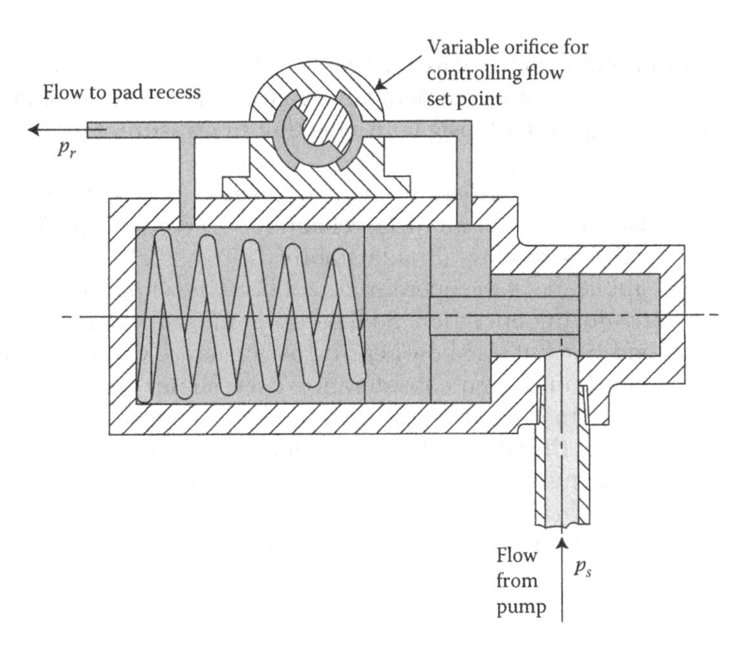

FIGURE 8.5
Adjustable pressure-compensated constant-flow control valve.

$$Q_{cv} = \text{Constant} \tag{8.13}$$

The following are example design computations for these three flow-compensating methods using the bearing design started in the previous section.

Basic system parameters are as follows. $p_r = 236$ psig, $\mu = 4.4 \times 10^{-6}$ lb-sec/in^2, Q = 0.385 in^3/sec = 0.10 gpm, and $P_L = 510$ psig. Assuming a supply pressure of 300 psig, the density of most petroleum oils is approximately 7.5 lb/gal, which converts to mass density as follows.

$$\rho = \frac{\gamma(\text{lb/gal})}{231(\text{in}^3/\text{gal}) \times 386(\text{in/sec}^2)} = 84 \times 10^{-6} \text{ lb-sec/in}^4$$

Step 1: *Sharp-edged orifice* exampled with assumed $C_d = 0.6$ and using Equation 8.11, the orifice diameter is as follows.

$$0.385 = 0.6 \times \frac{\pi d_o^2}{4} \left[\frac{2(300-236)}{84 \times 10^{-6}} \right]^{1/2} \text{ gives } d_o = 0.026 \text{ in.}$$

Step 2: For the *capillary tube* example, since the pressure difference $(p_s - p_r)$ and flow Q are known, the relationship between the capillary diameter d_c and length l_c are determined from Equation 8.12 as follows.

$$0.385 = \frac{\pi d_c^4 (300-236)}{4.4 \times 10^{-6} \times 128 l_c} \text{ gives } \quad l_c = 96.2\, d_c^4 \times 10^4$$

Step 3: For a constant-flow control valve compensator, there is no design computation needed since such valves are commercially available. The valve should be able to supply a flow of 0.1 gpm at 300 psig with a maximum pressure at liftoff of 510 psig.

The decision of which flow-compensation-type to select is based on considerations including (1) initial cost, (2) space required, (3) reliability, (4) accessibility, (5) serviceability, (6) useful life, (7) tendency to clog, (8) adjustability, (9) availability, and (10) type of application. In many applications the operating conditions, most prominently the applied load, vary significantly during operation. So the design must not only focus on the base nominal operating condition but also how bearing performance varies over the full range of operation. That important design consideration necessitates rigorous assessment of how flow control restrictors influence the bearing performance over the full range of operating conditions. From the *pressure-flow* characteristics of the compensating element, film thickness, flow, bearing power dissipation, and stiffness can all be expressed as functions of β, the *ratio of recess pressure to supply pressure*. For sharp-edged orifice or capillary compensation, from zero load to maximum load β varies from *zero to one*, respectively.

$$\beta \equiv \frac{p_r}{p_s} = \frac{W}{a_f A_p p_s} \tag{8.14}$$

By equating flow through the bearing pad to flow through the flow compensator, bearing performance can be analyzed as functions of dimensionless performance coefficients as follows (Rippel 1965). Accordingly for capillary *compensation*, performance coefficients can be expressed as follows.

Capillary Performance Coefficients:

Film thickness

$$h_c = k_h \left(\frac{k_c}{a_f q_f} \right) \tag{8.15}$$

$$k_h = \left(\frac{1-\beta}{\beta} \right)^{1/3} \tag{8.16}$$

Flow

$$Q_c = k_Q \frac{k_c}{\mu} p_s \tag{8.17}$$

$$k_Q = (1-\beta) \tag{8.18}$$

Power

$$H_{Bc} = k_H \frac{k_c}{\mu} p_s^2 \tag{8.19}$$

$$k_H = (1-\beta)\beta \tag{8.20}$$

Stiffness

$$S_c = k_s (a_f A_p p_s) \left(\frac{a_f q_f}{k_c} \right)^{1/3} \tag{8.21}$$

$$k_s = \frac{3W}{h_c}(1-\beta) \tag{8.22}$$

Malanoski and Loeb (1961)

$$S_c = \frac{3W}{h_c}(1-\beta) \tag{8.23}$$

The four performance coefficients for capillary compensation are graphed in Figure 8.6, consistent with which are the two optimum pressure ratio β values with a capillary; (1) for stiffness, $k_s = 0.866 @ \beta_{\text{opt-}S} = 0.67$ and (2) for power, $k_H = 0.26 @ \beta_{\text{opt-}H} = 0.5$. For applications

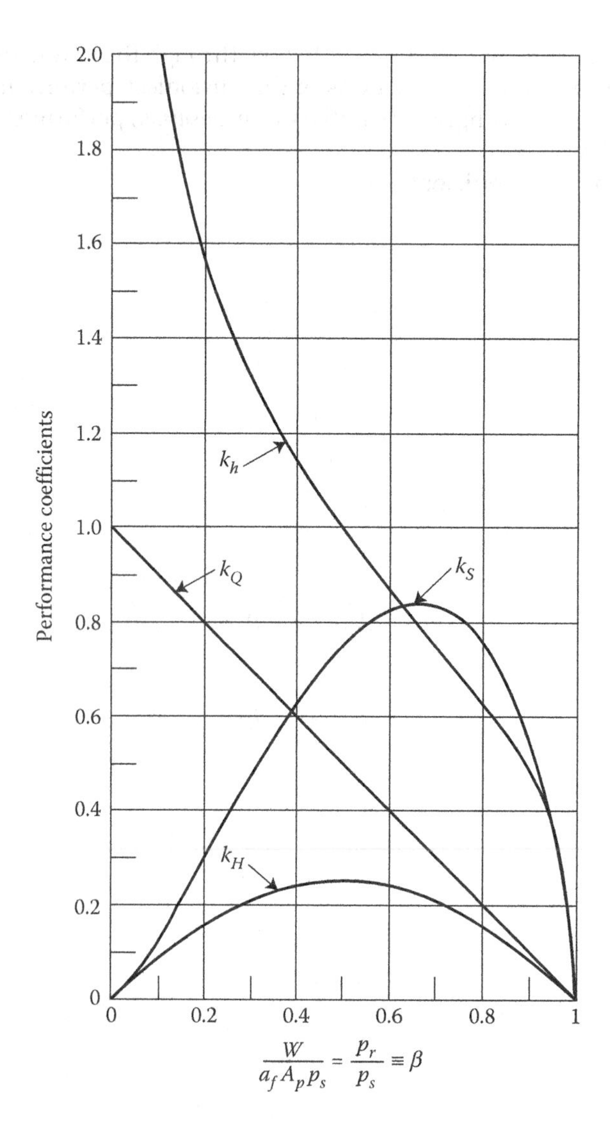

FIGURE 8.6
Performance coefficients for capillary-compensated pads.

with a significant range of operating parameter values like load, designing around an optimum value for β is obviously the preferred approach, especially regarding stiffness.

Orifice Performance Coefficients:

Film thickness

$$h_o = k_h \left(\frac{k_o \mu}{a_f q_f} \right)^{1/2} \left(\frac{1}{p_s} \right)^{1/6} \tag{8.24}$$

$$k_h = [2(1-\beta)]^{1/6} \left(\frac{1}{\beta} \right)^{1/3} \tag{8.25}$$

Flow

$$Q_o = k_Q k_o p^{1/2} \tag{8.26}$$

$$k_Q = [2(1-\beta)]^{1/2} \tag{8.27}$$

Power

$$H_{Bo} = k_H k_o p_s^{3/2} \tag{8.28}$$

$$k_H = [2(1-\beta)]^{1/2}\beta \tag{8.29}$$

Stiffness

$$S_o = k_s(a_f A_p p_s)\left(\frac{a_f q_f}{k_o \mu}\right)^{1/3}(p_s)^{1/6} \tag{8.30}$$

$$k_s = \frac{3\beta^{4/3}[2(1-\beta)]^{5/6}}{(2-\beta)} \tag{8.31}$$

Malanoski and Loeb (1961)

$$S_o = \frac{3W}{h_o}\left[\frac{2(1-\beta)}{2-\beta}\right] \tag{8.32}$$

The four performance coefficients for orifice compensation are graphed in Figure 8.7, consistent with which are the two optimum pressure ratio β values with an orifice; (1) for stiffness, $k_s = 0.94 @ \beta_{\text{opt-}S} = 0.69$ and (2) for power, $k_H = 0.54 @ \beta_{\text{opt-}H} = 0.68$. Again, for applications with a significant range of operating parameter values like load, designing around an optimum value for β is obviously the way to go, especially regarding stiffness.

Constant-Flow Valve Performance Coefficients:
Film thickness

$$h_v = k_h\left(\frac{k_o \mu}{a_f q_f}\right)^{1/2}\left(\frac{1}{p_s}\right)^{1/6} \tag{8.33}$$

$$k_h = \left(\frac{1}{\beta}\right)^{1/3} \tag{8.34}$$

Flow

$$Q_v = k_Q k_v \tag{8.35}$$

$$k_Q = 1 \tag{8.36}$$

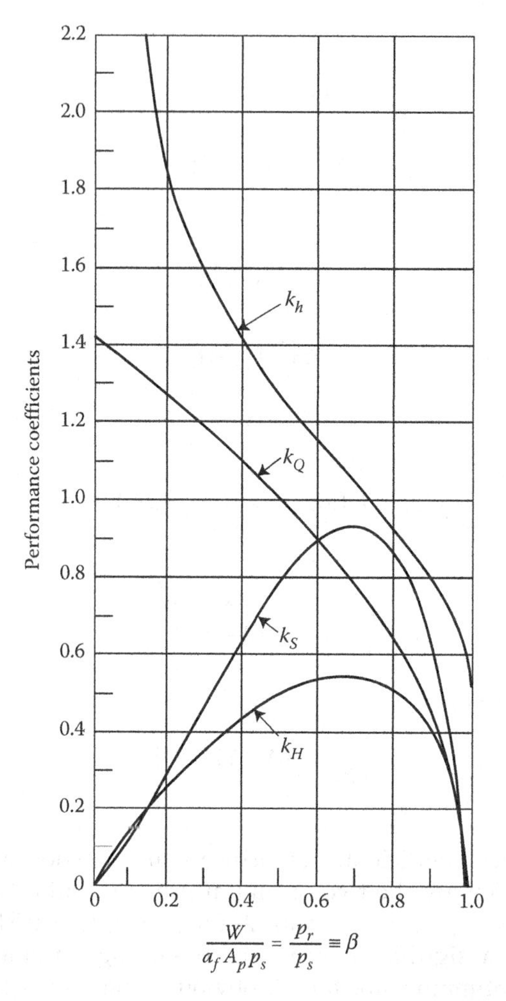

FIGURE 8.7
Performance coefficients for orifice-compensated pads.

Power

$$H_{Bv} = k_H k_v p_s \tag{8.37}$$

$$k_H = \beta \tag{8.38}$$

Stiffness

$$S_o = k_s (a_f A_p p_s) \left(\frac{a_f q_f}{k_o \mu} \right)^{1/3} (p_s)^{1/3} \tag{8.39}$$

$$k_s = 3\beta^{3/4} \tag{8.40}$$

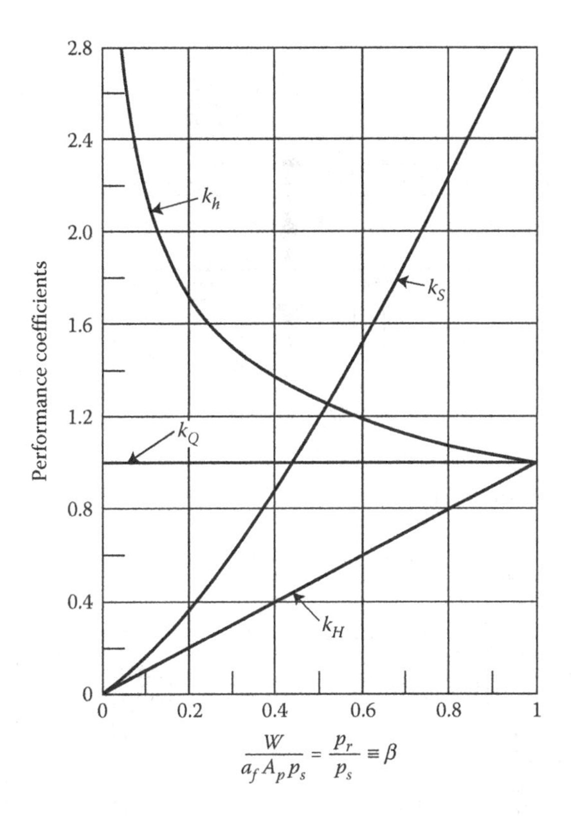

FIGURE 8.8
Performance coefficients for constant-flow-compensated pads.

Malanoski and Loeb (1961)

$$S_v = \frac{3W}{h_v} \tag{8.41}$$

The four performance coefficients for constant-flow compensation are graphed in Figure 8.8, and are significantly different in form than those in Figures 8.6 and 8.7 for capillary and orifice compensation, respectively. The stiffness coefficient continues to increase with increasing β and does not exhibit a maximum value.

8.4 Practical Flat-Pad Designs

As indicated in the previous section, hydrostatic bearings almost always consist of more than one pad. Except for the simplest cases of basic pads opposing other pads (e.g., double-acting axial thrust bearing, Figure 8.9), thorough design analyses require use of general purpose hydrostatic-bearing software employing a finite-difference numerical solution methodology that accommodates: (1) non-uniform film thickness, (2) multiple pads, and (3) multiple compensated recesses per pad, for example, Castelli and Shapiro (1967). The

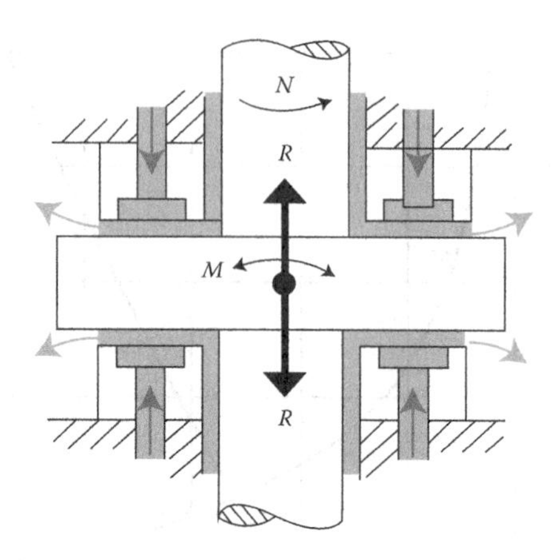

FIGURE 8.9
Multi-flat-pad double-acting thrust bearing.

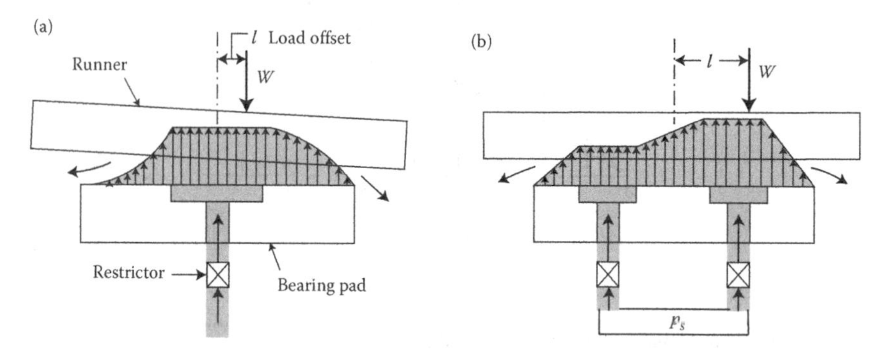

FIGURE 8.10
Effect of offset loads on flat-bearing pads produces restoring moment. (a) Single-recess pad, (b) multi-recess pad.

stiffness capacity of pads is not limited just to force, but also inherently resist moments from load offsets as illustrated in Figure 8.10. Figure 8.10a illustrates the inherent property of an applied moment to cause a non-uniform film thickness, altering the pressure distribution shapes that provide positive moment stiffness. Figure 8.10b shows how a two-recess pad can provide an even greater positive moment stiffness to handle moments from offset loading. Clearly, hydrostatic pads not only have force stiffness but also moment stiffness. Figure 8.11

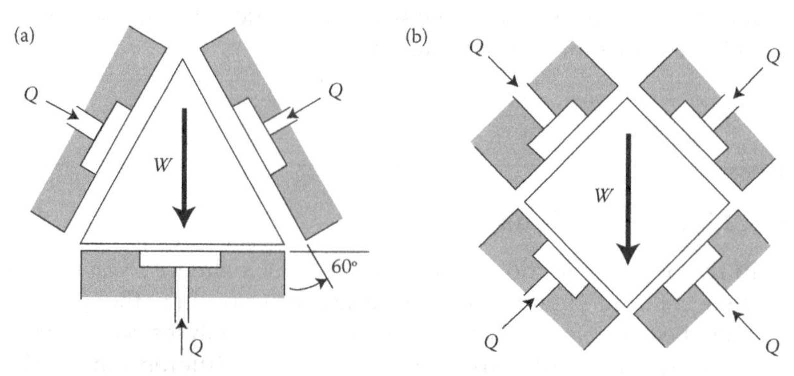

FIGURE 8.11
Planar containment of loaded component with flat pads.

illustrates two flat-pad hydrostatic-bearing arrangements for providing simultaneous *planar force and moment support* of a loaded member. An important breakthrough reported in Section 8.6 shows how the four-pad configuration in Figure 8.11b can virtually eliminate radiated *acoustic noise* inherent in high pitch-line velocity *gear sets.*

8.5 Configurations with Non-Flat Pads

Figure 8.12 illustrates common non-flat-pad configurations. While *preliminary design analysis* of such configurations can be performed using basic theory with flat-pad performance

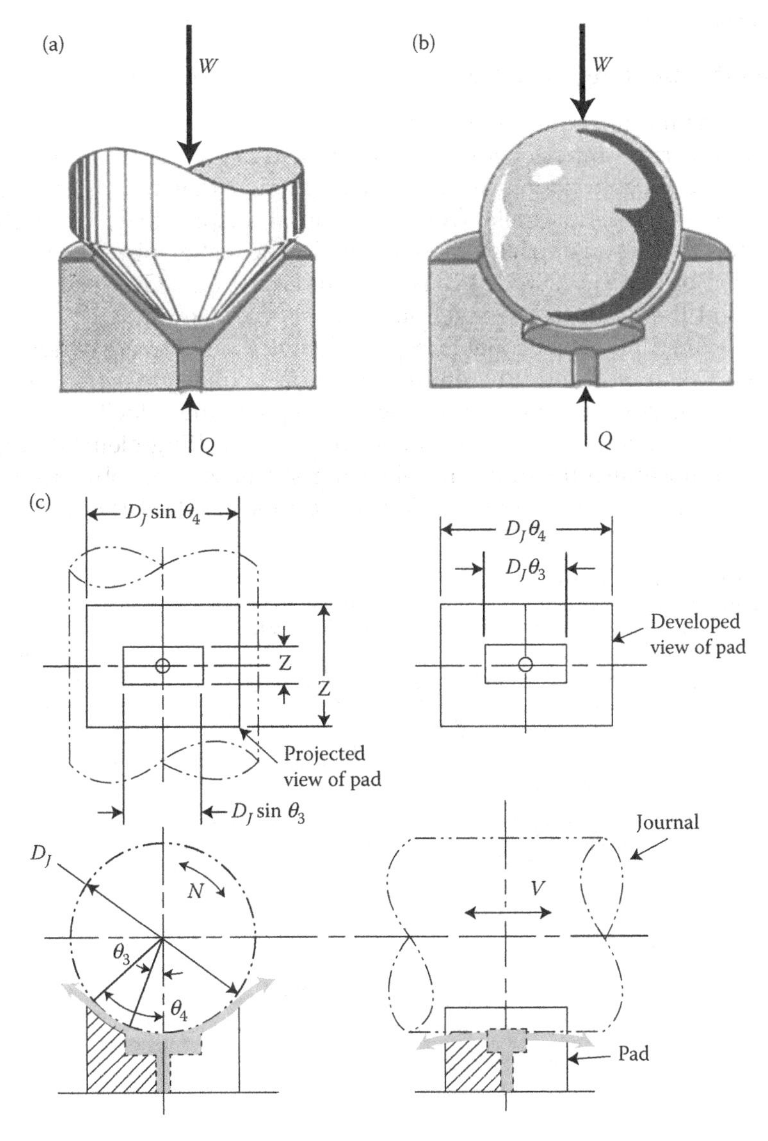

FIGURE 8.12
Hydrostatic non-flat pads, (a) conical, (b) spherical, and (c) cylindrical.

coefficients, final thorough design analyses require use of general purpose hydrostatic-bearing software employing a finite-difference numerical solution methodology that accommodates: (1) non-uniform film thickness, (2) multiple pads, and (3) multiple-compensated recesses per pad, as pointed out in the previous section for general flat-pad configurations. Figure 1.21 details the author's journal-bearing research test rig employing a multi-pad test bearing-loading system that includes extensive use of hydrostatic-bearing pads including flat pads, cylindrical pads, and one with a spherical seat like that illustrated in Figure 8.12b, to eliminate any test-bearing-to-journal misalignment as well as providing torque isolation for precise test-bearing friction torque measurement (Adams 2017).

8.6 Hydrostatic Bearings for Near Elimination of Gear Noise

The origin of gear mesh-frequency acoustic noise is well known. It is the ever-present residual manufacturing imperfections, tooth elasticity, and sliding fiction that always preclude perfect conjugate action. It is also well known that most gear-set-generated acoustic noise first passes as mesh-frequency vibration primarily through the shaft support bearings to the housing, and then radiated as acoustic noise to the environment. This is explained by Houser (2007) as illustrated in Figure 8.13. Gear-set-generated noise is concentrated at the mesh frequency, (rotational speed) × (number of teeth), which is typically quite acoustically objectionable, especially in the modern era of noise abatement. Gear teeth dynamic interaction inherent in non-perfect conjugate action produces more intense gear-mesh-generated vibration the higher the pitch line velocity.

Whether the gear shaft support bearings are of the rolling-element (typical) or a hydrodynamic fluid-film-type, quite high-bearing stiffness is an obvious requirement to maintain needed gear centerline positioning accuracy so that the level of precision

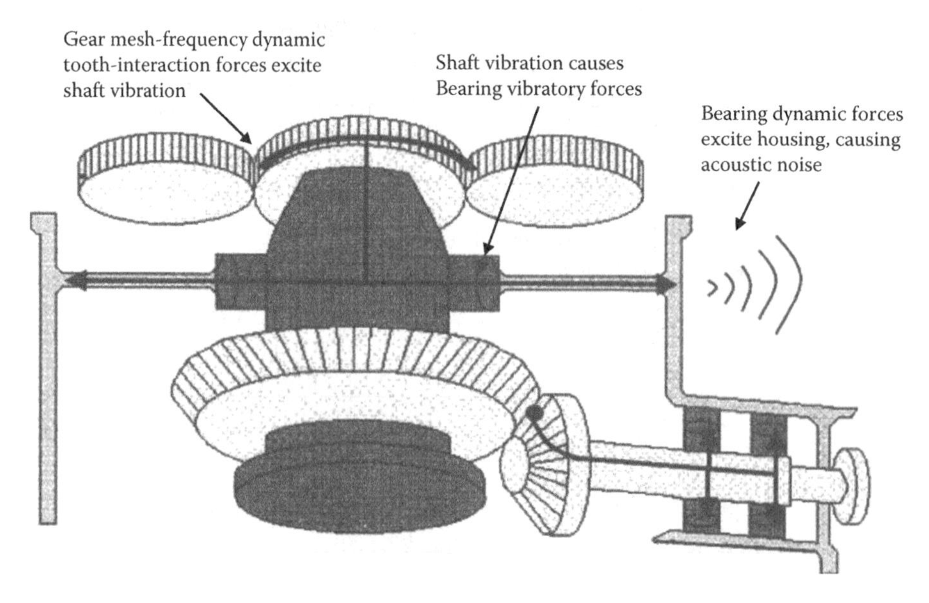

FIGURE 8.13
Gear-train schematic showing noise source and transmission path.

manufactured into the gear set is realized in its performance. So *therein is the dilemma* of attempts at gear noise attenuation measures, that is, high-bearing stiffness facilitates mesh-frequency vibration transmission to the housing. The concept presented here was developed by the author utilizing the small degree of compressibility of any liquid including lubricating oil in a hydrostatic-bearing backup in series load-wise with the primary shaft support bearings. At the same time, the hydrostatic-bearing backup is configured to ensure the very high overall static stiffness normally required for precision gear sets. However, the liquid compressibility in the hydrostatic-bearing backup will progressively increase the vibration filtering effect as vibration frequency is increased. This is clearly indicated by analysis as follows.

Employing hydrostatic bearings alone as the main bearings could accomplish the same objective, but with a potential for lower operational reliability. Because any interruption of pressurized supply flow to the hydrostatic bearings would interrupt operation of the gear set. But if incorporated as a bearing support component, interruption of pressurized supply flow would not have to interrupt operation of the gear set. In the event of a hydrostatic-bearing supply pump stoppage or other interruption of flow to the hydrostatic backups, the primary bearings continue to properly function as intended without service interruption, albeit temporarily without the noise filtration benefit. Furthermore, since hydrostatic backups do not take the rotational velocity of the shaft, their film thicknesses can be chosen relatively quite small, keeping the fluid flow requirement of the hydrostatic backups very small, thus requiring only a relatively small supply pump.

The analysis approach is to utilize the compressibility of the hydrostatic-bearing liquid, typically but not always oil, as quantified by a bulk modulus β, defined as follows.

$$\beta \equiv \frac{-\Delta p V}{\Delta V} \tag{8.42}$$

where p = pressure, V is volume. The analysis presented here is focused on understanding the performance of a single pad. The envisioned application configuration, illustrated in Figure 8.14, is based on the Figure 8.11b bearing where a main shaft rolling element bearing is backed up by a square four-pad hydrostatic bearing.

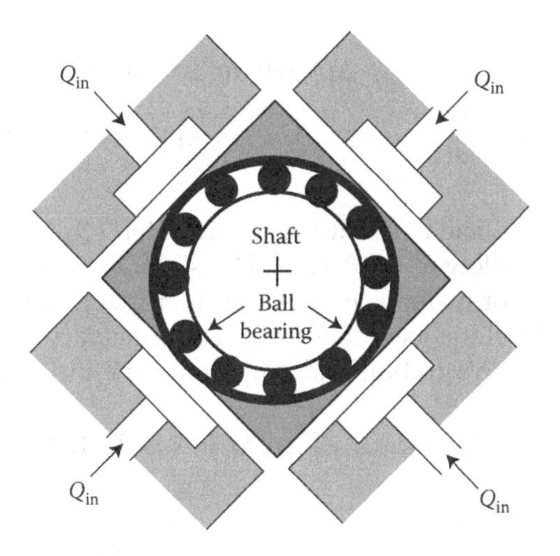

FIGURE 8.14
Example four-pad configuration of a hydrostatic backup bearing.

Simplifying assumptions made in this analysis include the following.

1. Because of the relatively high magnitude of acoustic or pressure propagation velocity in a typical liquid, the liquid pressure in the trapped volume V, though not constant in time, is assumed uniform throughout V at all times.
2. The designated trapped volume V of liquid is a constant-volume *control volume* defined to not include the film-thickness portion.
3. At any instant of time, V has a volumetric inflow and a volumetric outflow, equal to each other only under static load. But with shaft vibration added, inflow and outflow are not instantaneously equal.
4. Constant volumetric inflow is assumed since it presents the most stringent test of the concept developed here and is a little simpler to formulate than orifice or capillary compensation. Accordingly, the constant volumetric inflow condition, Q_{in} = constant, is imposed. The pressure within V for static-load conditions (i.e., no vibration present) is defined as P_0.

Since the control volume V is assumed constant here, the bulk modulus is defined consistent with that assumption, Equation 8.43. During vibration, the inflow and outflow are not constrained to be equal, so it is the incremental change in trapped liquid mass Δm, not ΔV in Equation 8.42, that must be used, as follows. ΔV is equivalent to $-\Delta m/\rho_0$ where Δm and ρ_0 are the incremental mass and nominal liquid density

$$\beta = \frac{\Delta p V \rho_0}{\Delta m} \tag{8.43}$$

The internal hydrostatic-bearing nominal pressure under static load alone is typically quite high. Consequently, vibration-induced time-varying incremental changes in liquid density are assumed to be much smaller than the nominal static-load value, ρ_0. That assumption justifies the following derivation that expresses the instantaneous mass flow rates entering and leaving V as follows, where Q is volumetric flow.

$$\dot{m}_{in} = \rho_0 Q_{in}$$

$$\dot{m}_{out} = \rho_0 Q_{out} = \rho_0 C(p_0 + \Delta p) \ \text{ where } C = \text{ outflow coefficient}$$

$$\therefore \dot{m}_{in} - \dot{m}_{out} = \frac{dm}{dt} = \rho_0(Q_{in} - Q_{out}) = \rho_0\left[Q_{in} - C(p_0 + \Delta p)\right]$$

The value of outflow coefficient at static-load condition ($Q_{in} = Q_{out}$) is $C = Q_{in}/p_0$. For a small step reduction in outflow at $t = 0$, C will be slightly smaller than at the static-load condition. And similarly for a small incremental increase in outflow at $t = 0$, C will be slightly larger than at the static-load condition. The time varying incremental dynamic change in pressure is Δp. Integration and employing the bulk modulus, Equation 8.43 yields the following.

Rearranging $\Delta m = \rho_0 Q_{in} t - C\rho_0 p_0 t - C\rho_0 \int_0^t \Delta p \, d\tau$ yields the following:

$$\frac{\Delta p V}{\beta} = Q_{in}\, t - C p_0 t - C\int_0^t \Delta p \, d\tau \tag{8.44}$$

Fluid inertia is neglected, assuming that viscous effects dominate. To track the time varying incremental change in pressure as a result of a prescribed vibration signal, a convolution integral is employed utilizing the pressure response to an instantaneous step change in outflow. The instantaneous load transmitted through the hydrostatic bearing film is instantaneously proportional to the hydrostatic-bearing pressure. So the transmitted dynamic load is proportional to Δp. Equation 8.44 is put into the standard form first order linear ordinary differential equation, as follows.

$$(\Delta p)' + \frac{\beta}{V} C\Delta p - \frac{\beta}{V}(Q_{in} - Cp_0) = 0 \tag{8.45}$$

The function shape in Figure 8.15 shows the expected shape of the response of dynamic pressure Δp within V resulting from a step decrease in outflow at time t = 0. A step increase in outflow would correspondingly be the reflection of the shown function about the time axis. When the incremental change in pressure reaches steady state, Δp_∞, the inflow and outflow must be equal, as follows.

$Q_{in} = Q_{out}$ so $Q_{in} = C(p_0 + \Delta p_\infty)$ That yields the following expression.

$$\Delta p_\infty = \frac{Q_{in}}{C} - p_0 \tag{8.46}$$

Dimensionless pressure and time are chosen as follows.

$$\Delta P \equiv \frac{\Delta p}{\Delta p_\infty} \quad T \equiv \left(\frac{Q_{in}}{V}\right) t \tag{8.47}$$

The non-dimensional form of differential Equation 8.45 is then as follows.

$$\frac{d}{dT}(\Delta P) + B\Delta P - B = 0 \quad B \equiv \beta C / Q_{in} = \beta / p_0 \quad \text{(static-load}\, C\, \text{used)} \tag{8.48}$$

With the physical insight embodied in Figure 8.15 to guide trial problem solutions, the following exact solution of Equation 8.48 "passes the insight test," perfectly matching Figure 8.15.

$$\Delta P = 1 - e^{-BT} \tag{8.49}$$

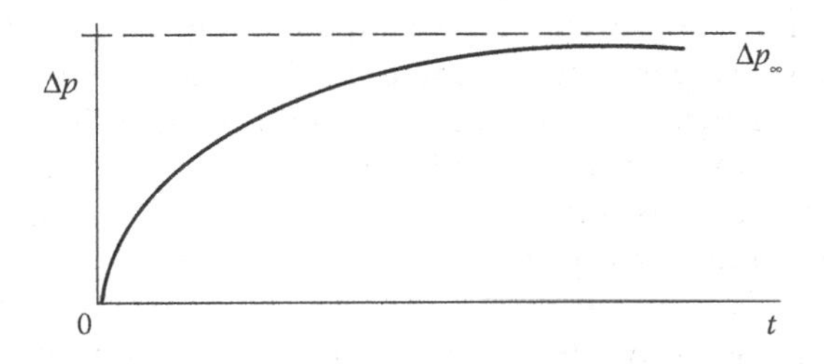

FIGURE 8.15
Expected pressure function shape for a step change in outflow.

The solution given by Equation 8.49 is the theoretical pressure response to a step change in hydrostatic recess outflow. So for a specified shaft vibration signal, a convolution integral can be formulated to give the theoretical time signal for dynamic pressure, and therefore dynamic force, transmitted through the hydrostatic fluid film. One can thereby predict the transmitted dynamic force caused by a harmonic input vibration. The dimensionless harmonic shaft vibration and dimensionless frequency are chosen as follows.

$$X = \sin \Omega T \quad \text{and} \quad \Omega = \frac{\omega V}{Q_{in}} \tag{8.50}$$

The convolution integral for transmitted dynamic pressure, ΔP_{Ω}, is then as follows.

$$\Delta P_{\Omega} = \int_0^T \frac{dX}{d\tau} \Delta P(T-\tau) d\tau = \Omega \int_0^T \left(1 - e^{-B(T-\tau)}\right) \cos \Omega \tau d\tau$$

This integrates to the following expression (Thomas 1956).

$$\Delta P_{\Omega} = \sin \Omega T - \frac{\Omega^2 \sin \Omega T + \Omega B \cos \Omega T}{B^2 + \Omega^2} - \frac{Be^{-BT}}{\Omega}$$

The steady state portion of this is the following.

$$\Delta P_{\Omega} = \sin \Omega T - \frac{\Omega^2 \sin \Omega T + \Omega B \cos \Omega T}{B^2 + \Omega^2} \tag{8.51}$$

The single peak amplitude of this harmonically varying dynamic pressure at frequency Ω is as follows:

$$|(\Delta P_{\Omega})| = \left\{ \left[1 - \frac{\Omega^2}{B^2 + \Omega^2}\right]^2 + \left[\frac{\Omega B}{B^2 + \Omega^2}\right]^2 \right\}^{1/2} \tag{8.52}$$

Clearly, as frequency is progressively made smaller, the amplitude of incremental transmitted pressure force will asymptotically approach that of the incremental change in static load, that is, $\Delta P_{\Omega} = 1$. And as frequency is progressively increased, the transmitted incremental dynamic force approaches zero. Figure 8.16 samples result from Equation 8.52 for three values of B. Those parametric sample results clearly show the *hydrostatic bearing to act as a low-pass filter.* Thus, while retaining the design high stiffness for static and low-frequency varying loads, the hydrostatic pad is shown to progressively filter out the higher-frequency load from being transmitted to the shaft, for example, in high-pitch-line-velocity gear sets. This low-pass filtering capability was experimentally verified by Zulkefli (2013).

To utilize Equation 8.52, the specified inputs are the amplitude of the dimensionless transmitted dynamic pressure ΔP_{Ω} (portion of dynamic force getting through the filtering), gear set mesh frequency ω, liquid bulk modulus β, and hydrostatic-bearing static-load parameters (p_0, Q_{in}). Back solving with Equation 8.52 then determines the solution for the

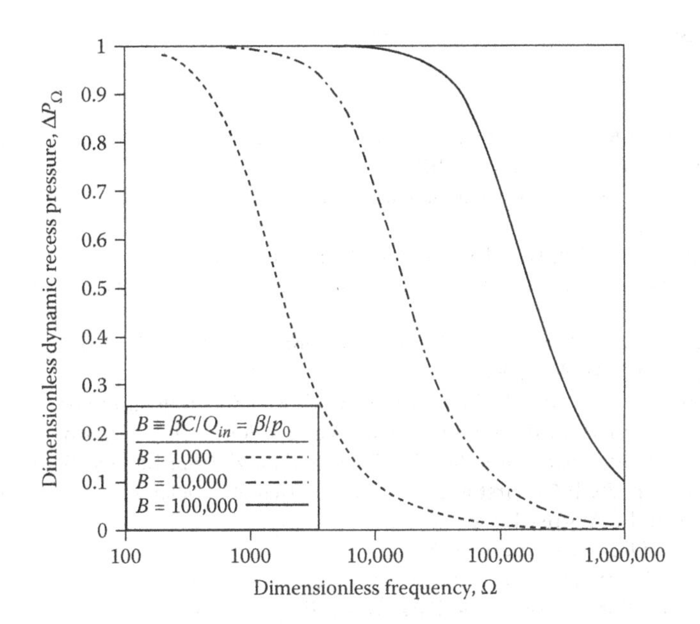

FIGURE 8.16
Transmitted dynamic pressure as a function of vibration frequency.

required liquid trapped volume V (mainly recess plus that upstream to restrictor) that satisfies the specified inputs.

The following is a sample computation.

Inputs: $\Delta P_\Omega = 0.1$ (90% amplitude filtering), $\beta = 250{,}000$ psi (2.46×10^9 N/m^2), $p_0 = 250$ psi (2.46×10^6 N/m^2), $Q_{in} = 0.1$ in^3/sec (1.64cm^3/sec), $\omega = 10{,}000$/sec = 1592 Hz (e.g., 53 toothed gear turning at 1800 rpm).

Outputs: $B = \beta/p_0 = 1000$ curve with $\Delta P_\Omega = 0.1$ yields $\Omega = 10{,}000$, as shown in Figure 8.16. From the definition for Ω this yields $V = 0.1$ in^3 = 1.639 cm^3.

For precision gear sets, the required quite high shaft support stiffness is preserved by suitable setting of the low-pass filter cut-off frequency range. In the case as backup support for rolling element gear set primary shaft support bearings (Figure 8.14), a secondary benefit of the low-pass filtering effect of the hydrostatic bearing is the potential for fatigue life extension of the rolling element bearings. Zaretsky (1999) relates how rolling-bearing life predictions that account only for bearing static load can be considerably optimistic by not including the additional contribution of vibration-induced dynamic loads. Since gear mesh frequency is a large multiple of the rotation frequency, the potential for rolling element-bearing fatigue life extension from the hydrostatic backup is suggested.

It is easy to appreciate that significant attenuation of mesh-frequency noise transmitted to the environment is a desirable objective. But the trade-offs in actual gear set designs probably will not generally facilitate making the elimination of mesh-frequency noise the only competing design objective. In that context, the compact analytical form of Equation 8.52 facilitates its incorporation into gear set design optimization analyses that have weighted objectives.

It is clear that gear mesh tooth dynamic force interactions feed torsionally into the rotor. But the fundamental character of rotor torsional vibration generally does not provide a mechanism to transmit rotor torsional vibration directly to the housing; see Chapter 3 of Adams (2010). While rotor radial and axial vibrations are significantly caused by torsional dynamic tooth interactions, their transmission to the housing is not a direct torsional mechanism.

References

Adams, M. L., *Rotating Machinery Vibration—From Analysis to Troubleshooting*, 2nd ed., CRC Press/Taylor & Francis, Boca Raton, FL, 2010.

Adams, M. L., *Rotating Machinery Research and Development Test Rigs*, CRC Press/Taylor & Francis, Boca Raton, FL, 2017.

Castelli, V. and Shapiro, W., Improved Method for Numerical Solutions of the General Incompressible Fluid Film Lubrication Problem, *ASME Journal of Lubrication Technology*, Vol. 89(2), pp. 211–218, 1967.

Houser, D. R., Gear Noise Vibration Prediction and Control Method, *Handbook of Noise and Vibration Control*, Wiley, pp. 847–856, 2007.

Malanoski, S. B. and Loeb, A. M., The Effect of the Method of Compensation on Hydrostatic Bearing Stiffness, AME Journal of Basic Engineering, Vol. 83, Series D, No. 2, June 1961.

Rippel, H. C., *Cast Bronze Hydrostatic Bearing Design Manual*, Cast Bronze Bearing Institute, Inc., Cleveland, OH, p. 75, 1965. First appeared as a series of articles in *Machine Design* magazine, beginning with the August 1, 1963 issue.

Thomas, G. T., *Table of Integrals*, Addison-Wesley, 1956.

Zaretsky, E. V. (editor), *Life Factors for Rolling Bearings*, 2nd ed., Society of Tribologists and Lubrication Engineers, STLE Publication SP-34, Park Ridge, IL, p. 314, 1999.

Zulkefli, Z. A. B., Mitigation of Gear Mesh-Frequency Vibrations Utilizing a Hydrostatic Bearing, *PhD Thesis*, Case Western Reserve University, Cleveland, OH, p. 184, January 2013.

9

Rolling Element Bearings

9.1 Overview

Chapter 2 plus this chapter combine to constitute a "book within a book" on rolling element bearings (REB). In sharp contrast to all the various types of sliding bearings, the machine designer has but one option when it comes to REBs. That is, perform the necessary analyses that lead to the selection of an appropriate bearing from an REB manufacturer's catalog. That is because (1) the manufacturing of REBs is a highly specialized endeavor, and (2) the life rating of REBs is based on the manufacturers' extensive in-house high-cycle subsurface-initiated-fatigue life testing, as treated in Chapter 2. Thus, this chapter basically presents the essence of what is typically summarized in manufacturers' catalogs as well as in mechanical engineering undergraduate machine design textbooks. However, referring to Figure 2.36 (Zaretsky and Poplawski 2009), REB high-cycle fatigue has the second lowest bearing rejection rate for life-ending use, with only *cage wear* lower, and both at less than 5%. In comparison, the *corrosion pitting* rejection cause has the highest rejection rate at over 30%.

Loads imposed on REBs are both *static* and *dynamic*. Depending on the type of machine application, dynamic loads can be from a variety of sources such as (1) rotor unbalance, (2) turbomachinery fluid dynamic unsteady flow (e.g., centrifugal pumps, compressors, turbines), (3) gear set mesh-frequency manufacturing imperfections, and (4) impact loads. To the extent that dynamic loads are a significant factor in a particular application, that aspect of bearing selection is primarily up to the machine designer/builder to assess. That is, it must be based on *intensity* and *exposure time* of the dynamic loads in the particular application. Given a significant degree of uncertainty for REB-life expectancy vis a vis dynamic loading, feedback on REB longevity from the field for a particular application is surely the best approach to optimize product integrity. Given all that, the initial bearing selection is based only on the bearing-static loads which are generally better defined than the dynamic loadings. Correspondingly, the REB manufacturers' life ratings are based only on static loading, so *Caveat Emptor* (buyer beware). For "run-of-the-mill" applications, selecting an REB from the catalog is a low-cost engineering task and can be the appropriate approach in many applications. But again referring to Figure 2.36, for a highly specialized application where the various REB-life-threating possibilities linger, if resources for advanced engineering are available then all those non-cookbook technological aspects of REBs can be properly assessed and engineered for specialized applications. In an ultimate pursuit of the right bearing in a special application, the highest possibility of success is to retain one of the non-OEM highest-credibility REB experts who have all the required advanced REB software (see Section 2.4) to reliably evaluate the various elusive REB-operating parameters reviewed in Chapter 2 such as (1) heat and thermal excess,

(2) dynamic loading, (3) excessive axial thrust, (4) roller-edge stresses, (5) skidding, (6) misalignment, and (7) cage fracture. One such independent organization devoted to such high-tech-bearing analysis and selection is J. V. Poplawski and Associates.

9.2 Bearing Static Loads

Using bearing static-load fatigue life as the initial selection criteria and entree to the REB-manufacturer's selection catalog requires primarily (1) bearing-static-load components, (2) rotational speed, and (3) bearing ID, OD, or length limitations. Figure 9.1 example typifies the process of determining bearing static loads. But as not uncommon, that example is deceptively simple in appearance. Although loading from rotor weight W is a quite reliable parameter to specify both in magnitude and direction (i.e., g = constant), other bearing-load sources often varying widely in *magnitude* and *direction* over a machine's full operating range and possibly fraught with considerable uncertainty for their magnitude and direction. Centrifugal pumps are a prime example, since both the axial and radial loads imposed by the impeller hydraulic forces (static and dynamic) are a strong function of the operating percent of best efficiency flow (Adams 2017), and are not nearly as well quantified as rotor weight. Similarly, the ship propulsion gear-set radial journal bearing example presented in Section 7.1.3 is a prime example of how bearing-load *magnitude* and *direction* can vary considerably in gear sets operating over a wide range of torque. Note that bearing 2 in the Figure 9.1 example is a duplex set, as illustrated in Figures 2.6 and 2.7,

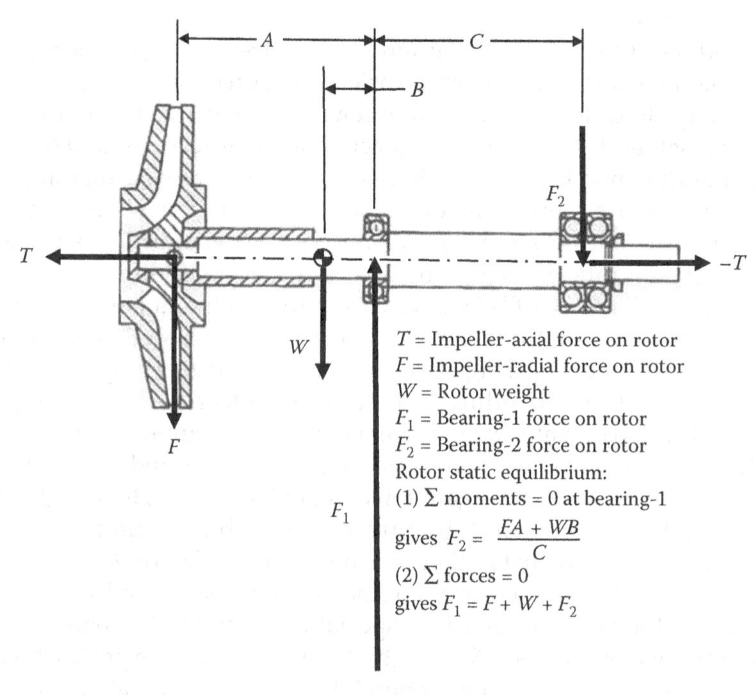

FIGURE 9.1
Equilibrium-static loads on a single-stage centrifugal pump rotor.

because it needs to carry not only radial load but also the substantial axial-thrust load from the centrifugal-pump-impeller hydraulic force.

9.3 Manufacturer Catalog Information and Bearing Selection

All REB manufacturers have catalogs that for each REB-type lists numerous sizes, each size identified by a *basic-bearing number* combined with complete dimensional information. These catalogs generally start with a primer on how to use the catalog to zero-in on the right bearing for a given application. For the more common REB types, this catalog information is provided in tabulations of all the available sizes for a particular bearing configuration, with tabulation headings as exampled in Table 9.1 (nomenclature defined in Figure 9.2).

To begin the process of selecting a bearing for a specific application, one specifies (1) the REB-type (see all types in Figure 9.3), like ball or roller, (2) grade of precision, like ABEC-1, (3) lubrication-type, like grease, liquid oil, fine oil mist, (4) closure, like open, shielded, or

TABLE 9.1

Typical Rolling-Element-Bearing Manufacturer Catalog Listing

Bearing Basic No.	Bore (mm)	OD (mm)	w (mm)	r (mm)	d_S (mm)	d_H (mm)
•	•	•	•	•	•	•
•	•	•	•	•	•	•
•	•	•	•	•	•	•

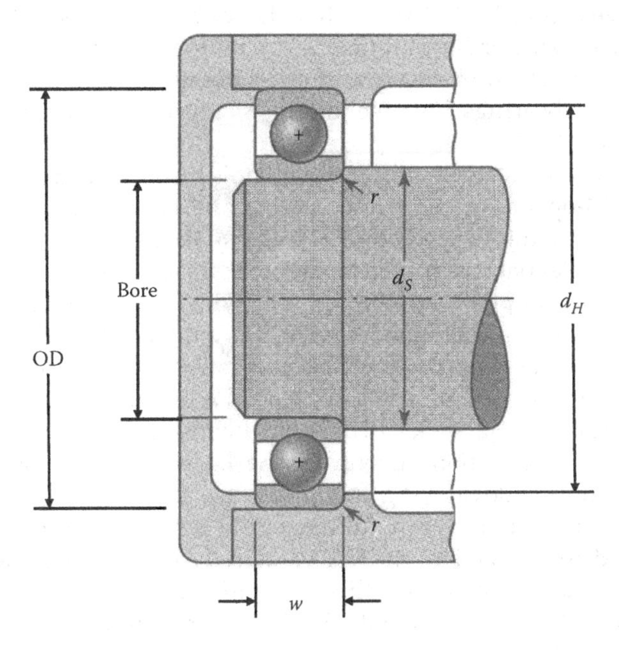

FIGURE 9.2
Shaft and housing shoulder dimensions.

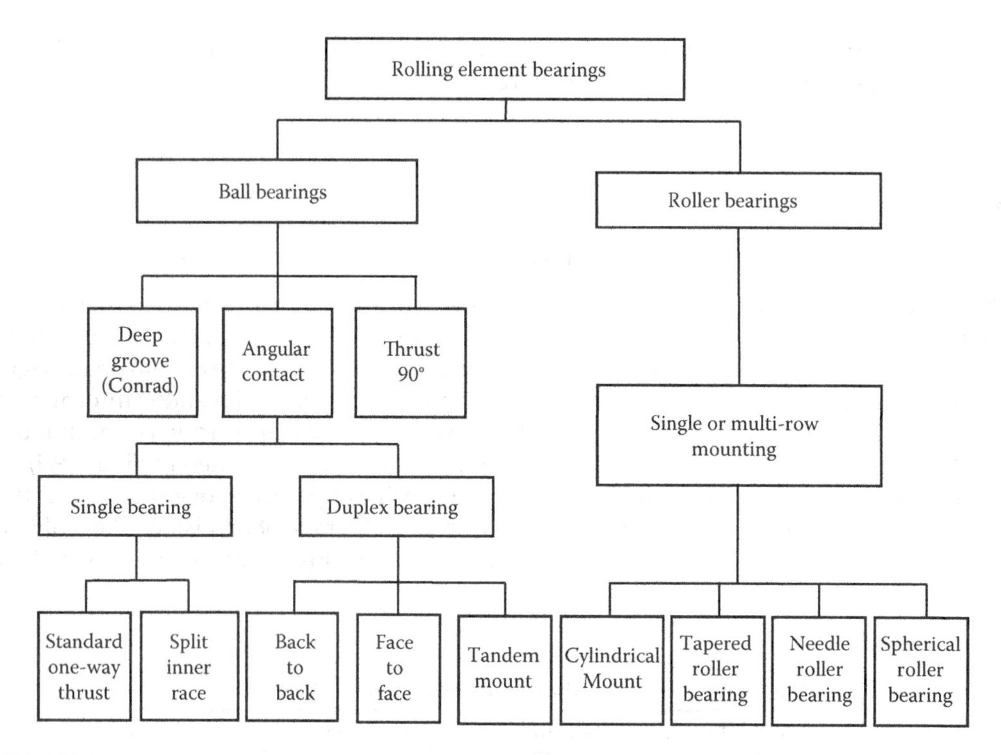

FIGURE 9.3
Categories of rolling element bearings for rotating machinery.

sealed, (5) the basic load rating, and (6) load capacity when not rotating. Excessive loads while not rotating can cause the rolling elements to impart small dents into the raceways, called brinelling, see Figure 2.38f, which will make the bearing noisy when rotating. As an example result, when transporting automobiles by railroad it is a common practice to shift the vehicle weight from the four wheels to support points on the underbody. That isolates the auto vehicle wheel bearings from feeling the relatively bumpy ride transmitted from the railroad car hard wheels.

REB-raceway maximum-surface speed, not rpm, is also a critical REB-selection factor that determines REB-operating temperature, wear and type of lubrication required, see Figure 2.33. Especially in high-speed applications, *fine oil mist* is the *best lubrication method,* since the oil-mist particles plus the copious continuous through-flow of the oil-mist-carrying air provide significant cooling of the bearing. A continuous flooding by a through-flow of liquid oil would not provide such good cooling since the churning viscous losses of the liquid oil would impart significantly more heat generation to the bearing elements than the added heat removal by the continuous through flow of liquid oil.

Referencing Juvinall and Marshek (2012), ABEC-1 ball bearings employing non-metallic separators and oil-mist lubrication can run at inner-raceway surface speeds up to 75 m/s with a life of 3000 hours, while carrying one-third of the rated-load capacity. This translates into a DN value (D is bore diameter in millimeters, N is speed in rpm) of about 1.25×10^6. With splash or oil drip lubrication, the DN value reduces by about one-third, and with grease lubrication by about two-thirds. Operating under ideal conditions, roller bearings can operate up to a DN value of about 450,000. Operation at significantly higher DN values is feasible, but the ABEC precision quality and other bearing particulars should be determined not from an REB catalog but should be determined in consultation with a recognized non-OEM REB specialist. Likewise, if the REB will be subjected to possible misalignment (see Figure 2.29) or temperature extremes, bearing selection should then also be made in consultation with a recognized REB expert.

REB size is often influenced by shaft diameter and space considerations. Furthermore, the bearing needs to have a load rating sufficiently high for *acceptable life and reliability* of the given application. REB life is quantified as the probability of survival for a given life at a given load. This reflects that REB failure from subsurface-initiated fatigue failure is viewed as a statistical phenomenon (Palmgren 1959). Figure 9.4 illustrates this with the percent survival rate as a function of the number of shaft revolutions to failure at a given load, based on numerous tests-to-failure for a bearing of specific *bearing basic number.* A typical REB-life rating is L_{10} which means a 90% chance of survival for the given load. To more insightfully explain the Figure 9.4 illustration, Figure 9.5 is constructed to introduce a 3D *xyz* picture that graphically relates REB (1) % *survival,* (2) *life,* and (3) *load* with a 3D surface. Note from Figure 9.5 that as the assumed load is increased, the % survival reduces, clearly illustrating how the load rating at a specified life decreases as load is increased, and vice versa.

Under ideal conditions (i.e., free of dirt, properly lubricated, properly installed, well matched to requirements of application, and free of chemical attack like corrosion), the following equation relates two different points on the life surface for a given bearing, as illustrated in Figure 9.5 (L is life, F is load),

$$\frac{L_1}{L_2} = \left(\frac{F_2}{F_1}\right)^a \tag{9.1}$$

where according to Palmgren is approximated by a = 3 for ball bearings and 10/3 for roller bearings, which does not reflect any applied mechanics fundamentals, only reflecting curve-fitting approximations over many thousands of REB life-test results.

With reference to Equation 9.1 and motivated by the simplification of user-bearing selection, REB-manufacturer catalogs may reference bearing-load ratings to a specific

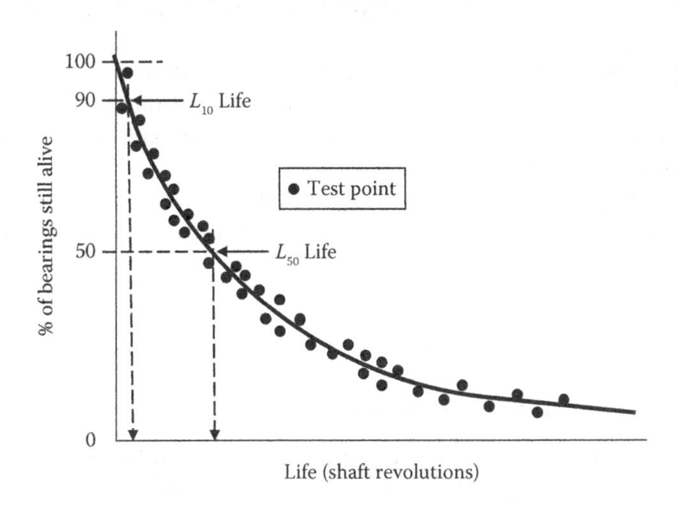

FIGURE 9.4
Life testing results for a specified REB at a given test load.

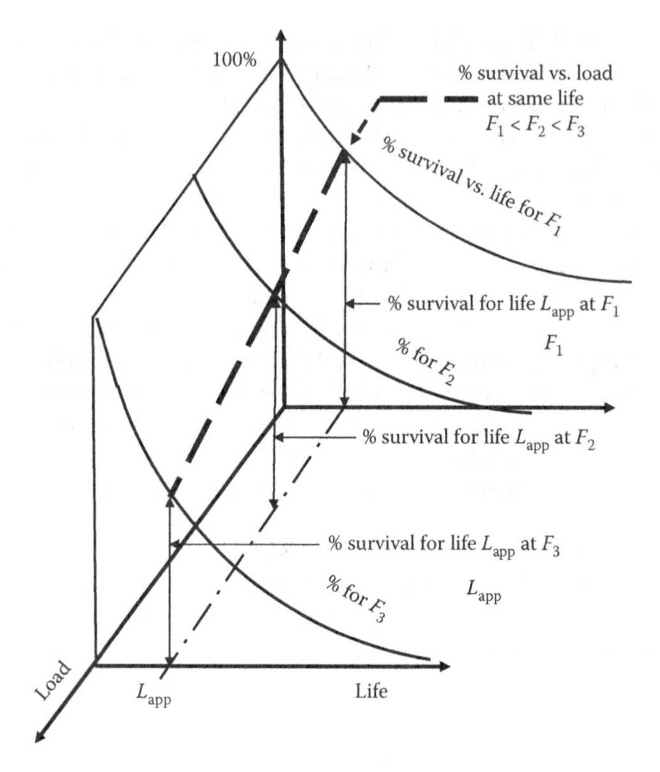

FIGURE 9.5
REB-survival surface graph of survival versus life as a function of load.

reference number of revolutions N_R like, for example, *one million* for ball bearings, which is an extremely small number of loading cycles. For example, at 3600 rpm the operating time accrued by one million cycles is only 4.6 hours. Therefore, the catalog reference load *is just a reference value* which when utilized with Equation 9.1 provides to the user the bearing life as a function of load for the bearing under consideration. In fact, the reference load may actually be a much larger-load value than what the REB elements would tolerate without gross plastic yielding deformation.

The following is an example utilizing Equation 9.1 for a Timken roller bearing, for which the reference life is 3000 hours $\times$ 500 rpm $=$ 90 million revolutions, the application life is 1200 hours $\times$ 600 rpm $=$ 43.2 million revolutions, and the application load is $F = 4$ kN. Equation 9.1 leads to the following.

$F_R = F[(L/L_R)(n/n_R)]^a = 4[(1200 \times 600)/(3000 \times 500)]^{3/10} = 3.21$ kN, which provides the load information to determine how big or robust a bearing to select from the REB catalog.

9.4 Differential Thermal Expansions

During operation, shaft and housing will generally not be at the same temperature. The rotor is typically warmer, depending on the type of machine, for example, jet engine, gear set, centrifugal pump. Therefore, axially fixing the shaft to the housing at both ends of the shaft could lead to excessive axial loads on both bearings due to rotor-to-housing

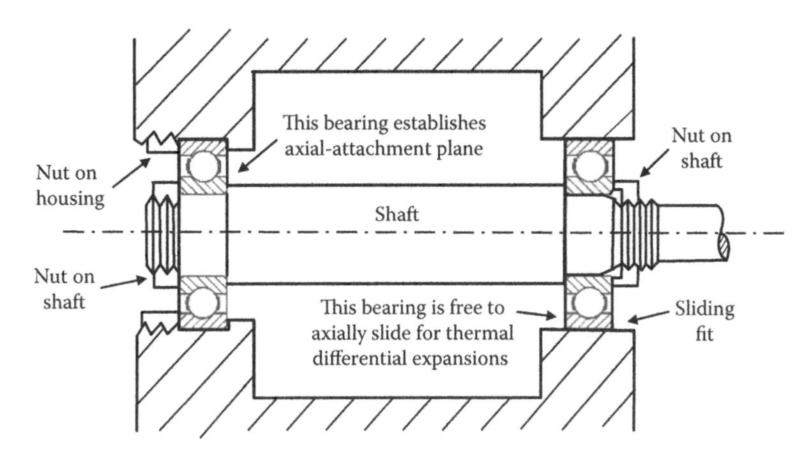

FIGURE 9.6
Axial positioning of shaft relative to housing.

differential thermal growth. Also, even without differential thermal expansion, axially fixing both bearings to the housing would produce a challenge on specified axial dimension tolerances stack-up of both shaft and housing. Figure 9.6 illustrates one of the typical ways of accommodating rotor-to-housing differential thermal growth by having one of the two bearings with a near-zero radial clearance sliding fit. Sliding fit can also be on the shaft instead of the housing.

References

Adams, M. L., *Power Plant Centrifugal Pumps Problem Analysis and Troubleshooting*, CRC Press/Taylor & Francis, Boca Raton, FL, 176 pp., 2017.

Juvinall, R. C. and Marshek, K. M., *Fundamentals of Machine Component Design*, 5th ed., Wiley, Hoboken, NJ, 2012.

Palmgren, A., *Ball and Roller Bearing Engineering*, SKF Industries, Inc., Philadelphia, 1959.

Zaretsky, E. V. and Poplawski, J. V., *Rolling Element Bearing Technology, Handout for 3-day short course*, Case Western Reserve University, Cleveland, Ohio, 973 pp., June 23–25, 2009.

10

Vibration Characteristics of Bearings

10.1 Overview

Lateral rotor vibration (LRV) analyses are critically dependent on reliable radial-bearing-rotor dynamic stiffness and damping coefficients that "connect" the rotor model to the stator (stator ≡ everything that does not rotate). In many rotating machinery types (e.g., turbomachinery) other liquid- and gas-filled internal close-clearance annular gaps, such as seals and turbomachinery bladed stages, are also of considerable LRV importance in connecting rotor and stator simulation models. However, this chapter is confined only to radial-bearing LRV characteristics. This chapter draws its contents from Adams (2010), which also covers LRV characteristics of radial seals and turbomachinery stages. Bearing LRV characteristics are highly sensitive to dimensional tolerances, in particular bearing-radial clearances. Thus, the vibration behavior of rotors is really somewhat *stochastic* rather than *deterministic*. The CD accompanying this text contains rotor vibration stiffness and damping tabulations for several journal-bearing configurations as well as the general rotor vibration analysis RDA software and the balancing software FLEXBAL software, with user instructions for both as given in Adams (2010).

10.2 Liquid-Lubricated Fluid-Film Journal Bearings

The starting point for this topic is Section 1.2, which is focused on a rigorous derivation of the Reynolds lubrication equation (RLE), Equation 1.10, which is also included here as follows:

Sliding-velocity term ↘ *Squeeze-film term* ↙

$$\frac{\partial}{\partial x}\left[\frac{h^3}{\mu}\left(\frac{\partial p}{\partial x}\right)\right]+\frac{\partial}{\partial z}\left[\frac{h^3}{\mu}\left(\frac{\partial p}{\partial z}\right)\right]=6U\frac{dh}{dx}+12\frac{dh}{dt} \quad (10.1)$$

where

$p = p(R\theta, z),$

$h = h(R\theta, z),$

$0 \leq \theta \leq 2\pi,$

$-L/2 \leq z \leq L/2,$

$\mu =$ viscosity.

From the solution of Equation 10.1 for the pressure distribution on the journal, the fluid-film force on the journal is computed by integration of the pressure distribution $p(R\theta, z)$ into x and y components to obtain the following equations for the x and y fluid-film radial-force components on the journal.

$$
\begin{aligned}
F_x &= -R\int_{-L/2}^{L/2}\int_{0}^{2\pi} p(R\theta, z)\cos(\theta)d\theta dz, \\
F_y &= -R\int_{-L/2}^{L/2}\int_{0}^{2\pi} p(R\theta, z)\sin(\theta)d\theta dz
\end{aligned}
\tag{10.2}
$$

Referring to Figure 7.4, the total instantaneous fluid-film force on the journal is composed of the static-load equilibrating force $-W$ and the instantaneous time-varying dynamic force f which is the bearing-film stiffness-and-damping response to rotor vibration. Assuming that the level of rotor vibration is sufficiently small, only the *linear portion* of the dynamic force f is assumed important, an engineering assumption that permeates nearly all standard vibration engineering analyses. Accordingly, the now standard linear vibration model for a journal bearing is as follows:

$$
\begin{Bmatrix} f_x \\ f_y \end{Bmatrix} = -\begin{bmatrix} k_{xx} & k_{xy} \\ k_{yx} & k_{xx} \end{bmatrix}\begin{Bmatrix} x \\ y \end{Bmatrix} - \begin{bmatrix} c_{xx} & c_{xy} \\ c_{yx} & c_{yy} \end{bmatrix}\begin{Bmatrix} \dot{x} \\ \dot{y} \end{Bmatrix}
\tag{10.3}
$$

As rigorously shown by Adams (2010), these four stiffness and four damping coefficients are strong functions of the static equilibrium position of the journal with respect to the bearing. Also, since the RLE (Equation 10.1) embodies only fluid viscosity but not fluid inertia, the model in Equation 10.3 contains only stiffness and damping. Furthermore, since fluid inertia is not included, the damping matrix is symmetric (Adams 1987, 2010).

Since solutions for the fluid-film radial-force components F_x and F_y are usually obtained through numerical integration of $p(\tau, z)$ as it is obtained from finite-difference numerical solution of the RLE, the partial derivatives of F_x and F_y that define the bearing stiffness and damping coefficients must also be numerically computed. This is shown by the following equations:

$$
\begin{aligned}
-k_{xx} &\equiv \frac{\partial F_x}{\partial x} \cong \frac{\Delta F_x}{\Delta x} = \frac{F_x(x+\Delta x, y, 0, 0) - F_x(x, y, 0, 0)}{\Delta x} \\
-k_{yx} &\equiv \frac{\partial F_y}{\partial x} \cong \frac{\Delta F_y}{\Delta x} = \frac{F_y(x+\Delta x, y, 0, 0) - F_y(x, y, 0, 0)}{\Delta x} \\
-k_{xy} &\equiv \frac{\partial F_x}{\partial y} \cong \frac{\Delta F_x}{\Delta y} = \frac{F_x(x, y+\Delta y, 0, 0) - F_x(x, y, 0, 0)}{\Delta y} \\
-k_{yy} &\equiv \frac{\partial F_y}{\partial y} \cong \frac{\Delta F_y}{\Delta y} = \frac{F_y(x, y+\Delta y, 0, 0) - F_y(x, y, 0, 0)}{\Delta y}
\end{aligned}
$$

$$
\begin{aligned}
-c_{xx} &\equiv \frac{\partial F_x}{\partial \dot{x}} \cong \frac{\Delta F_x}{\Delta \dot{x}} = \frac{F_x(x,y,\Delta\dot{x},0)-F_x(x,y,0,0)}{\Delta \dot{x}} \\
-c_{yx} &\equiv \frac{\partial F_y}{\partial \dot{x}} \cong \frac{\Delta F_y}{\Delta \dot{x}} = \frac{F_y(x,y,\Delta\dot{x},0)-F_y(x,y,0,0)}{\Delta \dot{x}} \\
-c_{xy} &\equiv \frac{\partial F_x}{\partial \dot{y}} \cong \frac{\Delta F_x}{\Delta \dot{y}} = \frac{F_x(x,y,0,\Delta\dot{y})-F_x(x,y,0,0)}{\Delta \dot{y}} \\
-c_{yy} &\equiv \frac{\partial F_y}{\partial \dot{y}} \cong \frac{\Delta F_y}{\Delta \dot{y}} = \frac{F_y(x,y,0,\Delta\dot{y})-F_y(x,y,0,0)}{\Delta \dot{y}}
\end{aligned}
\tag{10.4}
$$

The definitions contained in Equation 10.4 for the eight *stiffness and damping* coefficients are compactly expressible using subscript notation, as follows:

$$
k_{ij} \equiv -\frac{\partial F_i}{\partial x_j} \text{ and } c_{ij} \equiv -\frac{\partial F_i}{\partial \dot{x}_j} \tag{10.5}
$$

It is evident from the 8 Equations 10.4 that the journal radial-force components F_x and F_y are expressible as continuous functions of journal-to-bearing radial displacement and velocity components, as follows:

$$
\begin{aligned}
F_x &= F_x(x,y,\dot{x},\dot{y}) \\
F_y &= F_y(x,y,\dot{x},\dot{y})
\end{aligned}
\tag{10.6}
$$

It is also evident from Equation 10.4 that for each selected static equilibrium operating condition $(x, y, 0, 0)$, five solutions of the RLE are required to compute the eight *stiffness and damping* coefficients for each equilibrium journal position with respect to the bearing. Those five slightly different solutions are tabulated as follows:

$$
\begin{aligned}
&(x,y,0,0), \text{Equilibrium condition} \\
&(x+\Delta x,y,0,0), x-\text{displacement perturbation about equilibrium} \\
&(x,y+\Delta y,0,0), y-\text{displacement perturbation about equilibrium} \\
&(x,y,\Delta\dot{x},0), x-\text{velocity per turbation about equilibrium} \\
&(x,y,0,\Delta\dot{y}), y-\text{velocity perturbation about equilibrium}
\end{aligned}
$$

The perturbation sizes $(\Delta x,\Delta y,\Delta\dot{x},\Delta\dot{y})$ used must be appropriately sized (not too small, not too large) for the particular static equilibrium condition to ensure reliable results. That is, so as not to lose accuracy either from (1) round-off error if the perturbation is too small or from (2) the inherent nonlinearity of $p(R\theta,z)$ as a function of $(\Delta x,\Delta y,\Delta\dot{x},\Delta\dot{y})$ if the perturbation is too large. The proper approach for setting perturbation sizes is detailed by Adams (2010), showing an automated perturbation size adjustment utilizing the number of significant figures in the digital computation implicit in Equation 10.4.

Normalized (dimensionless) journal-bearing stiffness and damping coefficients are functions of the bearing dimensionless speed (Sommerfeld number, S), and defined as follows:

$$
\bar{k}_{ij} \equiv \frac{k_{ij}C}{W}, \bar{c}_{ij} \equiv \frac{c_{ij}\omega C}{W}, S \equiv \frac{\mu n}{P}\left(\frac{R}{C}\right)^2 \tag{10.7}
$$

where

C = radial clearance,
W = static load,
S = Sommerfeld number,
μ = lubricant viscosity,
$P = W/DL$ (unit load),
R = nominal radius,
$D = 2R$,
L = length,
n (revs/sec) $= \omega / 2\pi$.

The CD that accompanies this book contains extensive dimensionless tabulations of stiffness and damping coefficients for various journal-bearing configurations. The earliest major compendium of such journal-bearing rotor dynamic property coefficients was published by Lund et al. (1965), and it is still a significant resource for rotor vibration analyses. It contains the normalized stiffness and damping coefficients plotted against bearing dimensionless speed (Sommerfeld number) for several types of journal-bearing configurations, including (1) 360° cylindrical, axially grooved, (2) partial-arc, (3) lobe, and (4) tilting-pad, for both *laminar* and *turbulent* flow films. The most significant more recent compendium of journal-bearing rotor dynamic properties is provided by Someya et al. (1988). It is based on data contributed by technologists from several Japan-based institutes and companies participating in a joint project through the Japanese Society of Mechanical Engineers (JSME). It contains not only computationally generated data but also corresponding comparison data from extensive laboratory testing. Although many industry and university organizations now have computer codes that can generate bearing dynamic properties for virtually any bearing configuration and operating condition, urgent on-the-spot rotor vibration analyses in troubleshooting circumstances are more likely to necessitate the use of existing available bearing dynamic coefficient data. The published data, such as by Lund et al. (1965) and by Someya et al. (1988), are thus invaluable to successful trouble shooters and designers. For use with rotor vibration computer codes, tabulated bearing dynamic properties are more convenient than plotted curves since the bearing input for such codes is tabulated data at specific rotational speeds, as demonstrated with sample problems in Adams (2010) and contained on the CD accompanying this book. Furthermore, tabulations are more accurate than reading from plotted curves, especially semi-log and log-log plots spanning powers of 10. This accuracy issue is particularly critical concerning simulation predictions for *instability thresholds of self-excited rotor vibration.*

The fundamental caveat of lateral rotor vibration analyses stems from the *uncertainty* in the values of journal-bearing stiffness and damping coefficients. The primary source of uncertainty is the journal-bearing *radial clearance*, which is a relatively small number that is the difference between two relatively larger numbers. This is clearly demonstrated by the following realistic close-tolerance example.

$$\left.\begin{array}{ll}\text{Bearing bore diameter,} & 5.010^{\pm 0.001}\ \text{in} \\ \text{Journal diameter,} & 5.000^{\pm 0.001}\ \text{in}\end{array}\right\} \begin{array}{l}\text{Giving this} \\ \text{range of radial} \\ \text{clearance}\end{array} \rightarrow \begin{cases}0.004\ \text{in, min.} \\ 0.006\ \text{in, max.}\end{cases}$$

The result: $\dfrac{S_{max}}{S_{min}} = \left(\dfrac{0.006}{0.004}\right)^2 = 2.25$

This is more than a 2-to-1 range of dimensionless speed, which can be related to dimensional parameter ranges such as 2-to-1 in rpm or lubricant viscosity or static load. This 2-to-1 range involves a sizable variation in journal-bearing dynamic coefficients, to say the least. This is just one of the factors that prove that old power plant adage that *no two machines are exactly alike.* Other prominent factors which add uncertainty to journal-bearing characteristics include the following:

- Large variations in *oil viscosity* from oil temperature variations
- Journal-to-bearing angular *misalignment*
- Uncertainties and operating variations in bearing *static load*
- Bearing *surface distortions* from loads, temperature gradients, wear, and so on
- Basic simplifying assumptions leading to the RLE

These revelations are not intended to show *lateral rotor vibration analyses* (LRV) to be worthless because they most assuredly are of considerable value. But, the savvy analyst and trouble shooter must keep these and any other sources of uncertainty uppermost in their mind when applying LRV analyses. As described in Adams (2010, 2017), laboratory experimental efforts to measure bearing stiffness and damping coefficients are a corresponding challenge.

10.3 Squeeze-Film Dampers

Vibration-damping capacity of a rolling-element bearing (REB) is extremely small and therefore to measure it is virtually impossible since any test rig for this purpose would have its own damping that would mask that of a tested REB. As is well known and shown by Figure 10.1, the benefit of damping is in preventing excessively high vibration amplitudes at resonance conditions, $\omega/\omega_n = 1$. Thus, for the many machines running on REBs that have the maximum running speed well below the lowest critical speed resonance, the absence of any significant REB damping presents no problem.

Since a REB can usually operate for sustained periods of time after the normal lubrication supply fails, RCBs are usually a safer choice over fluid-film bearings in aerospace applications such as modern aircraft gas turbine jet engines. In such applications, however, the inability of the REBs to provide adequate vibration damping capacity to maintain tolerable unbalance vibration levels through critical speeds frequently necessitates the use of *squeeze-film dampers* (SFD). A SFD is formed by a cylindrical annular oil film within a small radial clearance between the OD cylindrical surface of an REB-outer raceway and the precision cylindrical bore into which it is fitted in a machine. The radial clearance of the SFD is similar to that for a comparable sized journal bearing, possibly a bit larger as optimized for a specific application. Figure 10.2 shows an SFD configuration employing *centering springs.* Since there is no journal rotation like in a journal bearing, centering springs establish a static equilibrium eccentricity. For the SFD, only the squeeze-film right-hand side term in Equation 10.1 is non-zero.

For a first order approximation to compute SFD damping coefficients, the perturbation approach given by Equation 10.4 may be used. However, the factors of *film rupture* and *dissolution of air* in the SFD-oil film produce considerably more complication and thus even

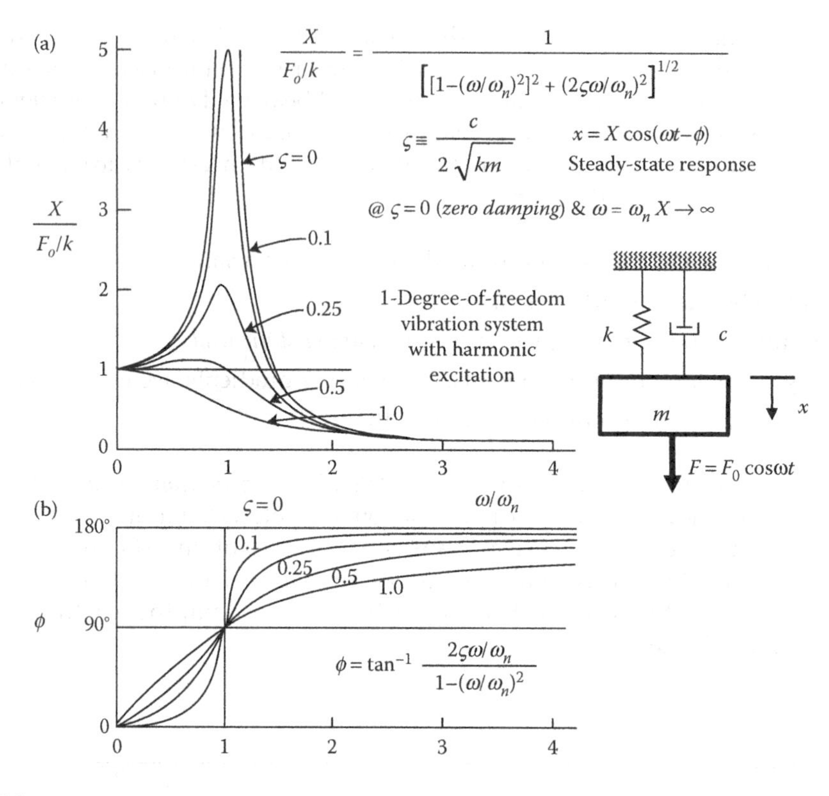

FIGURE 10.1
One-degree-of-freedom steady-state response to a sinusoidal force, (a) vibration amplitude versus speed, (b) phase-angle lag between harmonic force and steady-state vibration; $\varsigma = 1$ at critically damped.

more uncertainty of computational predictions for damping coefficients than for journal bearings. Also, the neglect of fluid inertia effects implicit in Equation 10.1 is not as good an assumption for SFDs as it is for journal bearings.

The SFD configuration shown in Figure 10.2 employs *centering springs* since there is no active *sliding velocity term* to generate static-load-carrying capacity in the hydrodynamic oil film. To create a static equilibrium position about which the vibration occurs and is damped by the SFD, centering springs may be employed to negate the bearing static load and maintain the damper approximately concentric. The radial stiffness of the centering springs

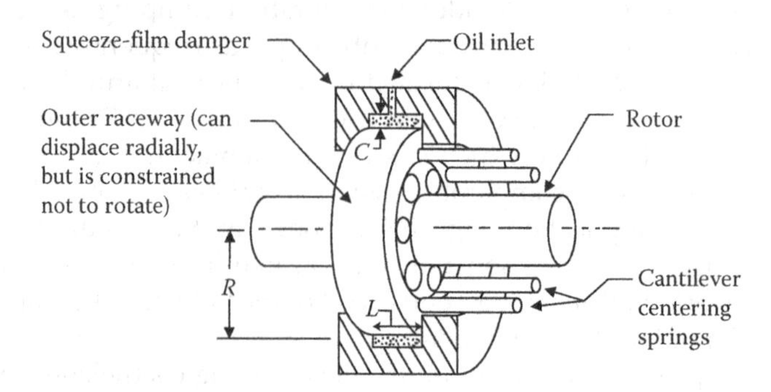

FIGURE 10.2
Squeeze-film damper concept with REB and centering springs.

is probably less than the radial stiffness of the REB, developed in the next section. Thus, the stiffness coefficient array is essentially the isotropic-radial stiffness of the *centering springs*, k_{cs}. Assuming validity of linearization as postulated for journal bearings in Equation 10.4, the radial-interactive force at a bearing station employing an SFD with centering springs is then expressible as follows:

$$\begin{Bmatrix} f_x \\ f_y \end{Bmatrix} = -\begin{bmatrix} k_{cs} & 0 \\ 0 & k_{cs} \end{bmatrix}\begin{Bmatrix} x \\ y \end{Bmatrix} - \begin{bmatrix} c_d & 0 \\ 0 & c_d \end{bmatrix}\begin{Bmatrix} \dot{x} \\ \dot{y} \end{Bmatrix} \tag{10.8}$$

where

c_d = damping coefficient for the damper film.

The SFD's length (L) is typically much smaller than its diameter ($D = 2R$). Consequently, it is customary to consider two cases: (1) SFD does not have end seals, and (2) SFD does have end seals, as illustrated in Figure 10.3 (Adams 2017). For Case 1, the supplied oil flow is continuously squeezed out the two axial boundaries of the damper film, and since typically $L/D < 0.2$, the solution to Equation 10.1 using the *short bearing* approximation is justified, see Figure 1.8. For Case 2, the use of end seals essentially prevents the significant axial oil flow encountered in Case 1 and thus using the *long bearing* approximation is justified. Also for Case 2, one or more drain holes are typically employed in the damper to maintain a specified oil through-flow to control damper oil temperature.

Postulating a concentric circular orbit for the rotor within the SFD, the solution of Equation 10.1 yields an instantaneous radial-plane force vector on the rotor which can be decomposed into its radial and tangential components. As an example of this, Vance (1988) lists these two force components based on the *short bearing* approximation which is appropriate for the above Case 1 (no end seals) and 180° cavitation zone trailing the orbiting line-of-centers (i.e., minimum film thickness), as follows:

$$\begin{aligned} F_R &= -\frac{2\mu R L^3 \Omega \varepsilon^2}{C^2(1-\varepsilon^2)^2} \\ F_T &= -\frac{\pi \mu R L^3 \Omega \varepsilon}{2C^2(1-\varepsilon^2)^{3/2}} \end{aligned} \tag{10.9}$$

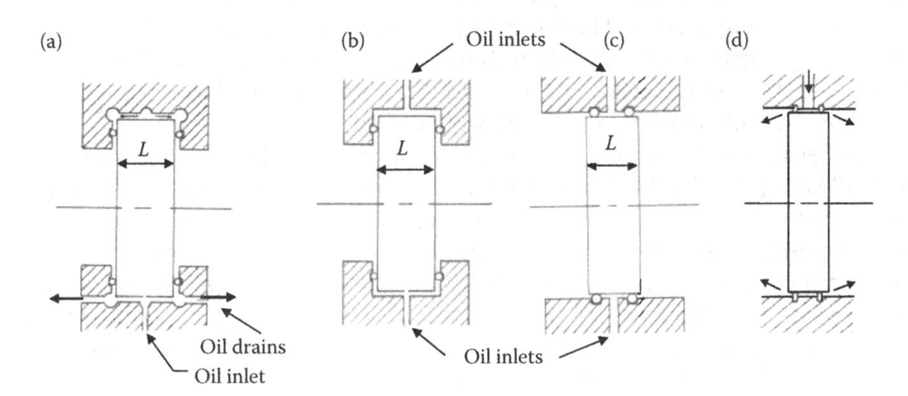

FIGURE 10.3
Squeeze-film damper configurations employing axial end seals, (a) and (b) O-rings with no centering spring, (c) O-rings acting as centering spring, (d) snap-ring end seals.

where

μ = viscosity,
R = damper radius,
L = damper length,
Ω = orbit frequency,
C = damper radial clearance,
ε = orbit eccentricity ratio e/C.

Radial and tangential force components can also be similarly derived using the *long bearing* solution of Equation 10.1, which is appropriate to the previous Case 2 (with end seals). It should be noted that the force components given by Equation 10.9 are obviously quite nonlinear functions of the orbit magnitude. However, they can be linearized for "small" concentric circular orbits. Equations 10.9 can be simplified for $\varepsilon << 1$ ($\varepsilon \to 0$) to the following.

$$F_R \cong -\left(\frac{2\mu RL^3\Omega\varepsilon^2}{C^2}\right)$$
$$F_T \cong -\frac{\pi\mu RL^3\Omega\varepsilon}{2C^2} \tag{10.10}$$

Since the radial force approaches zero one order faster than the tangential force (i.e., ε *vs* ε^2), the only non-zero coefficient retrieved is the diagonal damping coefficient. Thus, for the *short bearing* approximation with boundary conditions for the 180° cavitation zone trailing the orbiting line-of-centers, the Equation 10.8 damping coefficient is given as follows:

$$c_d \cong \frac{\pi\mu RL^3}{2C^3} \tag{10.11}$$

For a sufficiently high damper ambient pressure to suppress cavitation, the solution yields a damping coefficient that is twice that given by Equation 10.11.

Eliminating the centering springs makes the SFD mechanically simpler and more compact. Also, the possibility of centering spring fatigue failure does not need to be addressed if there are no centering springs. However, from a rotor vibration point of view, eliminating the centering springs makes the system dynamically less simple. The damper now tends to "sit" at the bottom of the clearance gap and it requires some vibration to "lift" it off the bottom. That is a quite nonlinear dynamics phenomenon.

Some modern aircraft engines are fitted with *springless* SFDs while some have spring-centered SFDs (Figure 10.4). Under NASA sponsorship, Adams et al. (1982) devised methods and software to retrofit algorithms for both types of dampers into the general purpose nonlinear time-transient rotor response computer codes used by the two major U.S. aircraft engine manufactures. Adams (2017) shows damper experimental results from a small table-top test rig that illustrate a family of nonlinear rotor vibration orbits that develop in a springless SFD as a rotating unbalance force magnitude is progressively increased (Figure 10.5). With a static decentering force effect (e.g., rotor weight) under small rotor unbalance force magnitudes, the orbit barely lifts off the bottom of the SFD, forming a small orbital trajectory that has been likened to a crescent moon. As unbalance magnitude is progressively increased, the orbit overcomes the static decentering force with the orbit progressing from the small crescent-moon trajectory, to a more elliptical shape. One could

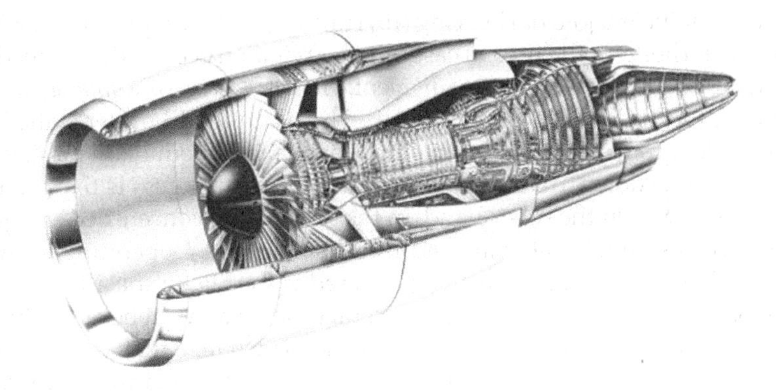

FIGURE 10.4
Modern double-spool-shaft high-bypass-flow gas-turbine jet engine.

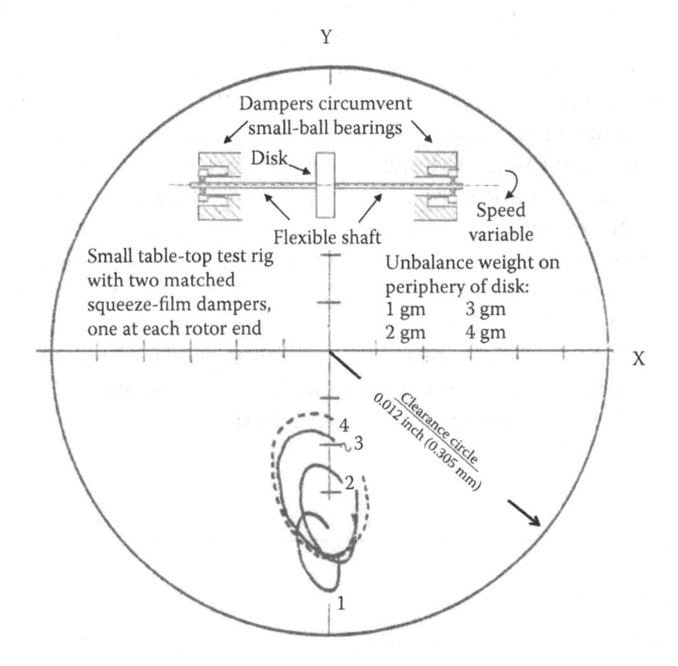

FIGURE 10.5
Rotor orbits with springless SFD at increasing rotor unbalance levels.

say that as the rotor vibration level needs attenuation, the springless SFD "rises to the occasion."

To provide a reasonable linear approximation to the nonlinear behavior of both springless and spring-centered SFDs, Hahn (1984) developed methods and results to approximate SFD dynamic characteristics with equivalent linearized stiffness and damping coefficients compatible with lateral rotor vibration analysis software. Such an approach appears to make sense when parametric preliminary design studies are conducted, leaving a full nonlinear analysis to check out a proposed and/or final prototype engine design.

In developing Equations 10.9 through 10.11, a concentric circular orbit trajectory is postulated. If one views the Reynolds equation solution for film pressure distribution in the SFD from a reference frame rotating at the orbit frequency (Ω), the pressure distribution is the same as in an equivalent journal bearing (JB) running at static equilibrium with the same concentric orbit of radius e. In the typical case where cavitation occurs, the respective SFD and JB Reynolds equation solutions are still equivalent. However, there is a *quite significant physical difference* between the SFD and its equivalent JB. That is, in the JB under static load there is typically an oil inlet groove near where the film gap starts its reduction (or "wedge" effect) and the cavitation zone downstream of the minimum film thickness is fixed in the JB space (see Figures 1.8 and 1.9). On the other hand, in the SFD with a concentric Ω-frequency orbit, the cavitation zone also rotates at Ω around the SFD annulus. Depending on whether end seals are used and on the through-flow of oil metered to the SFD, a specific "blob" of oil may be required to pass into and out of cavitation several times at a frequency of Ω during a single residence period within the SFD film. It is reasonable to visualize that as the orbit frequency is progressively increased, the SFD oil *refuses to cooperate* in that manner. Experiments have in fact shown that as orbit frequency is progressively increased, the SFD becomes an oil froth producer and its damping capacity falls far short of Reynolds equation-based predictions.

Hibner and Bansal (1979) provide the most definitive description on the failure of classical lubrication theory to reasonably predict SFD performance. They show with extensive laboratory testing at speeds and other operating conditions typical of modern aircraft engines that fluid-film lubrication theory greatly overpredicts SFD-film pressure distributions and damping coefficients. They observed a frothy oil flow out of their test damper. They suggest that the considerable deviation between test and theory stems from *gaseous cavitation*, greatly enhanced by air bubbles being drawn into the SFD to produce a *two-phase* flow which greatly reduces hydrodynamic pressures. The work of Hibner and other SFD researchers indicates that for low-speed applications, classical hydrodynamic lubrication theory can provide reasonable predictions for SFD performance. But at the quite high rotational speeds typical of modern aircraft engine rotors, classical hydrodynamic lubrication theory greatly over estimates SFD damping capacity. Therefore, thorough testing of specific SFD configurations is required to reliably determine actual SFD performance to enable reliable rotor-vibration design-analysis predictions.

10.4 Rolling-Element Bearings

As illustrated in Chapter 2, several different configurations of rolling element bearings (REB) are used in numerous applications. The most common RCB configurations are ball bearings which subdivide into the specific categories of *radial contact*, *angular contact*, and *axial contact*. Other commonly used REB configurations utilize *straight-cylindrical*, *crowned-cylindrical*, and *tapered-roller elements*. In many applications that employ REBs, the complete rotor-bearing system is sufficiently stiff (e.g., machine tool spindles) to operate at speeds well below the lowest critical speed resonance, so that the only rotor vibration consideration is proper rotor mass balancing. In flexible rotor applications, where operating speeds are above one or more critical speeds, the REBs often have sufficient internal preloading so that in comparison to the other system flexibilities (i.e., shaft and/or support structure) the REBs act essentially as rigid connections. In applications where the bearings have no internal preload and possibly some clearance (or "play"), the dynamic behavior can be quite nonlinear and thus standard linear analyses are potentially quite misleading.

REBs can readily be configured to achieve high stiffness, thus they are frequently used in applications where precision positioning accuracy is important, such as in machine tool spindles. In contrast to their high-stiffness potential, REBs have very little inherent vibration damping capacity, unlike fluid-film journal bearings. Also, unlike a fluid-film journal bearing which will fail catastrophically if its lubricant supply flow is interrupted, an REB can operate for sustained periods of time when the normal lubrication supply fails, albeit with a probable shortening of the REB's useable life. Thus, REBs are usually a safer choice over fluid-film bearings in aerospace applications such as modern aircraft gas-turbine engines. In such applications, however, the inability of the REBs to provide adequate vibration damping capacity to safely pass through critical speeds frequently necessitates the use of *squeeze-film dampers* (SFD) (Section 10.3) to support one or more of the machine's REBs. When an SFD is employed, its rotor vibration characteristics are usually the governing factor at the bearing, not the very high stiffness of the REB in series with the SFD.

As described in Chapter 2, *roller bearings* inherently possess much *higher-load capacity* and Hertzian contact stiffness than ball bearings, because a ball-load-supporting footprint is conceptually a *point* contact, whereas a roller-load-supporting footprint is conceptually a *line* contact, see Figures 1.25, 2.24, and 2.25. However, in roller bearings each roller has a single axis about which it must spin in proper operation. As rotational speed is increased for a roller bearing, the increased propensity for dynamic skewing of the rollers will impose a maximum useable rotational speed for the bearing. In contrast, a ball's spin may take place about any diameter of the ball. As a consequence, ball bearings have much higher maximum speed limits than roller bearings, given their inherent absence of dynamic skewing as explained in Chapter 2. Given the higher speed capability but inherently lower stiffness of ball bearings compared to roller bearings, it is far more likely that one would possibly need radial stiffness for a ball bearing than for a roller bearing when performing rotor vibration analyses. If one focuses on the load paths through an REB, two important factors become apparent.

1. Each contact between a rolling element and its raceways possesses a nonlinear load versus deformation characteristic (F vs. δ), for example, see Figure 2.6. That is, since the deformation footprint area between rolling element and raceway increases with load, the F versus δ characteristic exhibits a stiffening nonlinearity, that is, the slope of F versus δ increases with δ.
2. The total bearing load is simultaneously shared, albeit non-uniformly, by a number of rolling elements in compression as illustrated in Figure 2.9. Therefore, the contact forces taken by the rolling elements are *statically indeterminate*, that is, cannot be solved from force and moment equilibrium alone, but must include the flexibility characteristics of all elements simultaneously.

A suitable estimate of REB-radial stiffness for LRV analyses is obtained by assuming that the inner and outer raceways are both perfectly rigid. This simplifying assumption avoids employing the quite formidable and specialized complete static equilibrium-based solution described in Chapter 2 because it geometrically relates all the rolling elements' compressive deflections to a single bearing deflection. The load-versus-deflection for each rolling element is expressible from Hertzian elastic contact theory, such as in the following summary from Kramer (1993) (values for B based on steel):

$$F_i = \left(\frac{\delta_i}{B}\right)^n \tag{10.12}$$

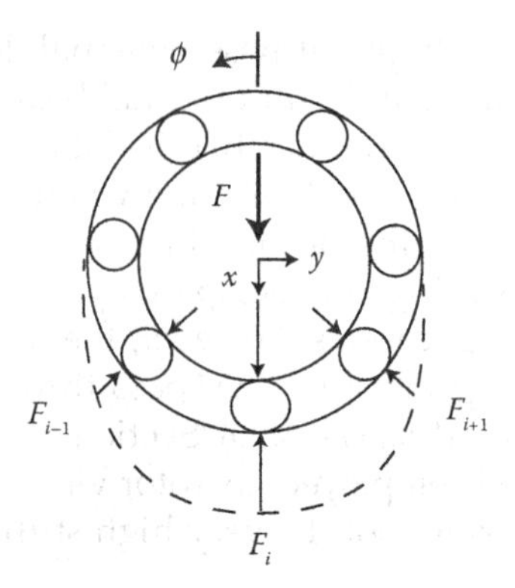

FIGURE 10.6
REB ball or roller-load arc, zero clearance and no preload.

where, F_i, δ_i are load and deflection of ith rolling element.

$$\left.\begin{matrix} n = 3/2 \\ B = 4.37\text{x}10^{-4} d^{-1/3} \end{matrix}\right\} \text{Ball} \qquad \left.\begin{matrix} n = 1/0.9 \\ B = 0.77\text{x}10^{-4} L^{-0.8} \end{matrix}\right\} \text{Roller}$$

Units: Ball diameter d (mm), roller length L (mm), contact force F_j (Newtons).

Contact forces occur only when a rolling element is in compression. It is implicit in the approximation here that the bearing has no internal preload and no play (clearance). Then the only source of contact loads is from the applied bearing load and the contact zone will be the 180° arc shown in Figure 10.6. The contact compressive deflection of each rolling element can then be expressed as follows, where x is the relative radial displacement between the raceways.

$$\delta_i = \begin{cases} x\cos\phi_i, & 90^\circ < \phi_i < 270^\circ \\ 0, & -90^\circ < \phi_i < 90^\circ \end{cases} \tag{10.13}$$

Clearance in a bearing tends to make the contact-load arc less than 180° and preload tends to make the contact-load arc greater than 180° as illustrated in Figure 2.9.

With ϕ referenced to the bearing load as shown, and with the bearing load equilibrated by the sum of the components of all the individual contact forces, that can be expressed as follows:

$$\begin{aligned} F &= \sum_{i=1}^{N} F_i \cos\phi_i = \sum_{i=1}^{N} \left(\frac{\delta_i}{B}\right)^n \cos\phi_i = \sum_{i=1}^{N} \left(\frac{x\cos\phi_i}{B}\right)^n \cos\phi_i = \left(\frac{x}{B}\right)^n S_x \\ S_x &\equiv \sum_{i=1}^{N} (\cos\phi_i)^{n+1} \end{aligned} \tag{10.14}$$

where N = number of rolling elements within the 180° arc of contact loading. From Equation 10.14, the bearing deflection is expressible as follows:

$$x = B\left(\frac{F}{S_x}\right)^{\frac{1}{n}} \tag{10.15}$$

Differentiating the radial-bearing force F by its corresponding radial-bearing deflection x, bearing x-direction stiffness is obtained as the following equation.

$$k_{xx} = \frac{dF}{dx} = \frac{nx^{n-1}}{B^n} S_x = \frac{n}{x}\left(\frac{x}{B}\right)^n S_x = \frac{n}{x} F \tag{10.16}$$

Visualize each loaded rolling element as a nonlinear-radial spring in compression. The stiffness of each rolling in the direction perpendicular to the bearing load can be obtained by projecting a y-direction differential deflection onto its radial direction and projecting its resulting differential-radial force back onto the y direction. The radial stiffness of an individual loaded rolling element is obtained by differentiating Equation 10.12 as follows:

$$k_i = \frac{dF_i}{d\delta_i} = \frac{n\delta_i^{n-1}}{B^n} = \frac{n(x\cos\phi_i)^{n-1}}{B^n} \tag{10.17}$$

Projecting the y-direction differential deflection onto the rolling element's radial direction and its resulting differential-radial force back onto the y direction yields the following.

$$d\delta_i = dy\sin\phi_i \text{ and } dF_{iy} = dF_i \sin\phi_i$$

Therefore,

$$dF_{iy} = dF_i \sin\phi_i = k_i d\delta_i \sin\phi_i = k_i dy(\sin\phi_i)^2$$

The y-direction stiffness for an individual loaded rolling element is thus obtained as follows:

$$k_{iy} = \frac{dF_{iy}}{dy} = k_i(\sin\phi_i)^2 \tag{10.18}$$

Summing all the rolling elements' y-direction stiffness yields the bearing's y-direction stiffness, as follows.

$$\begin{aligned} k_{yy} &= \sum_{i=1}^{N} k_i(\sin\phi_i)^2 = \frac{nx^{n-1}}{B^n}\sum_{i=1}^{N}(\cos\varphi_i)^{n-1}(\sin\phi_i)^2 = \frac{nx^{n-1}}{B^n}S_y \\ S_y &\equiv \sum_{i=1}^{N}(\cos\varphi_i)^{n-1}(\sin\phi_i)^2 \end{aligned} \tag{10.19}$$

Combining Equations 10.16 and 10.19 yields the stiffness ratio, as follows.

$$R_k \equiv \frac{k_{yy}}{k_{xx}} = \frac{S_y}{S_x} < 1 \tag{10.20}$$

This easily calculated ratio increases with the number of rolling elements in the bearing. Kramer (1993) provides values for the following example cases. Number of rolling elements in bearing = 8, 12, 16 gives the following corresponding values:

Ball bearing: $R_k = 0.46, 0.64, 0.73$ *Roller bearing*: $R_k = 0.49, 0.66, 0.74$.

The bearing stiffness coefficients given by Equations 10.16 and 10.19 are derived as though neither raceway is rotating. There are three cases of rotation one could encounter: (1) only the inner raceway rotates (most typical), (2) only the outer raceway rotates, and (3) both raceways rotate (e.g., inter-shaft bearings for multi-spool-shaft gas turbine jet engines). The cage maintains uniform spacing between the rolling elements, and when it rotates the bearing load and resulting deflection are perfectly aligned with each other only when the bearing load is either directly into a rolling element or directly between two rolling elements. At all other instances, bearing load and resulting deflection are very slightly out of alignment. This produces a very slight cyclic variation in the bearing stiffness coefficients at roller or ball passing frequency, and thus suggests the possibility of what is generically referred to as *parametric excitation*.

As an input into standard rotor-radial vibration-analysis software, such as the RDA code supplied on the CD accompanying this book, Equations 10.16 and 10.19 provide the following bearing interactive force with stiffness only, but zero damping coefficients.

$$\begin{Bmatrix} f_x \\ f_y \end{Bmatrix} = -\begin{bmatrix} k_{xx} & 0 \\ 0 & k_{yy} \end{bmatrix} \begin{Bmatrix} x \\ y \end{Bmatrix} \tag{10.21}$$

Of course, the chosen x-y coordinate system orientation in a given rotor vibration model may not align with the Equation 10.21 principal x-y coordinate system orientation, which is shown in Figure 10.6. However, as explained in Adams (2010), bearing and seal stiffness and damping coefficient arrays are *second rank tensors* and thus can be easily transformed to any alternate coordinate system orientation. Equation 10.21 transformed to a non-principal coordinate system yields non-zero off-diagonal stiffness terms which are equal, that is, the stiffness array is symmetric. Thus, this model for REB stiffness does not embody any destabilizing effect, in contrast to journal bearings.

10.5 Hydrostatic Bearings

From the derivations presented in Chapter 8, the film stiffness of a flat single-recess hydrostatic-bearing pad for the three types of flow compensation are summarized as follows.

$$\text{Orifice-flow compensation}: S_o = \frac{3W}{h_o}\left[\frac{2(1-\beta)}{2-\beta}\right] \tag{10.22}$$

$$\text{Capillary-flow compensation}: S_c = \frac{3W}{h_c}(1-\beta) \tag{10.23}$$

$$\text{Constant-flow compensation}: S_v = \frac{3W}{h_v} \tag{10.24}$$

As clear from the Chapters 1–8 references on hydrostatic bearings, *load capacity, film thickness, flow compensation,* and *stiffness* characteristics were well developed and published by the mid-1960s. In fact, hydrostatic bearing technology was already in hand for the support system of the Mount Palomar telescope (Figures 1.16 and 1.17), designed in the early post WW II era. However, much later in 1967 when the author began his employment as a junior research engineer at the Franklin Institute Research Laboratories (FIRL), he overheard FIRL bearing experts mentioning that someone has yet to develop vibration damping information for hydrostatic-bearing pads. That inspired the author to research the topic on his own time, utilizing then recently developed FIRL software for film bearings. That endeavor culminated in the then-young author's first published paper (Adams and Shapiro 1969). The treatment in this section is extracted from that paper.

To provide insight, the first step is to compare FIRL software output for a rectangular flat plate to earlier published closed-form-solution results for squeeze-film characteristics on a flat plate (Hays 1962). That exact comparison is replicated in Figure 10.7. The film pressure distribution shape for the flat plat case is illustrated in Figure 10.8a, for any arbitrary section perpendicular to an edge. Also shown (Figure 10.8b) is the effect of having a recess in a flat plate. Not that the pressure distribution over the small-clearance lands is the same in both Figures 10.8a and 10.8b. That is by virtue of continuity-of-flow the pressure gradient across the respective land regions must be identical. Last, consider the pressure distribution when the recess is connected to a flow restrictor (orifice or capillary) of flow resistance R, that is, giving the squeeze flow an additional "escape route" (Figure

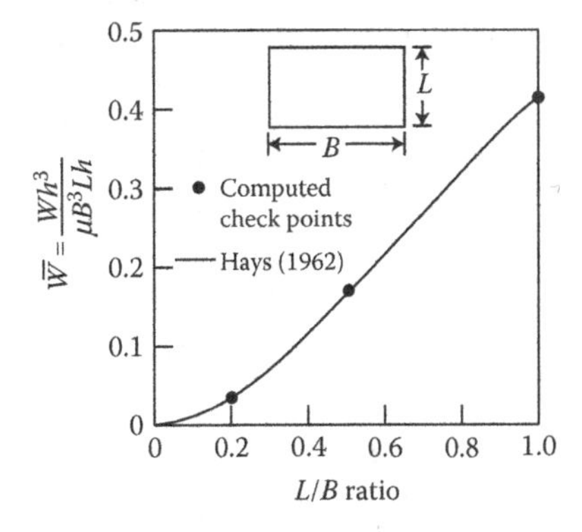

FIGURE 10.7
Squeeze-film-load capacity for a rectangular flat plate.

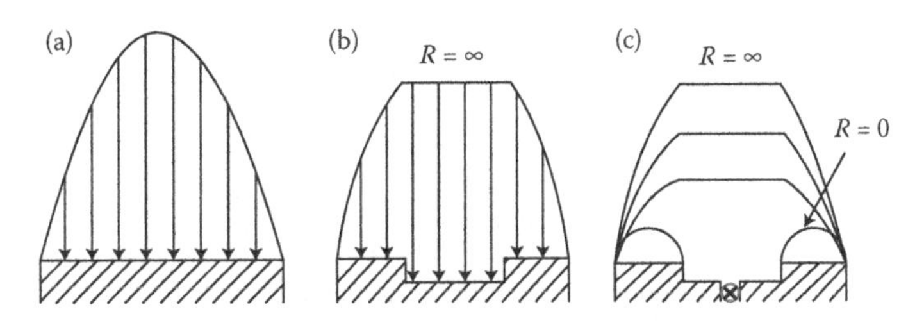

FIGURE 10.8
Squeeze-film pressure profiles for clearance decreasing in time; (a) Flat plate, (b) Flat plate with recess, (c) And flow restrictor.

10.8c). The implication here is that the damping capacity of a flat with recess(es) is always less than that for an equally sized flat plate of the same shape operated with the same parameters. Computed results show this to be true regardless of the static pressure ratio β, supply pressure, or flow-compensation-type.

Typical pressure profiles of a hydrostatic pad *with* and *without* squeeze-film velocity are illustrated in Figure 10.9. The squeeze-film force is the difference between the loads of two pressure distributions. When a *film-thickness-decreasing* squeeze action takes place in a hydrostatic pad (with either orifice or capillary compensation) one effect is to reduce or even reverse the instantaneous flow through the restrictor supplying the recess. This reduction in restrictor flow (constant supply pressure assumed) results from increased recess pressure. The increase in recess pressure, however, is not the sole contributor to squeeze load. While squeeze action takes place the pressure distribution across the sill is of a different form than under static conditions. This occurs because the total flow across the sills is the combination of flow from the recess plus the squeeze action on the sills.

When a *film-thickness-increasing* squeeze action takes place, the corresponding reduction of instantaneous pad pressure distribution is the negative of that for *film-thickness-decreasing* squeeze action. So the hydrostatic pad behaves similarly to a dashpot damper that gives the same damping force for both positive and negative velocity directions. That of course assumes that the *film-thickness-increasing* squeeze velocity is not high enough to cause pressures to drop to the fluid vapor pressure and then produce cavitation. In that case, the dynamic phenomenon would become highly nonlinear and not adequately modeled by a linearized damping coefficient. Summarizing, it can be seen that the damping-squeeze load is associated with two changes to the pad-pressure distribution, namely, (1) altered recess pressure and (2) change to the relative shape of pressure distribution across the sills.

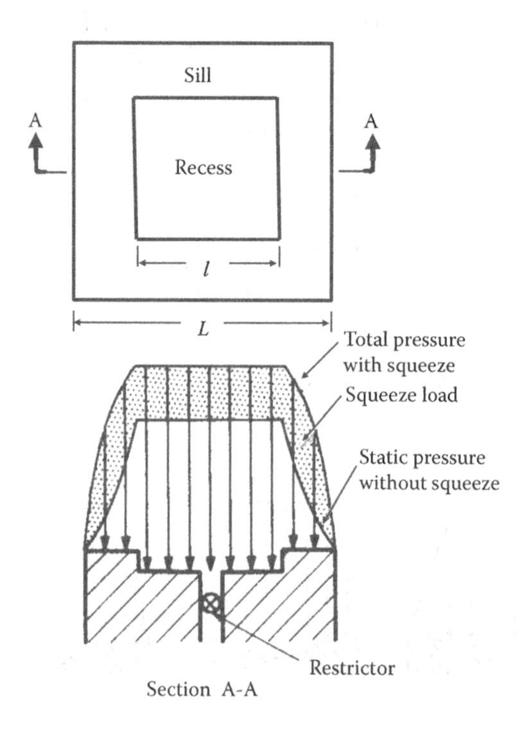

FIGURE 10.9
Hydrostatic pad with and without film-thickness-decreasing clearance.

One of the first findings established by Adams and Shapiro (1969) is the following relationship for squeeze film force for a plane rectangular recessed pad with flow compensation.

$$W \propto \frac{\mu B^3 L\dot{h}}{h^3} \tag{10.25}$$

This is the same general relationship that applies to the plane-rectangular flat-plate squeeze-film force. Furthermore, the Equation 10.25 relationship is similar to its counterpart for any flat hydrostatic recessed pad, for example, circular. Equation 10.25 only needs a dimensionless proportionality coefficient to specify the damping load, as follows:

$$W = \bar{W}\frac{\mu B^3 L\dot{h}}{h^3} \tag{10.26}$$

where, for orifice and capillary compensation, the dimensionless damping load is a function of (L/B, l/L, β). For constant flow compensation, the dimensionless damping load is a function of (L/B, l/L). The functional relationship for the dimensionless damping load cannot be expressed in closed mathematical form, except for a few pad configurations such as a circular pad with a concentric recess (Figure 8.2). Since damping-load equations for the *flat plate* and the same-shaped recessed *flat pad* have the same form, the damping force for the flat pad can be expressed as follows:

$$W(\text{for flat pad}) = D \times W(\text{for flat plate}) \tag{10.27}$$

where D is a proportionality coefficient.

Figures 10.10 and 10.11, replicated from Adams and Shapiro (1969), thus provide damping-load design charts for $L/B = 0.5$ and 1.0 in quite compact form. For specified operating parameters, those design charts actually provide a three-point curve fit against L/B, since

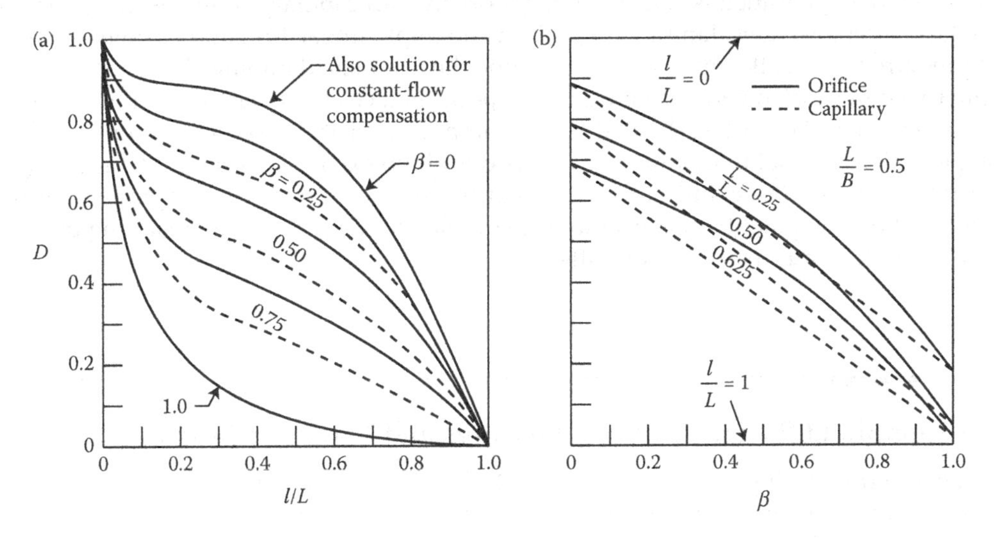

FIGURE 10.10
Hydrostatic damping D for single-recess flat pad, L/B = 0.5.

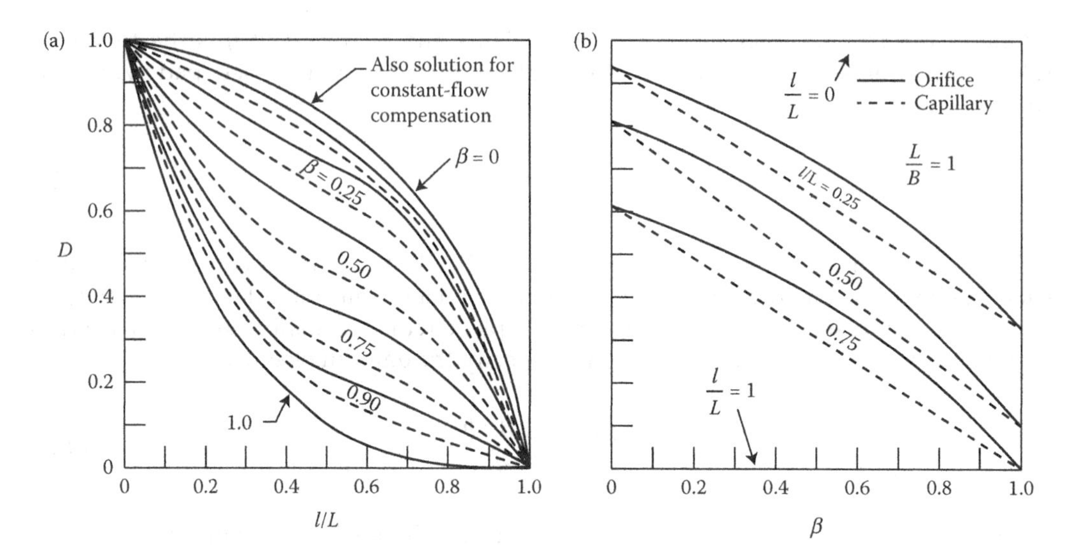

FIGURE 10.11
Hydrostatic damping ***D*** for single-recess flat pad, L/B = 1.0.

the L/D = 0.5 chart can also be used for $L/B = 2$, by interchanging the L and B roles. For vibration modeling purposes, the damping coefficient c is simply the ratio of *damping load* to *squeeze velocity*, as follows.

$$c = \frac{W}{\dot{h}} = W\frac{\mu B^3 L}{h^3} \tag{10.28}$$

On close examination of the design charts in Figures 10.10 and 10.11, certain generalizations become obvious: (1) the damping capacity of a given pad is considerably higher with constant-flow compensation than with either orifice or capillary compensation, (2) damping with orifice compensation is somewhat higher than with capillary compensation, (3) with capillary compensation, damping capacity is a simple linear function (straight line) of the pressure ratio β, allowing one to determine the complete damping characteristics of a capillary-compensated pad by computing its performance at only any two values of static pressure ratio β, like $\beta = 0$ and 1, and (4) the damping capacity of a pad with constant-flow compensation is equal to that of the same pad with no external supply to the recess (i.e., like with flow restrictor of infinite flow resistance). In other words, the damping load is not dependent on the quantity of constant flow supplied to the recess, only resulting because all squeeze flow must pass over the sills.

10.6 Compliant Surface Foil Gas Bearings and Magnetic Bearings

Stiffness and damping characteristics for (1) *compliance surface foil gas bearings* and (2) *magnetic bearings* are briefly treated in Chapters 1 and 3, respectively, where the difficulties of specifying and/or determining their stiffness and damping properties are explained. In consequence, neither of those two bearing types lends themselves to straightforward treatments on stiffness and damping as provided for the other bearing types covered in this chapter.

References

Adams, M. L., Insights into Linearized Rotor Dynamics, Part 2, *Journal of Sound and Vibration*, Vol. 112(1), 1987.

Adams, M. L., *Rotating Machinery Vibration—From Analysis to Troubleshooting*, 2nd ed., CRC Press/Taylor & Francis, Boca Raton, FL, 2010.

Adams, M. L., *Rotating Machinery Research and Development Test Rigs*, CRC Press/Taylor & Francis, Boca Raton, FL, 2017.

Adams, M. L. and Shapiro, W., *Squeeze Film Characteristics in Flat Hydrostatic Bearings with Incompressible Flow, Trans. American Society of Lubrication Engineers* (ASLE), July 1969.

Adams, M. L., Padovan, J., and Fertis, D., Engine Dynamic Analysis with General Nonlinear Finite-Element Codes, Part 1: Overall Approach and Development of Bearing Damper Element, ASME, *Journal of Engineering for Power*, Vol. 104(3), 1982.

Hahn, E. J., Equivalent Stiffness and Damping Coefficients for Squeeze-Film Dampers, *Proc., 3ed IMechE International Conference on Vibrations in Rotating Machinery*, York, England, 1984.

Hays, D. F., Squeeze Films for Rectangular Plates, *Trans. ASME Journal of Basic Engineering*, Paper No. 62-Lub S-9, 1962.

Hibner, D. and Bansal, P., Effects of Fluid Compressibility on Viscous Damper Characteristics, *Proc., Conference on the Stability and Dynamic Response of Rotors with Squeeze-Film Bearings*, University of Virginia, 1979.

Kramer, E., *Dynamics of Rotors and Foundations*, Springer-Verlag, Berlin, p. 383, 1993.

Lund, J. W., Arwas, E. B., Cheng, H. S., Ng, C. W., Pan, C. H. T., and Sternlicht, B., Rotor-Bearing Dynamics Design Technology, Part-III: Design Handbook for Fluid Film Type Bearings, *Wright-Patterson Air Force Base*, Technical Report AFAPL-TR-65-45, 299 p, 1965.

Someya, T., Mitsui, J., Esaki, J., Saito, S., Kanemitsu, Y., Iwatsubo, M., Tanaka, M., Hisa, S., Fujikawa, T., and Kanki, H., *Journal-Bearing Databook*, Springer-Verlag, Berlin, 323 pp, 1988.

Vance, J. M., *Rotordynamics of Turbomachinery*, Wiley, New York, p. 388, 1988.

Section III

Troubleshooting Case Studies

11

Turbine-Generator Destruction

11.1 Overview

Figure 11.1 shows photos from two major massive failures of two large 600 MW steam turbine-generator sets at (1) the Kinan Plant in Japan and (2) the Porsheville Plant in France. Motivated by these failures, the author was assigned the project to develop new general-purpose software to analyze time-transient nonlinear-rotor vibration of multi-bearing flexible rotors, to simulate operating with *very large rotor mass unbalance* as occurs with the detachment of large turbine blades. The development and example use of that then-first-of-its-kind software is documented in Adams (1980). Further revealing simulation results are provided by Adams and McCloskey (1984) from research funded by the Electric Power Research Institute (EPRI).

From the Figure 11.1a photograph it is clear where low-pressure (LP) turbine-released-blades at running speed burst through the LP outer housing. The resulting very large rotor unbalance force magnitude was approximately the same as, probably larger than, the LP rotor weight (85,000 lb, 38,600 kg). Figure 11.2 illustrates a double-flow LP steam turbine rotor that is the same configuration and with about the same blade sizes as in the two destroyed machines pictured in Figure 11.1.

Although the initiating root cause of the two failures pictured in Figure 11.1 was detachment of LP turbine blades, the ensuing dramatically high-vibration levels essentially destroyed both units end-to-end, for example, destroyed the generator shown in Figure 11.1c. Anyone who remembers their strength-of-materials course may recollect that shear stress from beam bending forces maximizes at the center of the beam. The destroyed generator rotor shown in Figure 11.1(c) experimentally confirmed that theory, big time!

Under normal shutdown of such large steam turbine generator sets, it typically takes about one-half hour or more for the rotor to completely coast down. Because such a large end-to-end rotor spinning at 3600 rpm (two-pole machine on 60 Hz AC system) possesses more kinetic energy than a 100-car freight train travelling at 70 mph, that is, $KE = \frac{1}{2} mv^2$. So when failures such as that pictured in Figure 11.1 occur, the rotor certainly does not quickly stop rotating. For example, if there is a sudden emergency automatic total cut-off of steam to the turbine rotors and the rotor only takes about 10 minutes to stop rotating, the operators already know that substantial internal unit damage has occurred even if all the shrapnel remains inside the housings.

An additional consequence of massive failures as pictured in Figure 11.1 stems from the fact that many of such large-unit generators are hydrogen gas cooled, not air cooled. This is because the windage power losses in the generator rotor-stator gap is smallest with H_2 gas which has the lowest molecular weight of any gas (remember the Hindenburg zeppelin?). In such units, the out-of-control rotor-vibration levels will usually wipe out

(a)

(b)

FIGURE 11.1
Destroyed 600 MW steam turbine-generator sets; (a) LP turbines, (b) brushless-exciter shaft. *(Continued)*

the generator hydrogen shaft seals, resulting in a massive hydrogen explosion and fire, putting any plant personnel near the generator at that point in considerable jeopardy. One such failure event occurred in the 1980s on the 350 MW Unit-1 (similar to unit in Figure 7.2) at the Lake Hubbard Plant near Dallas. Fortunately no one was injured, but the resulting intense fire heat not only deformed major I-beam support structures but also wiped out the control room for both Units 1 and 2. An afternoon Dallas rush-hour traffic-helicopter monitor announced over the radio to commuters that "There is an enormous fire at Lake Hubbard. Avoid the area." A few days later as the power company's consultant, the author saw firsthand the Lake Hubbard LP turbine casing looking almost identical to that in the Figure 11.1a picture, clearly exhibiting LP turbine blade detachment as the failure root cause. Peering into the large hole in the Lake Hubbard LP turbine housing, the author saw where on the rotor LP blades had detached. A similar failure also occurred in the

(c)

(d)

FIGURE 11.1 (Continued)
Destroyed 600 MW steam turbine-generator sets; (c) generator, (d) LP turbine last stage.

early 1980s at the American Electric Power's Amos Plant on an 800 MW unit. Such massive failures, although rare, are not one-of-a-kind occurrences. So such failures will surely occur somewhere else in the future, just like airliner crashes.

11.2 Simulation of Rotor Vibration from Very Large Unbalance

The unbalance force is modeled as a synchronous once-per-revolution co-rotating force on the rotor, placed at a last-stage LP blade row. The initial large-unbalance simulation results of the LP turbine in Figure 11.2 are shown in Figures 11.3 and 11.4, for two different journal-bearing types, respectively: (1) actual partial-arc journal bearings (see configuration, Figure 7.20), and (2) four-pad pivoted-pad journal bearings (see Figure 7.2c). These simulations were

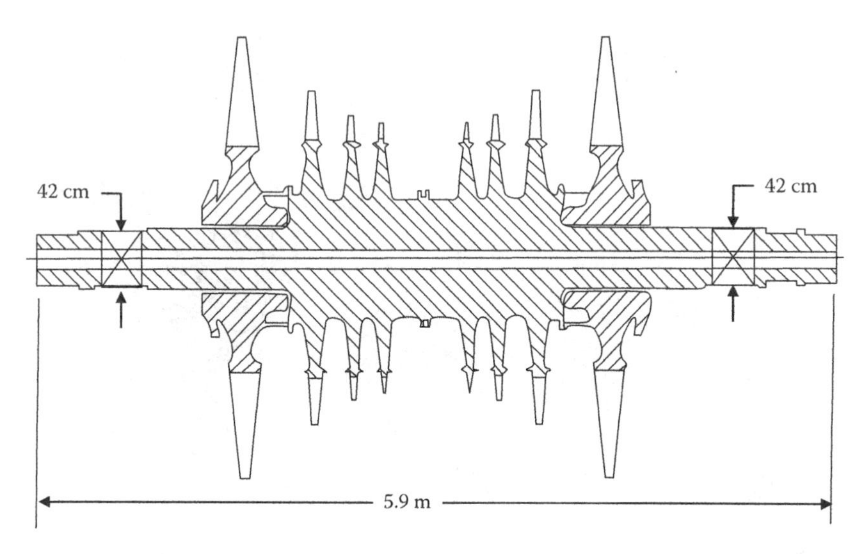

FIGURE 11.2
Low-pressure steam turbine rotor of a 3600 rpm 700 MW unit.

obtained with a time-transient dynamic integration of the FEA rotor model (160 degrees-of-freedom) with the journal bearing oil-film forces updated at each time step by resolving the Reynolds lubrication equation for each updated journal position and velocity relative to the respective journal bearing. The bearing support structures were also modeled as flexibly supported masses, so the bearings as well as the journals moved dynamically in the simulated response (Adams 1980).

The time-transient simulation is numerically integrated forward in time to determine if a steady-state periodic response results following an instantaneous last-stage blade detachment at running speed. In these simulations, a steady-state periodic response was predicted, with a period of three shaft revolutions. With the high degree of dynamic nonlinearity in journal-bearing oil-film force characteristics under such large dynamic motions, converging to a steady-state periodic motion is not automatically guaranteed because of the possibility of chaotic motion in the presence of such substantial dynamic nonlinearity. In both simulation cases (Figures 11.3 and 11.4), it required approximately 60 shaft revolutions following the simulated blade detachment to converge to the period-three steady-state motions shown, that is, 60 revolutions at 3600 rpm requires one second in real-time. So in such a real failure, anyone in the plant would first hear a *large bang* follow by a one-second build-up of substantial *noise and floor vibration*, and would then instinctively head out the door to a potentially safer location!

The explanation for the period-three motion is that the lowest LP critical speed resonance is near 1150 rpm, a mode lightly damped at 3600 rpm as evidenced by occasional dynamic-instability self-excited oil-whip vibration of this unit design in the field. When a dynamic system is linear, its steady-state vibration after the start-up transient dies out will contain only the excitation frequency, as rigorously proven by elementary ordinary differential equation mathematics. But when a system becomes highly nonlinear, that "opens the door" for excitation-force energy to "leak" into harmonics (i.e., 1/2, 1/3, 1/4 …, and 2, 3, 4 …) of the excitation frequency. In this case, that strongly energized the very lightly damped and quite receptive 1150 cpm harmonic, close to 1/3 the excitation frequency.

The long-time standard journal bearing configuration for the LP turbine illustrated in Figure 11.2 is the 160° partial-arc viscosity-pump design illustrated in Figure 7.20. It is the

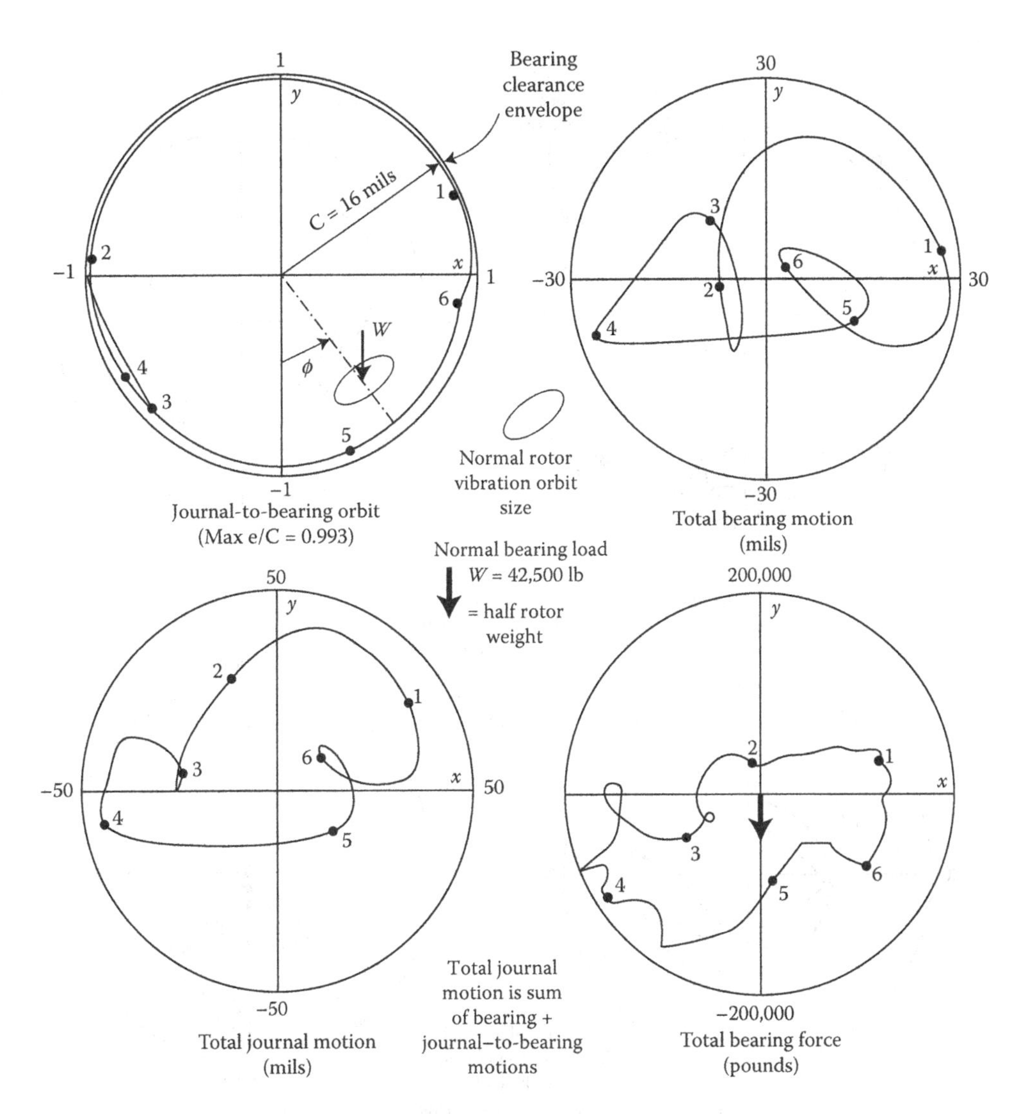

FIGURE 11.3
Steady-state periodic response at bearing nearest the unbalance with force magnitude of 100,000 lb (445,000 N); rotor supported on two identical fixed-arc journal bearings modeled after the actual two journal bearings. Timing marks at each one-half revolution, that is, 3 revs shown. (From Adams, M. L., *Rotating Machinery Vibration–From Analysis to Troubleshooting*, 2nd ed., CRC Press/Taylor & Francis, Boca Raton, FL, 2010.)

journal-bearing design modeled for the simulation results in Figure 11.3. Realizing that pivoted-pad journal bearings do not inherently possess dynamic destabilizing characteristics that fixed-arc bearings have, a follow-up second simulation study with the two *partial-arc journal bearings replaced by equally sized pivoted-pad journal* bearings was undertaken. The steady-state results for that simulation are shown in Figure 11.4. The difference between the two simulations (Figures 11.3 and 11.4) is truly dramatic. That dramatic difference is made quite clear when the peak-to-peak journal vibration is fast Fourier transformed (FFT), as illustrated in Figure 11.5. The FFT clearly shows that while the steady-state period vibration is period-three in both simulations, the pivoted-pad bearings do not dramatically amplify the 1/3 subharmonic.

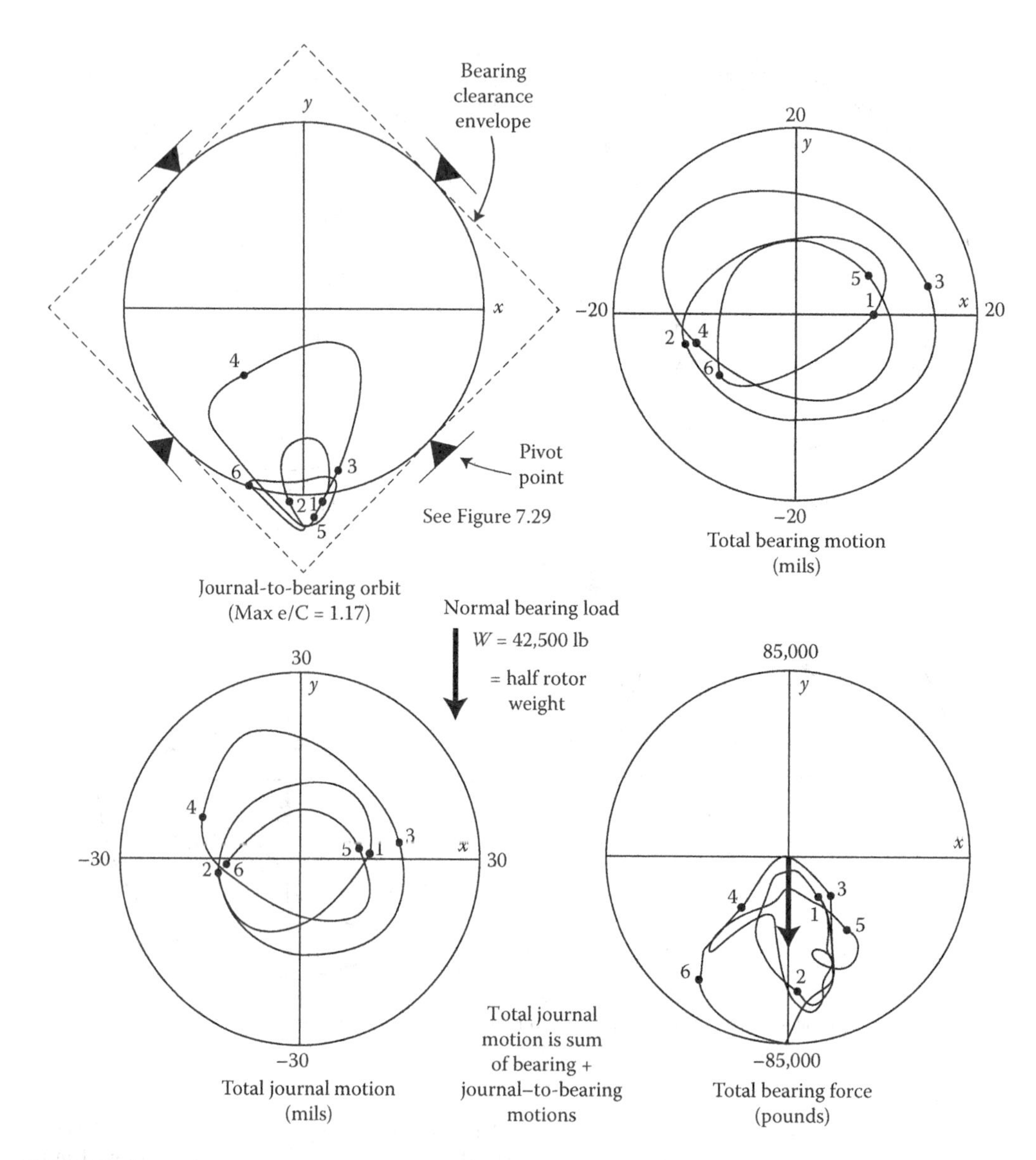

FIGURE 11.4
Steady-state periodic response at bearing nearest unbalance with force magnitude of 100,000 lb (445,000 N), rotor supported on two identical four-pad pivoted-pad bearings with the gravity load directed between the bottom two pads. Bearings have the same film diameter, length, and clearance as the actual fixed-arc bearings. Timings mark each one-half revolution, that is, 3 revs shown. (From Adams, M. L., *Rotating Machinery Vibration–From Analysis to Troubleshooting*, 2nd ed., CRC Press/Taylor & Francis, Boca Raton, FL, 2010.)

11.3 Simulation of a European Unit

First publication of these results (Adams 1980) led to the author's full-time 1984-summer consulting job at the Swiss company Bown-Boveri Corporation (BBC) in Baden near Zurich. The author used his software to model the resulting vibration of a large BBC LP turbine, in BBC journal *Bearings*, subjected to detachment of a full LP last stage blade. The two

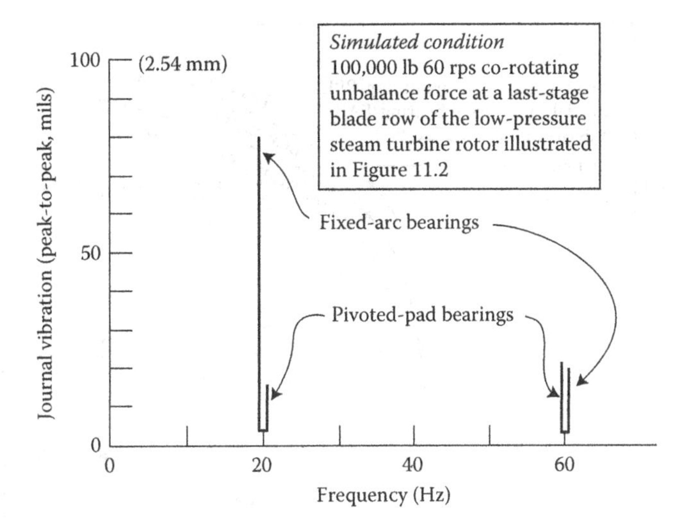

FIGURE 11.5
FFT of peak-to-peak journal vibration displacement amplitudes.

major U.S. manufacturers at that time employed many more but lighter blades on their double-flow LP last-stages, with wire-lashed-together blade groups to detune blade natural frequencies. In stark contrast, the European LP turbines employed substantially fewer but much more massive *free-standing* last-stage LP turbine blades, reflecting their approach to controlling LP last-stage blade vibration. As a consequence, the unbalance force from a completely BBC detached free-standing blade yielded an unbalance force magnitude of about 600,000 lb (2.7 million N), six times the unbalance force magnitude for the simulations shown in Figures 11.3 and 11.4. The BBC simulations were of course quite revealing to say the least, but did not exhibit a greatly amplified subharmonic as exhibited in the Figure 11.3 simulation results. That is because the BBC machine running at 3000 rpm (two-pole machine on 50 HZ AC system) was not close to a threshold speed for oil-whip self-excited vibration, that is, it did not have a very lightly damped subharmonic of the 3000-rpm spin speed.

11.4 Rotor Vibration as a Function of Unbalance Magnitude

The EPRI-sponsored research reported by Adams and McCloskey (1984) uncovered the sudden transition from the *linear* vibration zone to the *nonlinear* zone, by parametrically studying how the steady-state rotor vibration varies as a function of the unbalance force magnitude for the rotor-bearing systems in the Figure 11.3 and 11.4 simulation results. The quite revealing finding from that study is shown in Figure 11.6, for unbalance magnitudes from quite small up to the 100,000 lb (445,000 N) unbalance magnitude of the simulation results in Figures 11.3 and 11.4.

For the partial-arc bearing case, the transition from the linear zone to the nonlinear zone is quite sudden, at about 35,000 lb unbalance force magnitude, remarkably exhibiting a *classical nonlinear jump phenomenon* from a numerical time-transient integration algorithm of a multi-degrees-of-freedom system. As clearly shown, once the transition from journal-bearing

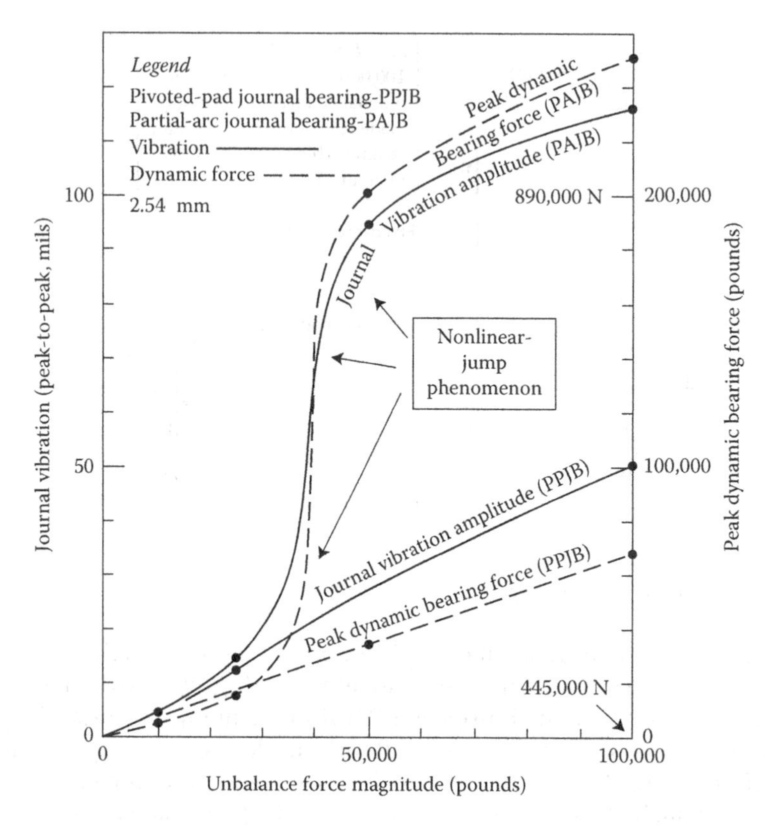

FIGURE 11.6
Comparison between partial-arc and pivoted-pad journal-bearing vibration-control capabilities under large unbalance operating conditions of an LP steam turbine rotor at 3600 rpm; steady-state journal motion and transmitted-peak dynamic bearing force over a range of unbalance magnitudes (data points mark computed simulation cases).

linear to nonlinear behavior takes place (in the 35,000 to 40,000 lb zone), further increases in the vibration-amplitude and total-bearing-force magnitude are relatively small. That is, the journal orbits are already filling up the bearing clearances, so further unbalance magnitude increases only reflect bearing support structure increased responses. Of course, the comparison in Figure 11.6 between the *partial*-arc bearing and *pivoted-pad* bearing cases is striking, exhibiting no nonlinear-jump phenomenon with the pivoted-pad bearings. If the author owned a 3600 rpm large steam turbine generator unit with fixed-arc journal bearings in the LP turbine(s), he would have the bearings replaced by pivot-pad journal bearings, primarily for the sake of power plant personnel safety.

For the case employing the actual partial-arc journal bearings, the author has a more insightful intuitive explanation of the sharp increase in vibration amplitude around the 40,000 lb unbalance force level (Figure 11.6) than just academically recognizing it as a "nonlinear-jump phenomenon." It is well known that for any linear vibration model with only harmonic excitation forcing functions imposed, the steady state vibration response can contain only the frequency(s) of the excitation force(s). This comes directly from elementary ordinary differential equation mathematics. The *particular solution* for vibration is the *steady-state solution* and the *homogeneous solution* is the *transient response*. However, when a system becomes nonlinear, steady-state response can contain integer super and subharmonics of the forcing frequency(s), (i.e., 2, 3, ... and 1/2, 1/3, ...). So the presence of nonlinearity can be thought up as a "valve" opening up to allow energy to invade the harmonics. In the Figure 11.6 example, as postulated unbalance force is increased, the degree of nonlinearity increases, progressively opening up that "valve," which is totally "closed" when the system is linear. If any of the harmonics are lightly damped, the overall vibration amplitude will be accordingly increased. And that vibration increase will cause the degree of nonlinearity to increase, which will in turn allow significantly more energy into any lightly damped harmonics, resulting in further increase in the degree of nonlinearity, and so on. That synergistic effect is what the academics call a "nonlinear-jump phenomenon," taking them less effort than acquiring a physical understanding of it. It's kind of like when physicians come up with a fancy name of a disease or disorder they don't understand and don't know how to treat.

References

Adams, M. L., Nonlinear Dynamics of Flexible Multi-Bearing Rotors, *Journal of Sound and Vibration*, Vol. 71(1), pp. 129–144, 1980.

Adams, M. L., *Rotating Machinery Vibration—From Analysis to Troubleshooting*, 2nd ed., CRC Press/Taylor & Francis, Boca Raton, FL, 2010.

Adams, M. L. and McCloskey, T. H. Large Unbalance Vibration in Steam Turbine-Generator Sets, *Proceedings, 3rd International Conference on Vibration in Rotating Machinery, Institution of Mechanical Engineering*, York, England, September 1984, I Mech E Conference Publication 1984-10, pp. 491–497.

12

Tilting-Pad Journal-Bearing Pad Flutter

12.1 Overview

Figure 12.1 illustrates the observed damage to the babbitt white metal surface of an unloaded top pad of a four-pad tilting-pad journal bearing of a high-pressure (HP) turbine in a large steam turbine generator unit such as those illustrated in Figures 7.2 and 7.3. The nominal bearing-journal diameter is approximately 12 in. (30.5 cm). Employing similar time-transient nonlinear dynamics motion simulation described in Chapter 11, the author undertook an investigation to discover the cause of this top-pad damage that had been observed on a number of similar units during major shutdowns of turbine generator units in the field for internal unit component inspections. Such major unit inspection shutdowns were then typically conducted at 5-year intervals, but are typically now problematically conducted at much longer time intervals, motivated by plant-owner cost-cutting MBA mentality. Although the top pads are essentially not load-carrying during normal operation, top pads or some fixed-arc upper bearing half is needed to contain the rotor under emergency conditions such as the topic of Chapter 11. The work reported in this chapter was first reported in Adams and Payandeh (1983).

12.2 Modeling of Single-Pad Dynamics

Figure 12.2 illustrates the two-degrees-of-freedom dynamics model used to simulate the behavior of a statically unloaded-tilting pad. The two-degrees-of-freedom include the pad angular pitch angle δ and the radial coordinate y as illustrated. Thus, there are two equations of motion, one for each degree of freedom as follows.

Pad pitch motion

$$I_p\ddot{\delta} = M(y,\dot{y},\delta,\dot{\delta}) \tag{12.1}$$

and

Pad-radial motion

$$m_p\ddot{y} = F(y,\dot{y},\delta,\dot{\delta}) \tag{12.2}$$

As detailed by Adams and Payandeh, in the time-transient marching algorithm the Reynolds lubrication equation is resolved each time-step based on updated displacement and velocity values for each motion coordinate. The instantaneous fluid-film moment M

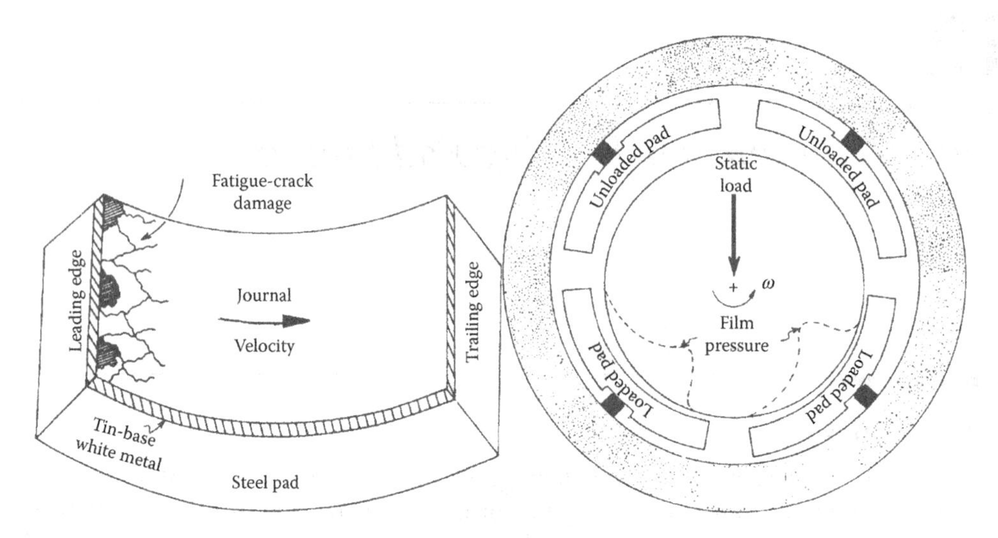

FIGURE 12.1
Four-pad tilting-pad journal-bearing loaded between bottom pads.

and radial force F on the pad are obtained by numerically integrating the pad-film pressure distribution $p(\theta, z)$ as follows.

$$M = -(R - d_R)R \int_{-L/2}^{L/2} \int_0^{\beta} p(\theta, z)\sin(\theta - \alpha)d\theta dz \tag{12.3}$$

and

$$F = -R \int_{-L/2}^{L/2} \int_0^{\beta} p(\theta, z)\cos(\theta - \alpha)d\theta dz \tag{12.4}$$

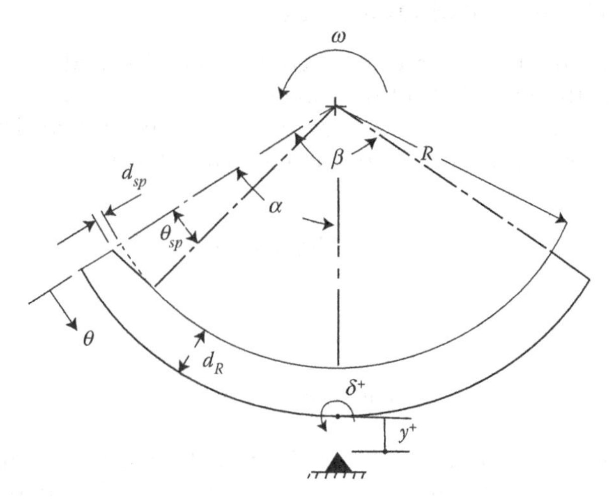

FIGURE 12.2
Single pivoting-pad two-degrees-of-freedom dynamics model.

Details for achieving stable time-step integration of Equations 12.1 and 12.2, given the relatively small pad inertia I_p and mass m_p, are provided by Adams and Payandeh. They present results from a broad parametric study that demonstrates for several parameter combinations whether an unloaded pad will exhibit self-excited pad flutter vibration. Those factors include (1) pad arc β, (2) pivot angular location α/β, (3) ratio of pivot clearance to ground clearance C_p/C (preload factor), and (4) Sommerfeld number parameters, that is, size, clearance, speed, and viscosity.

12.3 Simulation and Explanation of Unloaded Tilting-Pad Flutter

A representative sample of simulation results from Adams and Payandeh exhibiting unloaded-pad self-excited pad flutter are replicated in Figure 12.3. These simulation results

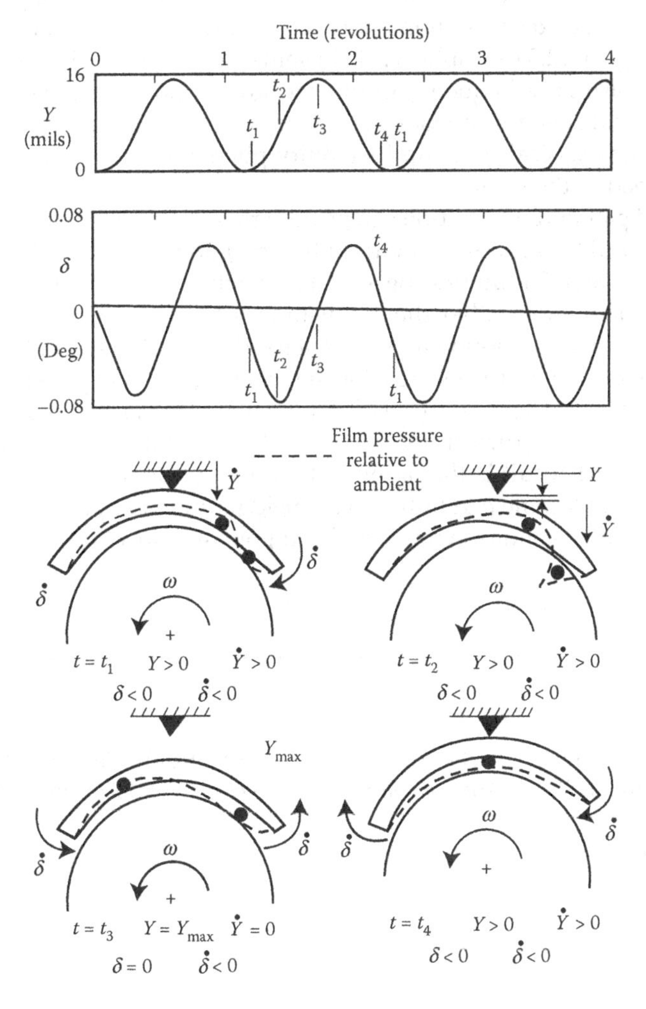

FIGURE 12.3
Numerical simulation of dynamic response of unloaded pad.

display the y and δ time-base signals and four corresponding synchronized snapshots of the pad in self-excited pad-flutter motion. The pad in this simulation has so-called *spragg relief* on the leading edge of the pad which is just visible on the Figure 12.3 illustration. Spragg relief is a slight 10° chamfer-like ramp on the pad surface at the leading edge, so when/if the pad tries to close the film at the leading edge it touches the pad not exactly at the leading edge but slightly downstream of the leading edge. So then even with the pad momentarily touching the journal near its leading edge, there is still a short converging film-thickness section at the pad leading edge to generate a hydrodynamic wedge effect to initiate pad pitching back in the positive direction. The damaged pad illustrated in Figure 12.1 had leading-edge spragg relief.

To their knowledge, Adams and Payandeh were the first to report on this self-excited pad flutter vibration phenomenon. Field experience suggests that it is probably a common occurrence in many machines employing tilting-pad journal bearings, although it may not always result in bearing damage like illustrated in Figure 12.1. The self-excited motion appears to occur in nearly the same fashion, regardless of the parameter combination which allows it to occur. Generally, the pad floats back and forth between the pivot point and the journal. In some simulated cases, momentary contact with the pivot occurs once per cycle as shown in the Figure 12.3 example, while in some cases the pad does not quite touch the pivot at all. In all cases, the frequency of the periodic motion is between 0.4 and 0.5 times the journal rotational speed frequency. In this regard, the motion appears to be a similar phenomenon to the classical oil-whip instability phenomenon, except that the unloaded pad vibrates instead of the rotor.

The self-excited pad flutter vibration can be thought of as simply the result of the absence of a stable static equilibrium position within the (δ, y) domain allowed by the kinematical constraints of the system. The interesting aspect is that the absence of an equilibrium position produces a periodic motion rather than a "wandering" type motion. In fact, if one plots the pad center-of-curvature motion relative to a fixed point, it traces out a subsynchronous orbit like the rotor does with respect to a journal bearing when experiencing oil-whip self-excited vibration. That says pad flutter is fundamentally the oil whip phenomenon except that the pad center-of-curvature orbits relative to the rotor rather than vice versa. The most useful troubleshooting result is that a modest amount of preload (see Figure 7.24) can be used to suppress troublesome pad flutter in the field if unloaded-pad leading-edge damage is observed during major shutdown internal component inspections.

Reference

Adams, M. L. and Payandeh, S., Self-Excited Vibration of Statically Unloaded Pads in Tilting-Pad Journal Bearings, *ASME, Journal of Lubrication Technology*, Vol. 105(3), pp. 377–383, 1983.

13

Earthquake-Induced-Bearing Damage

13.1 Overview

The *hysteresis loop* associated with the journal-bearing-caused dynamic instability self-excited vibration phenomenon called *oil whip* was for a long time an interesting topic for the academics. It did not attract the close scrutiny of rotating machinery development engineers. However, in the seismically active region of Japan, a team headed by Professor Y. Hori at the University of Tokyo brought the practical importance of the journal-bearing hysteresis loop to the wider engineering community. In the paper by Hori and Kato (1990), the distinct possibility of an earthquake-initiated high-amplitude sustained self-excited rotor vibration in large steam turbine generator sets is addressed. That work motivated subsequent research by the author and his team, reported in the paper by Adams et al. (1996).

13.2 Journal-Bearing Hysteresis Loop Simulations

A generic illustration from Adams et al. (1996) of the journal-bearing hysteresis loop is shown in Figure 13.1. Their illustration encapsulates the imbedding of the classical oil-whip phenomenon within an expanded view that shows *two stable solutions* at speeds below the classical oil-whip threshold speed ω_{th} (Hopf bifurcation) and *one unstable solution* which is bounded between the two stable solutions. Employing a time-transient forward-marching simulation algorithm as applied for the topics in Chapters 11 and 12 for tracking nonlinear dynamic motions, they simulated journal-bearing hysteresis loop data points. The simulation model is illustrated in Figure 13.2. Each data point is obtained by simulating the journal orbit transient forward from time = 0 to a convergent steady-state response, illustrated in Figure 13.3 for convergence to a nonlinear limit-cycle.

13.3 Danger of the Stable Nonlinear Limit-Cycle

Figure 13.4 shows steady-state hysteresis-loop vibration simulation results for a series of four ascending bearing dimensionless static loads, with simulation data points marked. The oil-whip threshold speed ω_{th} is at the small-perturbation Hopf bifurcation and is the

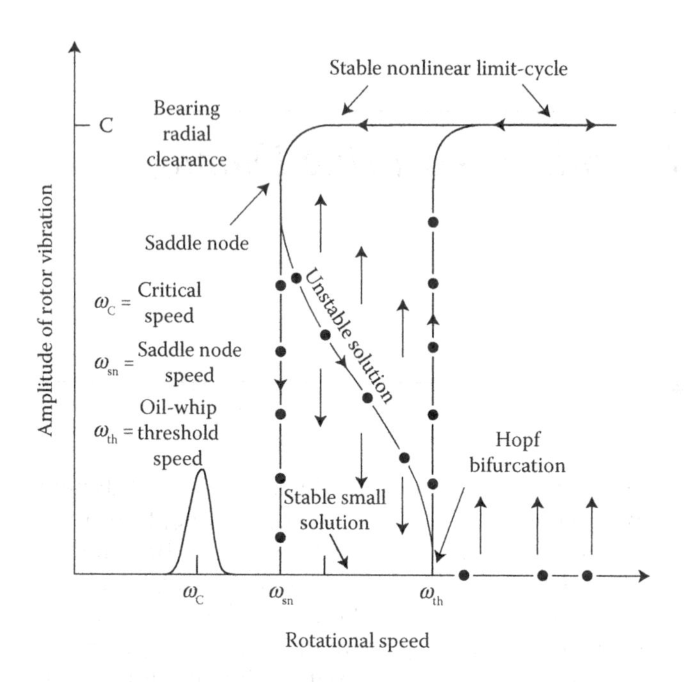

FIGURE 13.1
Journal-bearing hysteresis loop.

instability threshold speed predicted from linearized rotor-bearing equations of motion (Adams 2010). The Figure 13.4 simulation results show how the *hysteresis loop opens up* with increased bearing-static load. This encompasses the well-known fact that the oil-whip threshold speed ω_{th} is increased as journal-bearing-static load is increased. In other words, oil whip occurs when the bearing is too lightly loaded or oversized for the application load. With little or no bearing load, oil whip occurs at near twice the lowest frequency forward-whirl resonance mode critical speed ω_C. And the oil-whip orbital-vibration frequency transitions from $\omega_{th} \cong \omega_c$ at incipience to larger than ω_c as the effective stiffness of the oil film increases considerably as the journal orbit grows to nearly fill up the bearing clearance.

Again, the Figure 13.4 results confirm as is well known the favorable result that the oil-whip threshold speed increases significantly with increased bearing-static loads. But the other *not well-known unfavorable result* is that the hysteresis-loop saddle-node speed

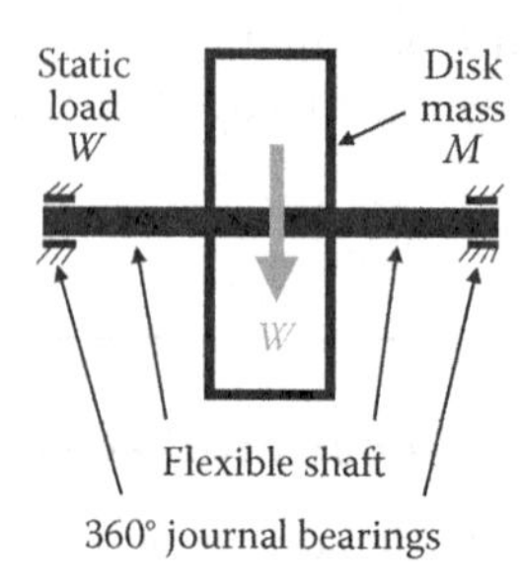

FIGURE 13.2
Rotor-bearing simulation model.

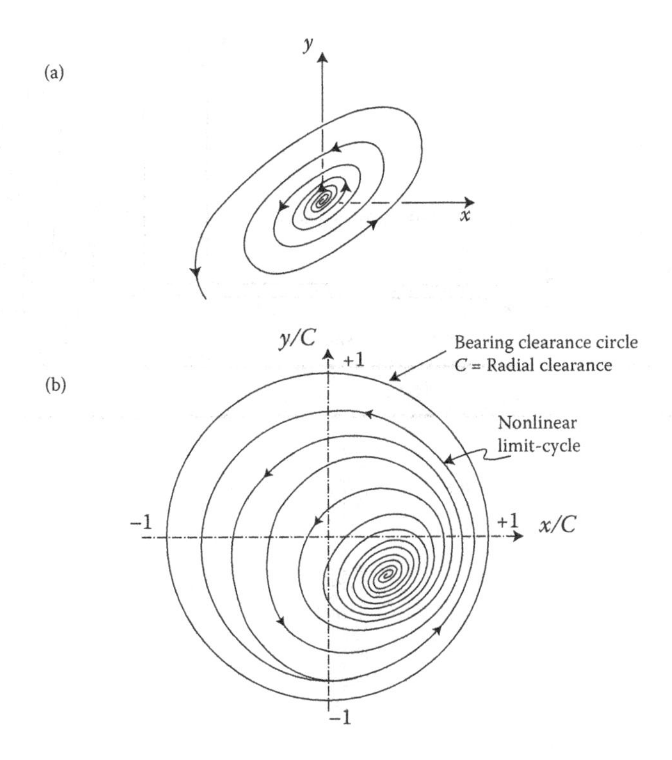

FIGURE 13.3
Simulated time-transient motion to nonlinear limit-cycle; (a) beginning of transient, primarily linear response, about static-equilibrium point, (b) orbit referenced to bearing center.

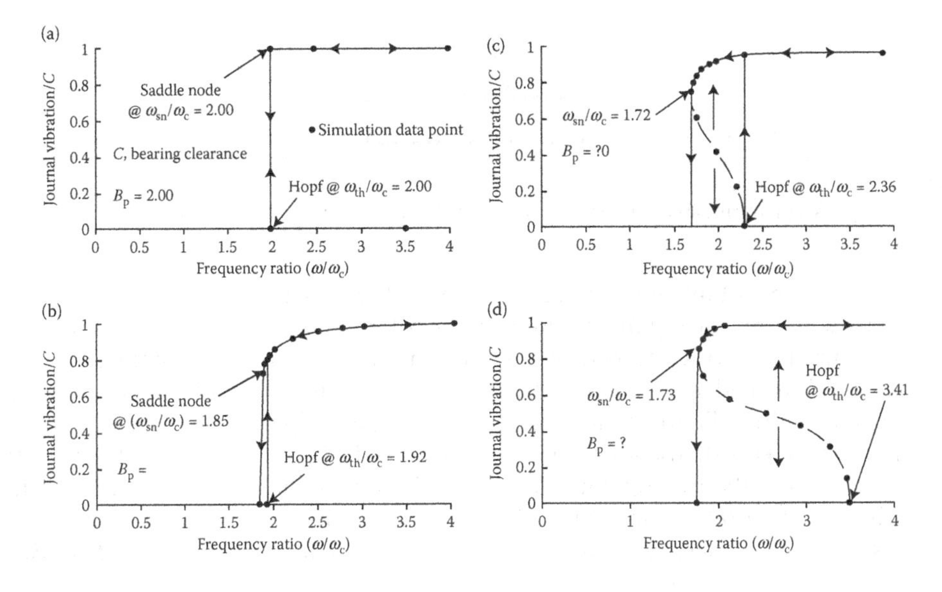

FIGURE 13.4
Hysteresis rotor response for ascending bearing static load; $B_p = S\omega_n/(\Omega_n\omega)$, where ω_n = natural frequency, $\Omega_n = \omega_n\sqrt{C/g}$, S = bearing Sommerfeld number; thus $1/B_p$ is a dimensionless load; nomenclature per Hori, Y. and Kato, T., Earthquake Induced Instability of a Rotor Supported by Oil Film Bearing. *ASME Journal of Vibration and Acoustics*, Vol, 112, pp. 160–265, 1990.

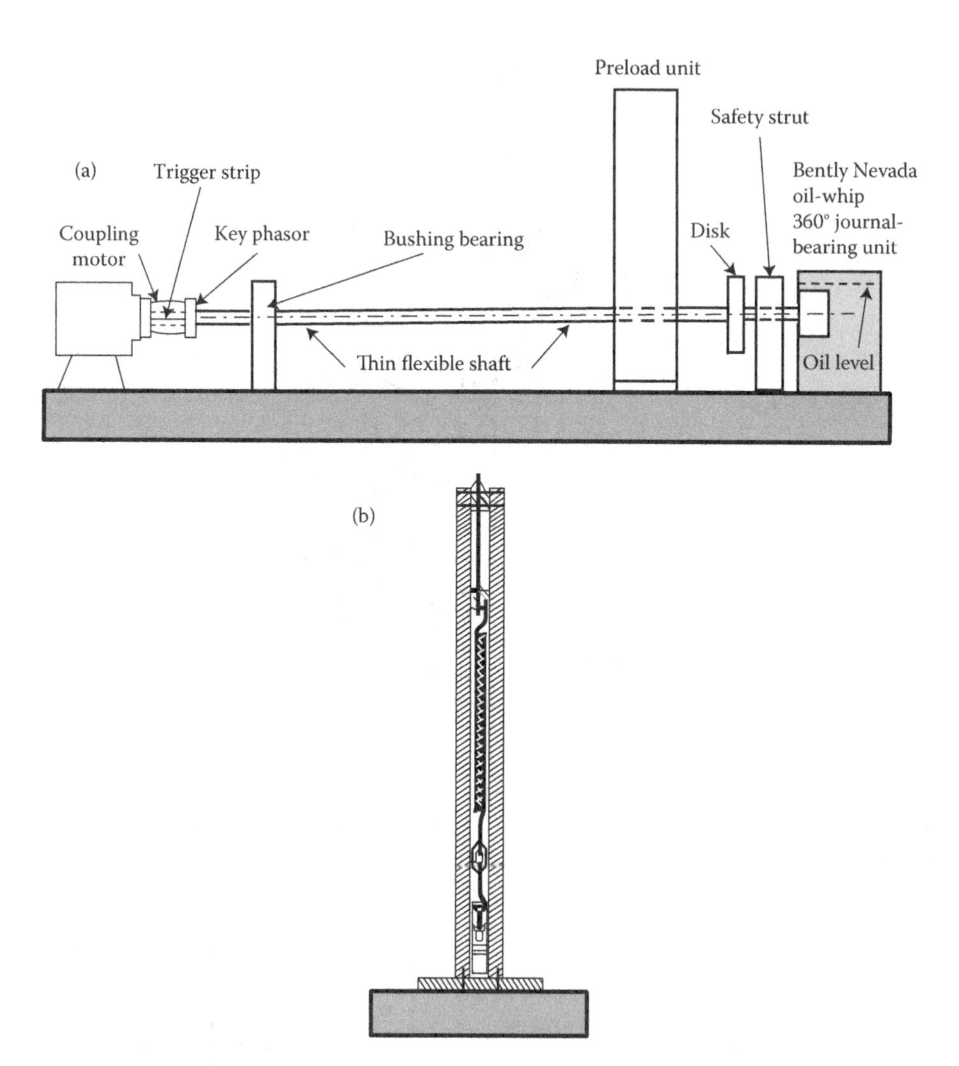

FIGURE 13.5
Hysteresis-loop test apparatus; (a) modified Bently Nevada kit, (b) detail of preload unit.

is reduced. The important practical consequence of this is that a large-enough motion disturbance such as an earthquake could push the rotor vibration up to the stable nonlinear limit-cycle amplitude which would persist even after the earthquake stops shaking the machine. And threshold spin speed for this is even lower than ω_{th} with the static load near zero. For example, from Figure 13.4d, even though the oil-whip threshold speed is 3.4 times ω_c (the first critical speed), the saddle node speed is only 1.73 ω_c (Figure 13.4d). So a large ground-motion disturbance might kick the rotor vibration up to the stable nonlinear limit-cycle, with journal-vibration amplitude filling up the bearing clearance. *Such an occurrence would possibly result in massive vibration-caused life-threatening machine destruction like that shown in the Figure 11.1 photographs.*

TABLE 13.1

Summary of Experimental Results from Test Rig in Figure 13.5

Preload lb	B_p	K_s lb/in	ω_n Hz	Ω_n	W lb	ω_{th} Hz	ω_c Hz	ω_{sn} Hz	ω_{sn}/ω_c
3.46	0.94	232	41.1	0.98	0.97	70.3	35.2	62.5	1.78
3.90	0.70	199	38.1	0.91	1.30	72.3	36.1	64.5	1.78
4.35	0.56	175	35.7	0.85	1.64	80.1	39.1	66.1	1.69
4.79	0.46	155	33.7	0.81	1.97	84.0	41.0	68.0	1.66

13.4 Hysteresis Loop Experiments

The research results reported by Adams et al. (1996) also include corresponding experimental results. A Bently Nevada rotor kit was modified to facilitate hysteresis loop speed bounds, that is, Hopf-bifurcation and saddle-node speeds (Adams 2017). The table top test rig is illustrated in Figure 13.5. The static load on the 360° journal-bearing load was applied on the flexible shaft in the upward direction through a relatively soft spring so that shaft-radial motion would not change the applied static journal-bearing load. Results from tests are summarized in Table 13.1, and are consistent with the simulation results.

References

Adams, M. L., *Rotating Machinery Vibration—From Analysis to Troubleshooting*, 2nd ed., CRC Press/Taylor & Francis, Boca Raton, FL, 2010.

Adams, M. L., *Rotating Machinery Research and Development Test Rigs*, CRC Press/Taylor & Francis, Boca Raton, FL, 2017.

Adams, M. L., Adams, M. L., and Guo, J. S., Simulations and Experiments of the Non-Linear Hysteresis Loop for Rotor-Bearing Instability, *Proceedings, 6th IMechE International Conference on Vibration in Rotating Machinery*, Oxford University, pp. 309–319 September 1996.

Hori, Y. and Kato, T., Earthquake Induced Instability of a Rotor Supported by Oil Film Bearing. *ASME Journal of Vibration and Acoustics*, Vol, 112, pp. 160–265, 1990.

14

Refrigerator Compressor Wrist-Pin Bearing

14.1 Overview

The piston and connecting rod sub-assembly shown in Figure 14.1 is from a single-piston reciprocating refrigerant compressor that was designed for use in both a home refrigerator and a window air conditioner, both produced by the same major U.S. home appliance manufacturer. Several months after some design modifications to this compressor were released into the market place (e.g., to lessen bearing loads by reducing the mass of the reciprocating parts and through flapper valve modifications), the refrigerator application started to show four times the rate of the 5-year warranty compressor failures as the window air conditioning application, that is, 20% versus 5%. This resulted in an annual $20 million loss (circa 1970) on that single refrigerator product from warranty replacements. That therefore quickly got onto the top corporate management's "radar screen." Close study of several of the failed compressors revealed that it was the wrist-pin bearing that was the failing part. The wrist pin is press-fitted into the piston to be surrounded by the wrist-pin sleeve bearing which is press-fitted into the connecting rod. The compressor was hermetically sealed within the refrigerant loop to eliminate any refrigerant leakage, so the bearing lubricant was mixed with the refrigerant.

That the refrigerator compressor warranty failure rate was four times that of the window air conditioner compressor mystified the refrigerator division's top compressor engineers because the wrist-pin peak load in the air conditioner was approximately 25% higher than in the refrigerator. The wrist-pin-bearing radial load versus crank angle is illustrated for both applications in Figure 14.2. The nominal size of the wrist-pin bearing was $D = L = 0.375$ in. $= 9.5$ mm. Therefore, the peak unit cyclic loads for the refrigerator and air conditioner bearings ($P = W/DL$) were 1387 psi (957 N/cm^2) and 1778 psi (1227 N/cm^2), respectively. In an attempt to uncover the root cause for the relatively large warranty failure rate in the refrigerator application, many different analyses and tests were conducted at the corporation R&D center, sort of a "fishing expedition" to placate the corporate top management and hopefully to get the refrigerator division engineers "out of hot water." The R&D center task force assembled for this troubleshooting "adventure" had been active for nearly two years when the author (at 31) commenced his new job at the corporate R&D center with his new boss, an internationally known fluid-film bearing expert on that task force, and the top management having just dictated that the task force had two months left to solve the problem or else.

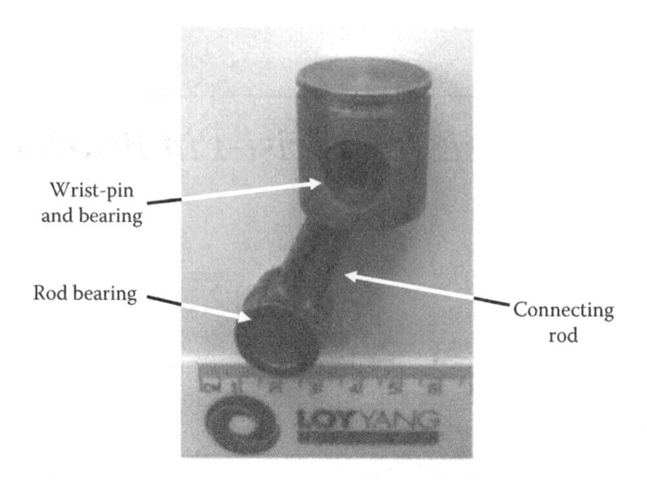

FIGURE 14.1
Piston and connecting rod of a small reciprocating compressor.

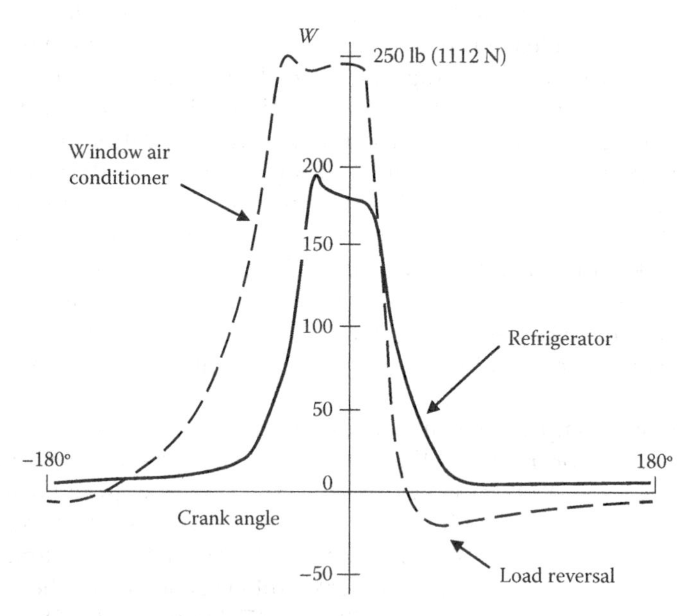

FIGURE 14.2
Wrist-pin-bearing load *W* curves versus crank angle.

14.2 Wrist-Pin Orbit Trajectories Solve Problem

One of the many analyses ordered was computation of the wrist-pin-bearing minimum film thickness within the 0–360° crank cycle. The author was assigned that task. Unlike the calculation of journal-bearing minimum oil-film thickness under bearing steady static load, it required the nonlinear dynamical orbit of the wrist pin relative to the bearing be modeled in order to predict the transient minimum oil-film thickness. The author had performed similar analyses, on different equipment, at his previous place of employment. This is now a well-established computational approach for reciprocating compressors and internal combustion piston engines, but not then in 1971. Both the air conditioner and refrigerator time-varying wrist-pin-bearing loads were entered into a new computer code and then quickly written by the author for this purpose, employing a nonlinear time-transient marching algorithm as for the topics in Chapters 11, 12, and 13.

The computed wrist-pin orbits relative to the bearing, not the transient minimum film thickness, provided the answer to the bearing failure root cause, much to the author's benefit being only a couple of months into his new job at the corporate R&D center. For his success in solving this corporate high-profile problem, the author was treated to a free dinner by the corporate vice president of engineering. That was when the author decided that after completing his PhD locally at Pitt, his next job would be as a professor somewhere.

Two of the many simulated wrist-pin orbits from this analysis are illustrated in Figure 14.3. They clearly show the root cause of the refrigerator compressor's higher warranty failure rate. The load curves illustrated in Figure 14.2 show that the air conditioner-loading curve goes slightly negative, while the refrigerator-loading curve does not. This was a feature not previously noted when the task force focus was entirely on the maximum peak loads. In the air conditioner application, just a slight amount of load reversal caused the wrist pin to substantially separate away from the oil-feed hole that channels oil from the rod bearing through a connecting hole in the rod, for example, see Figure 7.1b. In contrast, for the refrigerator application there was no load reversal and thus the wrist pin did not lift off the oil-feed hole as its oscillatory trajectory clearly shows. Subsequent endurance tests completely confirmed what the computed nonlinear dynamic orbits implied. That is, the refrigerator wrist pin continuously rubs on the bearing over the oil hole and thereby does not separate to allow enough oil to enter to adequately accommodate the next squeeze-film action in each loading cycle. With the root cause uncovered by the author's nonlinear analysis, modifications were then implemented to ensure that the refrigerator compressor-load curve included some load reversal. Ironically, it was the prior design modifications to reduce bearing cyclic peak load that eliminated a bit of load reversal that caused the fundamental problem. The design defect was eliminated and the compressor high failure rate ceased once those units still "in the pipeline" cleared the retailers' stocks.

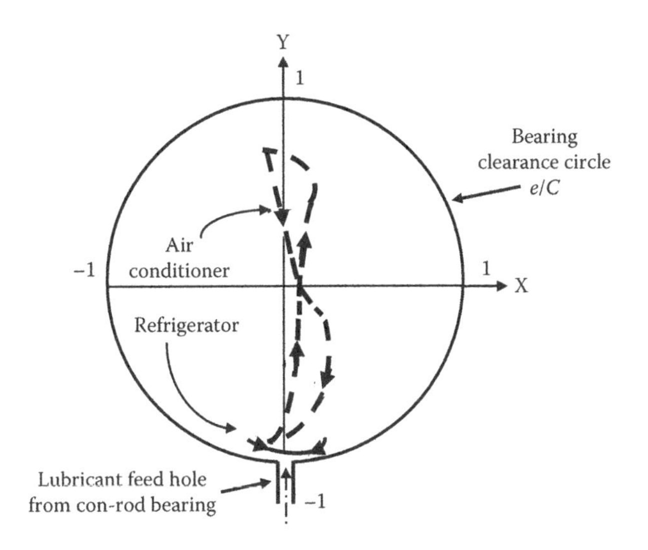

FIGURE 14.3
Numerical simulation of compressor piston wrist-pin cyclic motions.

15

Vertical Rotor Random-Bearing Loads

15.1 Overview

Both pressurized water reactor (PWR) primary reactor coolant pumps (RCP) and boiling water reactor (BWR) primary circulating pumps employ vertical rotational centerlines. RCPs have three journal bearings supporting a motor-pump rigidly coupled rotor configuration, such as shown in Figure 15.1. The two top radial bearings, oil-lubricated hydrodynamic tilting-pad journal bearings, are within the drive motor. The third lowest-of-the-three radial bearings is a hydrodynamic 360° cylindrical sleeve journal-bearing lubricated by PWR primary loop water at primary-loop ambient pressure (153 bar) but at a much lower temperature by virtue of the shown thermal barrier isolating the PWR-water-lubricated bearing directly from the PWR primary loop reactor core high temperature. This is the rotor-bearing configuration most frequently employed thus far in the United States. In the early generation PWR commercial nuclear power plants, the canned-motor-type pump was employed, resulting from a scaled-up version of what the RCP manufacturer had already developed for U.S. Navy nuclear powered submarines. However, switching to the shaft-sealed RCP in the early 1960s was strongly motivated by (1) a 10% increase in pump efficiency plus the valuable flexibility of both (2) pump internal component inspection and (3) the ability to make pump repairs in the field. Whereas canned-motor RCPs have to be shipped back to the factory for any internal inspections or repair work because of penetration and subsequent resealing the hermetically contained canned-motor pump rotor.

15.2 Random-Bearing-Loads Explain Random Vibration

As summarized in Figure 15.2, RCP radial journal-bearing loads are statically indeterminate since there are more than two radial bearings on a rigidly coupled rotor, necessitating that both the *rotor and bearing bi-planar static radial deflections* be appropriately taken into account in the journal-bearing load predictions, to correctly compute radial-bearing static loads as correctly as possible. The author is the one who enlightened the Figure 15.1 RCP manufacturer of this technological fact, which as a result is now a longstanding design analysis for these RCPs. Correspondingly, RCP-journal-bearing static loads are also significantly affected by manufacturing and assembly tolerances of radial misalignment offsets between the three journal bearings' centerlines, which are naturally random within the specified manufacturing and assembly tolerances. The author is the one who additionally enlightened the Figure 15.1 RCP manufacturer of these technological facts.

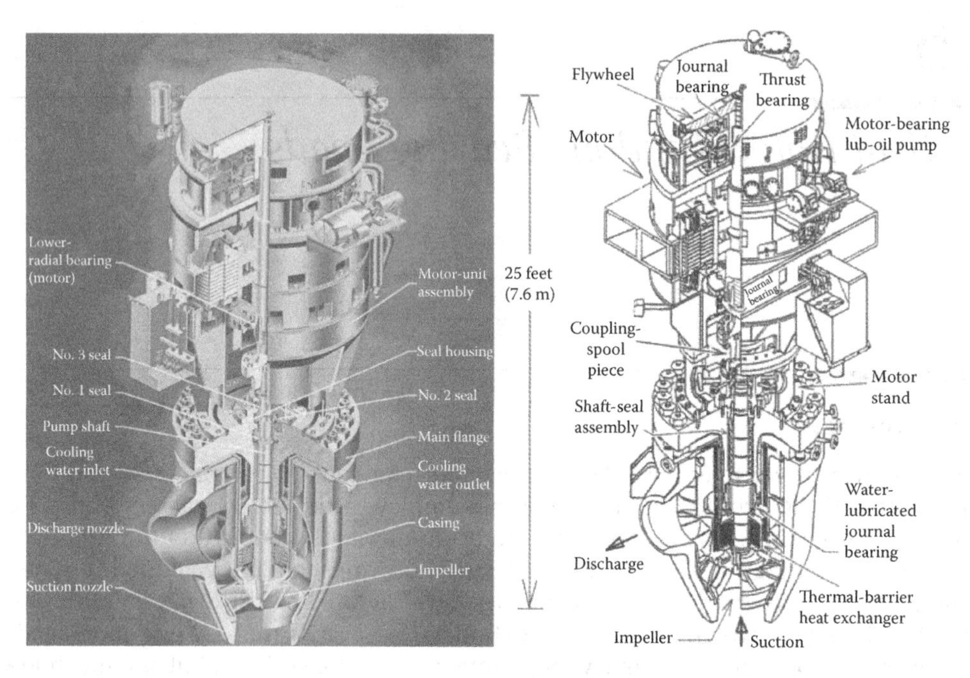

FIGURE 15.1
Shaft-sealed PWR primary reactor coolant pump.

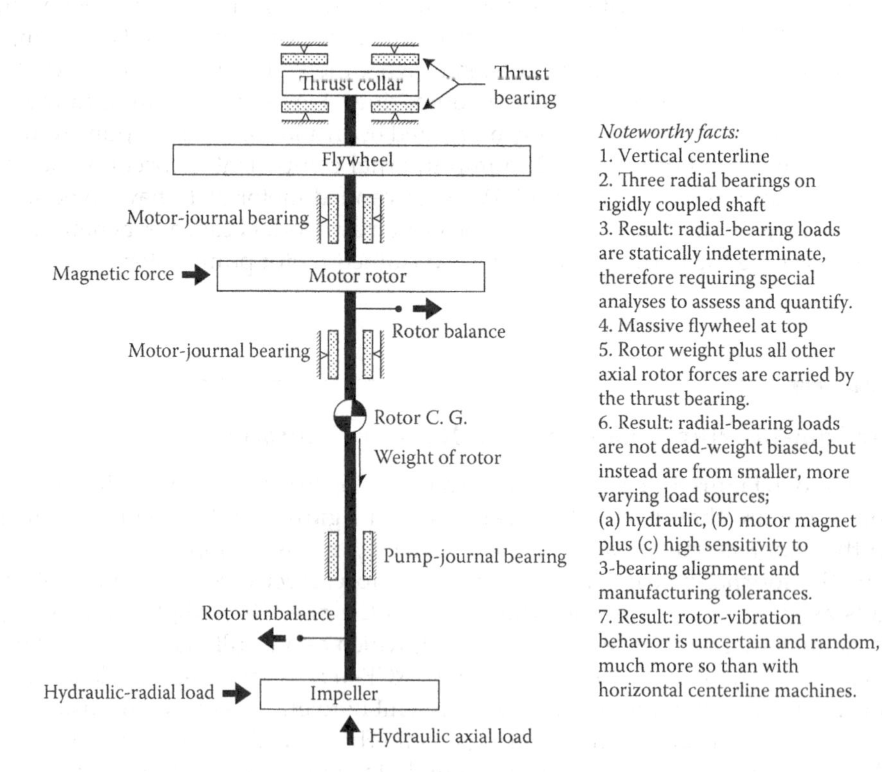

Noteworthy facts:
1. Vertical centerline
2. Three radial bearings on rigidly coupled shaft
3. Result: radial-bearing loads are statically indeterminate, therefore requiring special analyses to assess and quantify.
4. Massive flywheel at top
5. Rotor weight plus all other axial rotor forces are carried by the thrust bearing.
6. Result: radial-bearing loads are not dead-weight biased, but instead are from smaller, more varying load sources; (a) hydraulic, (b) motor magnet plus (c) high sensitivity to 3-bearing alignment and manufacturing tolerances.
7. Result: rotor-vibration behavior is uncertain and random, much more so than with horizontal centerline machines.

FIGURE 15.2
Summary of PWR-reactor coolant-pump-bearing loads.

Specialized computer iterative analyses are required as is now done to properly predict as correctly as possible the three radial-bearings' static loads.

Concerning rotor-bearing mechanics in general, vertical machines are fundamentally far more complex to analyze and understand than horizontal machines, primarily because the radial-bearing loads are not dead-weight biased like horizontal machines, the entire rotor weight being carried by the axial thrust bearing. The net result of all these complicating factors is that the radial-bearing static loads in RCPs are significantly less well defined and randomly variable much more from machine-to-machine than radial-bearing static loads in horizontal-rotor machines. Given the strong dependence of journal-bearing rotor vibration characteristics on bearing-static loads (Adams 2010), the rotor-vibration behavior of RCPs is more uncertain and randomly variable compared to horizontal machines.

To encapsulate all this, RCPs are among the most extreme examples that no two machines are ever exactly alike. Consequently, from a rotor-vibration perspective, RCPs are among the most challenging types of rotating machinery on which to make analysis-based predictions. Thus, what might be ascertained by successfully reducing a vibration problem on one specific RCP cannot be relied on to help solve a similar problem on another RCP, even one of the exact same configuration.

Jenkins (1993) attests to the considerable challenge in assessing the significance of monitored-vibration signals from RCPs, and focuses on possible correlation of vibration-signal content and equipment malfunction as related to machine age. He presents the "Westinghouse approach" in identifying vibration-problem root causes and corrective changes for Westinghouse RPC machines. In one case study presented by Jenkins, one of three "identical" RCP pumps for a specific PWR unit developed an excessive half-rotational-speed ($N/2$) frequency rotor vibration whirl. With a cavitation-free (i.e., no film rupture) full-film hydrodynamic water-lubricated sleeve bearing (153 bar ambient pressure) on a vertical centerline, there is nearly always some $N/2$ rotor-vibration content observed in the monitored rotor-vibration signals of these pumps. But this is typically at tolerable rotor-vibration levels when the pumps are operating normally. In this case study by Jenkins, the drastically increased level of $N/2$ vibration led to an investigation to determine the likely root cause(s) and the proper corrective action(s). Based on (1) the drastic increase in the monitored $N/2$ vibration component (changed from 2 to 6 mils p.p. at coupling) and (2) on a definitive shift in rotor static centerline position as indicated by the coupling proximity probes' DC voltages, it was diagnosed that the water-lubricated carbon graphite journal sleeve-bearing clearance had significantly worn open. Some motor-bearing trial-and-error alignment adjustments allowed the $N/2$ vibration component to be reduced within levels deemed operable, pending a replacement of the pump bearing at the next refueling, or sooner if the monitored rotor vibration developed a subsequent upward trend.

References

Adams, M. L., *Rotating Machinery Vibration—From Analysis to Troubleshooting*, 2nd ed., CRC Press/Taylor & Francis, Boca Raton, FL, 2010.

Adams, M. L., *Rotating Machinery Research and Development Test Rigs*, CRC Press/Taylor & Francis, Boca Raton, FL, 2017.

Jenkins, L. S., Troubleshooting Westinghouse Reactor Coolant Pump Vibrations, *EPRI Symposium on Trouble Shooting Power Plant Rotating Machinery Vibrations*, LaJolla, CA, May 19–21, 1993.

16

Steam Turbine Tilting-Pad-Bearing Retrofit

16.1 Overview

This case study involves the high-pressure-intermediate-pressure (HP-IP) portion of a 240 MW, 60 Hz, 3600 rpm coal-fired steam turbine generator driveline of an electric generating unit illustrated in Figure 16.1. Excessive synchronous (at rotational speed frequency) rotor vibration was consistently experienced on the unit's HP-IP rotor during load following in the 145–180 MW range. This characteristic was documented to exist from the unit's original commissioning several years earlier. But having been initially commissioned as a base-load unit before the electric company's nuclear units were operational, this vibration was not an operating problem since the unit was operated at full-rated capacity most of the time. After the commissioning of the electric company's nuclear powered units, this 240 MW fossil fired unit became used primarily for load following. The resulting frequent-load transitioning through the 145–180 MW range then made this vibration a constant problem for the operators, sufficiently problematic to necessitate root cause identification and substantial attenuation. The author undertook vibration measurements of the entire driveline. Examination of these measurements suggested the root cause to be the HP-IP rotor's *2nd critical speed,* as evidenced from the corresponding vibration peak as shown in Figure 16.2, accompanied by a sharp change in phase angle measurement. This critical speed shifted up or down across 3600 rpm operating speed as power output of the unit cycled during power demand following. The author's preliminary assessment was as follows. The root cause appeared to be the HP-IP journal-bearing static-load changes commensurate with the variable partial emission of the impulse turbine control stage during power cycling (Adams 2010).

16.2 Simulation of Vibration as a Function of Bearing Loads

The author developed a rotor vibration simulation model of the unit's entire driveline shown in Figure 16.1. This model was used to parametrically study rotor unbalance vibration during power output variation as influenced by journal-bearing hydrodynamic at-speed oil-film stiffness variations at 3600 rpm that result from variations in net impulse turbine nozzle radial force when transitioning through power changes. The computer-model simulations verified the preliminary diagnosis that the root cause was the HP-IP *rotor's shifting 2nd critical speed.* A composite of those simulation results are plotted in Figure 16.3,

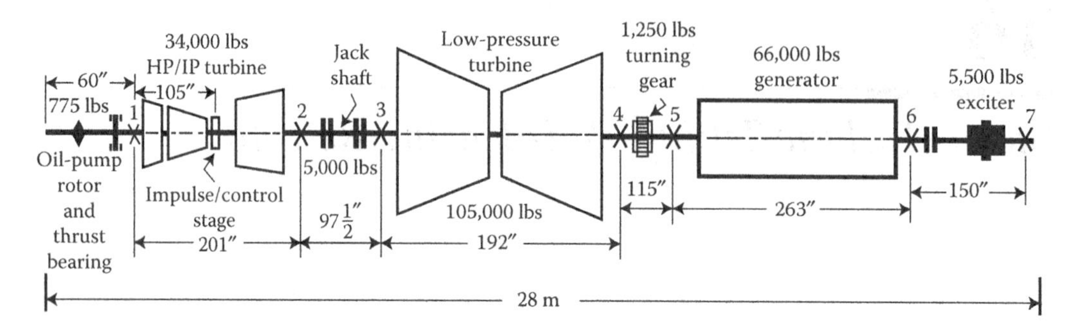

FIGURE 16.1
Large steam-turbine configuration.

showing the influence of journal oil-film radial stiffness variation with unit power output variation.

The turbine generator original equipment manufacturer's (OEM) HP-IP-bearing configuration (Figure 16.4a) was a four-pad pivoted-pad configuration with the rotor-weight load between the bottom two pads. As was expected, and as the Figure 16.3 simulations show, the computer model predicted the 2nd critical speed transitions up through the 3600-rpm operating speed as journal-bearing input stiffness is set at progressively higher values, from 0.3 million lb/in. (53 million N/m) up to 1.4 million lb/in. (245 million N/m). Figure 16.3 also shows the unbalance vibration response predicted with an author-recommended six-pad non-OEM journal-bearing retrofit at bearings #1 and #2.

16.3 Bearing Retrofit Eliminates Vibration Problem

The retrofit six-pad pivoted-pad-bearing design, with weight centered on the bottom pad, has substantially higher radial stiffness than the OEM design, and its stiffness is much less sensitive to bearing-load changes during power following of the unit. Based on the drastic

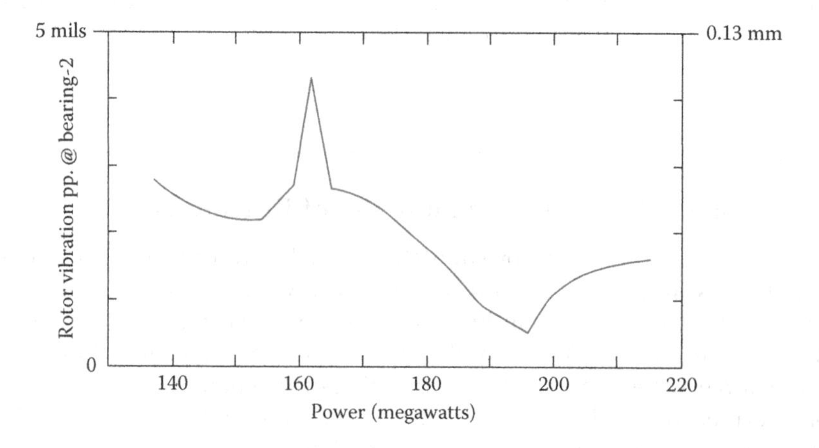

FIGURE 16.2
Measured shaft vibration versus power at HP-IP bearing #2.

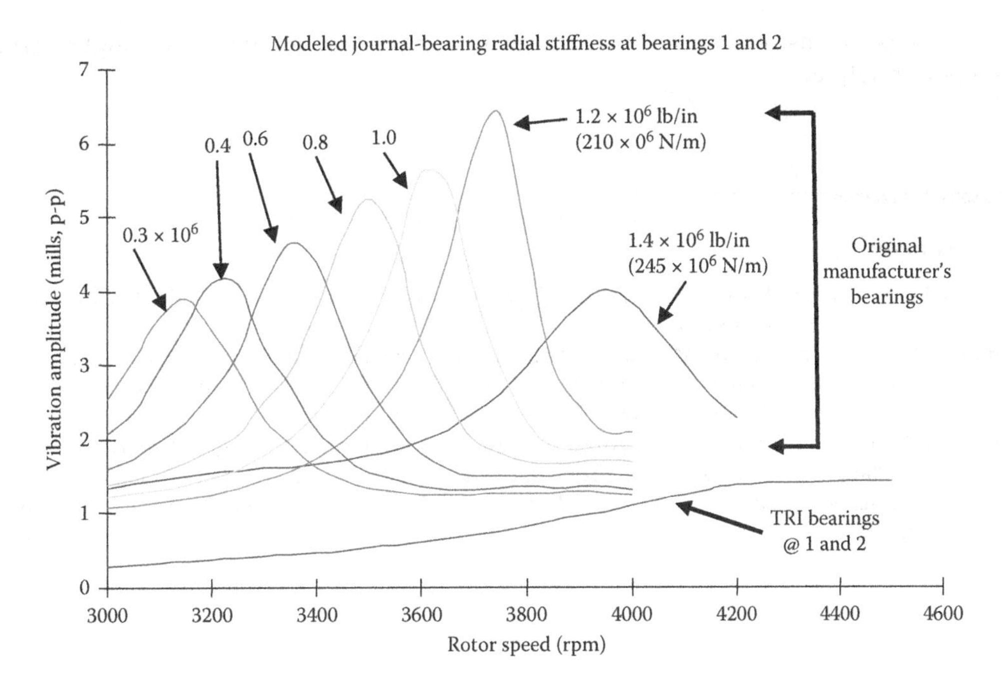

FIGURE 16.3
Predicted shaft unbalance vibration amplitude at bearing #1. At the nominal-bearing unit load of 200 psi (13.6 bar), OEM radial-bearing stiffness K is computed to be 1.98×10^6 lb/in. (256×10^6 N/m).

vibration attenuation predicted by the computer model with the six-pad Turbo Research Inc. (TRI) bearing retrofit, the author recommended this retrofit, supplied by TRI. A sketch of this bearing configuration is shown in Figure 16.4b and described in detail by Giberson (1993).

This bearing retrofit was purchased and installed by TRI at bearings #1 and #2 shortly thereafter by the power company owner of this unit during the next scheduled outage. Following that scheduled outage and bearing retrofit, the author again analyzed shaft vibration measurements on the complete machine's driveline, reported by Adams and Adams (2006). Those vibration measurements showed a more than twofold reduction in peak-vibration levels. Removal of the HP-IP 2nd critical speed from within the operating

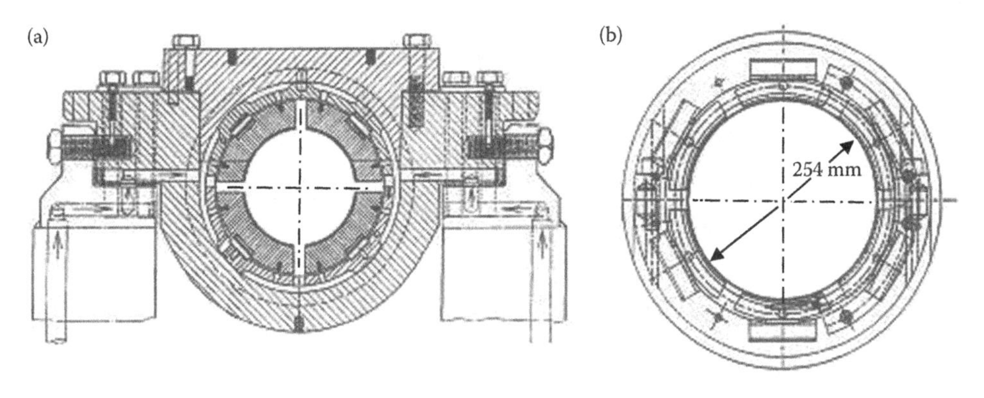

FIGURE 16.4
Turbine pivoted-pad-bearing configurations, (a) OEM, (b) TRI.

power range was also evidenced from a gradual change rotor vibration phase angle versus unit power output.

References

Adams, M. L., *Rotating Machinery Vibration—From Analysis to Troubleshooting*, 2nd ed., CRC Press/Taylor & Francis, Boca Raton, FL, 2010.

Adams, Maurice L. and Adams, Michael L., On the Use of Rotor Vibration Analysis and Measurement Tools to Cure Power Plant Machinery Vibration, *Proceedings, IFToMM 7th International Conference on Rotor Dynamics*, Vienna, Austria, September 25–28, 2006.

Giberson, M. F., Evolution of the TRI Align-A-Pad® Tilt-Pad Bearing Through 20 Years of Solving Power Plant Machinery Vibration Problems, *Proceedings, Electric Power Research Institute (EPRI) Symposium on Trouble Shooting Power Plant Rotating Machinery Vibrations*, San Diego, CA, May 19–21, 1993.

17

Ineffective Journal-Bearing Pinch

17.1 Overview

The rotor sectional view in Figure 17.1 is from a four-stage boiler feed water pump (BFP). In the power plant of this case study, the BFPs are installed as variable speed units with operating speeds from 3000 to 6000 rpm, each with an induction motor drive through a variable speed fluid coupling. In this plant, the BFPs are all 50% pumps, which means that when a main stream turbo-generator is at full load 100% power output, two such pumps are operating at their nominal design operating condition. This plant houses four 500 MW generating units, each having three 50% BFPs installed (i.e., one extra 50% BFP on standby), for a total of 12 BFPs, all of the same configuration. The full-load operating speed range for each 50% BFP is 5250–5975 rpm.

The BFPs at this plant had experienced a long history of failures, leading to operating times between overhauls under 10,000 hours, with the significant attendant monetary cost. Based on the operating experience at other power plants employing the same BFP configuration with quite similar operating ranges, these BFPs should have been running satisfactorily for over 40,000 hours between overhauls. Using vibration-velocity peak monitored at the outboard-bearing bracket as the indicative parameter, these BFPs were usually taken out of service for overhaul when that vibration level exceeded 15 mm/sec (0.6 in/sec). To wait longer, the plant experienced that significantly increased damage to pump internals occurred, significantly increasing the overhaul rebuild cost. The dominant vibration frequency was at rotational speed.

17.2 Initial Diagnosis and Analysis

The author's preliminary diagnosis was that these pumps were operating quite near a critical speed and that the resonance vibration resulting from this worked to accelerate the wearing open of inter-stage sealing-ring radial clearances. As these inter-stage clearances wear open and the overall vibration damping capacity diminishes significantly, this accelerates a continuous growth of vibration levels and associated internal wear. To confirm this preliminary diagnosis, the author developed a computer simulation model for this pump configuration to compute lateral-rotor-vibration unbalance response versus rpm. The manufacturer (OEM) of the pump provided a nominally dimensioned layout of the assembled pump, including weights and inertias for concentrated masses (impellers, balancing disk rotor, thrust-bearing collar, coupling piece, and shaft sleeves). The pump

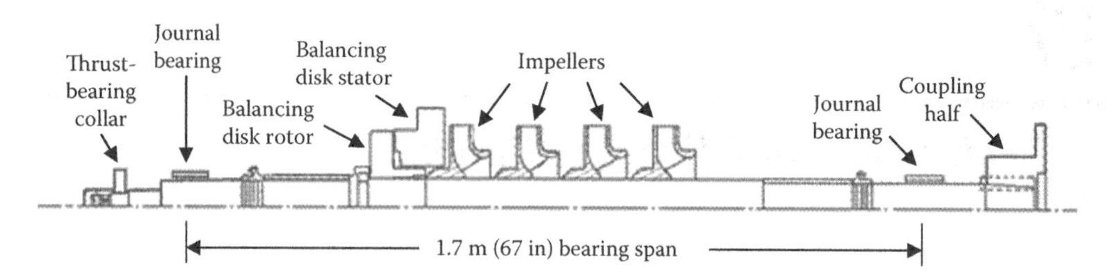

FIGURE 17.1
Rotor sectional view of a four-stage boiler feed pump.

OEM also provided detailed geometric dimensions for the journal bearings, inter-stage radial seals, and other close-clearance radial-annular gaps. This cooperation by the pump OEM greatly expedited the development of the computer simulation model, eliminating the need to take extensive dimension measurements from one of the BFPs at the plant or repair shop, which were a considerable distance outside the United States (Australia).

The radial-annular gaps have clearance dimensions that are quite small and are formed by the small difference between a bore (ID) and an outside diameter (OD), each with tolerances. The size of each of these small-radial-clearance gaps is very influential on the respective bearing or seal stiffness, damping, and inertia coefficients (Adams 2010) and thus very influential on the computed results for predicted rotor-vibration response. However, these small radial gaps vary percentage-wise significantly and randomly because of their respective ID and OD manufacturing tolerances plus any wearing open due to in-service use. BFPs are thus one of the most challenging rotating machinery types to accurately model and analyze for rotor vibration. The net result is that even in the easiest of cases, a realistic rotor vibration analysis for troubleshooting purposes, as opposed to design purposes, requires several trial input cases to get the model predictions to reasonably portray the vibration problem that the machine is exhibiting. By iterating the model inputs per radial-clearance manufacturing tolerances and allowances for wear, a set of inputs is sought that produce rotor-vibration response predictions that concur with the machine's vibration behavior. When a good-agreement model is so achieved, it is referred to as the *calibrated model*. Through computer simulations, the calibrated model can then be used to explore the relative benefits of various fix or retrofit scenarios, as was done in the successful steam turbine case studies presented here in the Chapter 16 steam turbine case.

A calibrated model was not initially achieved for this pump-vibration problem in that all reasonable model variations for input dimensions failed to produce predicted unbalance responses having a resonance peak below 8000 rpm, which was considerably above the BFP operating speed range. Since the power plant in this case was in Australia, a visit to the plant had not initially been planned. However, given the failure of all initial model variations to replicate or explain the pump-vibration problem, a trip to the plant was undertaken to study the pumps firsthand.

17.3 Hidden Flaw of Ineffective Bearing-Shell Squeeze

Poor hydraulic conditions in BFPs, such as from inaccurate impeller castings, can produce strong synchronous rotor vibrations (Adams 2010). Therefore, several of the BFP impellers were inspected for such inaccuracies. In the course of further searching for the vibration problem root cause, a number of serious deficiencies were uncovered in the local BFP overhaul and repair shop's methods and procedures, all of which collectively might have accounted for the vibration problem. Luckily, on the last day of the planned one-week visit to the plant, Loy Yang Plant south of Melbourne, *the root cause was discovered*, but it could have been easily overlooked.

Needing a break from number crunching, the author took a break to look at one of the recently overhauled BFPs being reinstalled, just when the journal bearings were being reinstalled by the mechanic. The author noticed the insertion of thin gasket-material strips being inserted between the inner bearing half shells and the outer bearing halves prior to the bolting together of the bearing housing halves. This use of thin gasket-material strips was discontinued many years earlier in most U.S. power plants. The net result of the interposed gasket strips was to greatly reduce the effective bearing dynamic stiffness to a value significantly below the range that the author had reasonably assumed in the initial unsuccessful attempts to develop a calibrated rotor-vibration simulation model. When the relatively soft gasket effect was incorporated into the computer simulation model-bearing stiffness inputs, a resonance peak showed up right in the normal operating speed range.

An analysis study was made to compute critical speed at which unbalance-excited vibration response peaks as a function net-bearing stiffness, using a stiffness value range consistent with the interposed gasket strips. A summary of the results for that analysis is shown in Figure 17.2. A bearing stiffness value of 100,000 lb/in. (17.5×10^6 N/m) places the critical speed right in the BFP normal full-load operating-speed range. The inherent variability of gasket material stiffness also explained the plant's experience with the excessive vibration fading in-and-out over time.

The gasket stiffness was essentially in-series with the bearing oil film's in-parallel stiffness and damping characteristics. Since the gasket was much less stiff than the journal-bearing oil film, the gasket also reduced considerably the damping action of the journal-bearing oil films. That was the worst-case scenario, a resonance with very little damping, as clearly illustrated in the "textbook" Figure 10.1. The use of thin gasket-material strips between the bearing inner half shells and outer housing was clearly the *"smoking gun,"* placing the critical speed within the normal full-load operating speed range, while at the same time

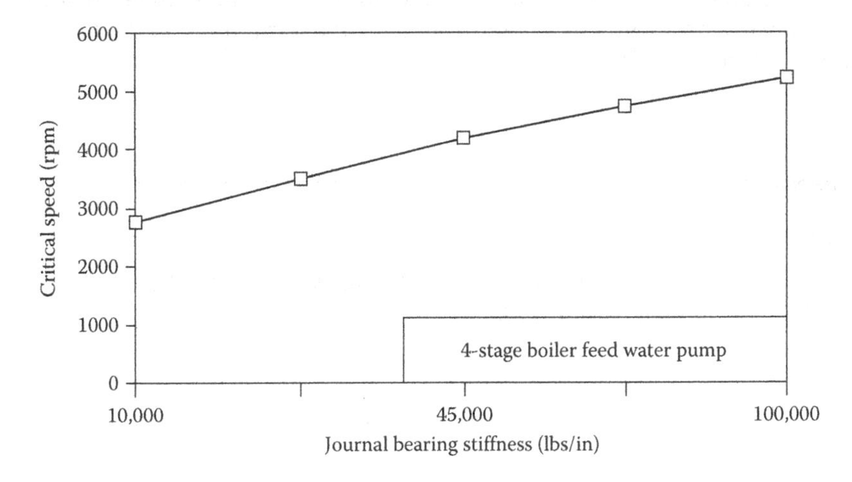

FIGURE 17.2
Computed critical speed versus interposed gasket stiffness.

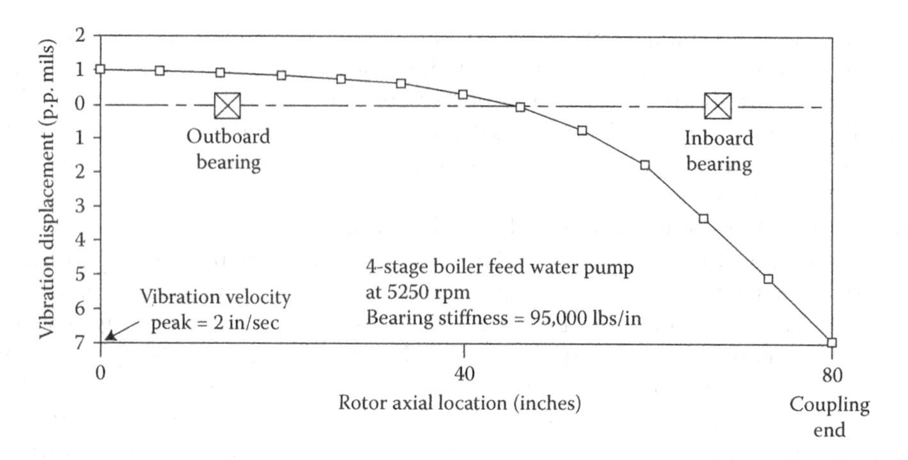

FIGURE 17.3
Critical speed rotor response mode shape with typical unbalances.

depriving the attendant resonance of reasonable damping. All bearings were reinstalled with thin metal shim strips, at the 45° and 135° angular locations relative to horizontal, providing a *bearing pinch* of about 0.001 in. (0.025 mm), which is pretty standard in normal assemblies, that is, with no gaskets interposed. Bearing pinch ensures stiff metal-to-metal contact between the inner-bearing half shells and the two outer-bearing housing halves without over compressing the inner-bearing half shells into unwanted distortion. That allows the bearing oil film to set the stiffness and damping characteristics that dynamically "connect" the rotor to the non-rotating part of the machine (Adams 2010). This fix saved the power plant several thousand dollars annually by extending the operating duration between pump major overalls from 10,000 hours to over 40,000 hours.

A view of the BFP non-planar critical-speed response mode shape from the simulation analysis with the interposed gasket strips is shown in Figure 17.3. The obvious conclusion drawn from this computed unbalance response shape was that coupling unbalance contributed significantly to this excessive vibration problem because the repair shop's rotor balancing procedure as witnessed by the author was inadequate in several areas. A supplementary benefit to the plant was the referral to a highly qualified non-OEM U.S. pump overhaul facility in Chicago.

Reference

Adams, M. L., *Rotating Machinery Vibration—From Analysis to Troubleshooting*, 2nd ed., CRC Press/Taylor & Francis, Boca Raton, FL, 2010.

18

Nuclear Plant Pump Rubber Bearings

18.1 Overview

The vertical centerline pump illustrated in Figure 18.1 is an Emergency Service Water (ESW) pump for a nuclear power plant on Lake Erie. The pump had just been reinstalled and returned to service after a complete overhaul during a refueling outage of the nuclear-powered generating unit. It was test run to confirm it being fully operational. It was then shut down to rest mode and drained. On its restart some days later, it began to show symptoms of internal damage based on monitored-vibration signals. It was then disassembled at the plant for inspection of pump internals. It was thereby discovered that the roll pins connecting journal-bearing support brackets to the housing were heavily damaged, that is, nearly sheared off. The author was retained to provide guidance in quickly determining the root cause of the failure in time to restart the plant on schedule at the conclusion of the 6-week refueling cycle.

18.2 Diagnosis and Fix

Figure 18.2 illustrates the radial-bearing arrangement with a detailed view of a polymer staved water-lubricated bearing in Figure 1.36a. The author's initial diagnosis was that the journal-bearing polymer liners had mysteriously swelled onto their respective journals, tight enough to mangle the roll pins that anchored the bearings to the pump housing (Figure 18.2). The author contacted one of the leading producers of polymer-bearing liners, Duramax Marine, coincidently located quite near the author's residence. They then inspected the disassembled pump and examined the polymer-bearing liners. They took samples of the polymer liner material to their laboratory and determined that the material was a neoprene rubber known for its considerable swelling property after being drained of water, as was the case at the plant when the pump drains down at shutdown.

The pump refurbishment vender had apparently chosen that neoprene rubber material because it was a lower cost alternative to the original equipment manufacturer (OEM) material. Duramax Marine supplied retrofitted bearings with staves of a standard 70 durometer Nitrile rubber known to have minimum swell behavior. The problem was thereby eliminated. Naturally, the power company owner of the nuclear power plant justifiably took the pump refurbishment vender to task for their mess-up in choosing

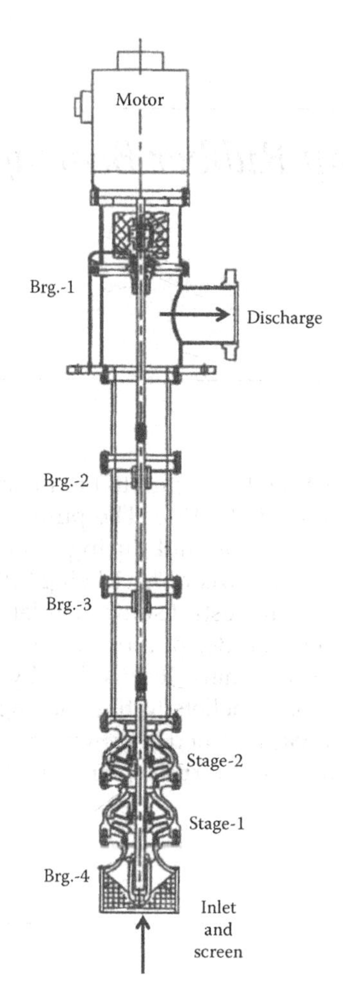

FIGURE 18.1
Nuclear-plant ESW pump, with bearings 1 through 4 so marked.

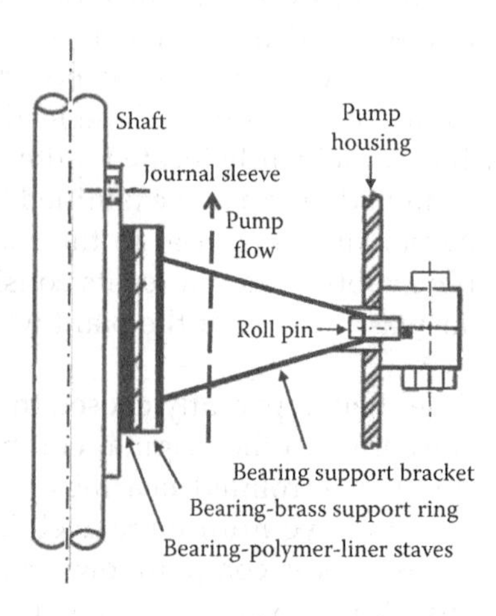

FIGURE 18.2
Details of journal-bearing support connection to pump housing.

the swell-prone-bearing-liner polymer material. Had the fundamental problem not been quickly diagnosed and corrected, the nuclear power generating unit would not have been able to power up as scheduled, with the consequence of a $2 million/day loss in generating revenue. A technology review of water-lubricated polymer-lined bearings is the topic of Section 1.11, presenting polymer bearing fundamentals from engineering information generously provided by Duramax Marine.

Index

D

I

J

K

L

M

T